Design of Steel-Concrete Composite Structures Using High-Strength Materials

WOODHEAD PUBLISHING SERIES IN CIVIL AND STRUCTURAL ENGINEERING

Design of Steel-Concrete Composite Structures Using High-Strength Materials

J.Y. Richard Liew
Department of Civil and Environmental Engineering, National University of Singapore, Singapore

Ming-Xiang Xiong
School of Civil Engineering, Guangzhou University, Guangdong, China

Bing-Lin Lai
School of Civil Engineering, Southeast University, Nanjing, China

ELSEVIER
WP
WOODHEAD PUBLISHING
An imprint of Elsevier
elsevier.com/books-and-journals

Woodhead Publishing is an imprint of Elsevier
The Officers' Mess Business Centre, Royston Road, Duxford, CB22 4QH, United Kingdom
50 Hampshire Street, 5th Floor, Cambridge, MA 02139, United States
The Boulevard, Langford Lane, Kidlington, OX5 1GB, United Kingdom

British Library Cataloguing-in-Publication Data
A catalogue record for this book is available from the British Library

Library of Congress Cataloging-in-Publication Data
A catalog record for this book is available from the Library of Congress

ISBN: 978-0-12-823396-2

For Information on all Woodhead Publishing publications visit our website at https://www.elsevier.com/books-and-journals

Publisher: Matthew Deans
Acquisitions Editor: Glyn Jones
Editorial Project Manager: Devlin Person
Production Project Manager: Surya Narayanan Jayachandran
Cover Designer: Greg Harris

Typeset by Aptara, New Delhi, India

Contents

Foreword

High-strength steel and concrete materials are typically used in the construction of high-rise structures. These materials have great advantages when they are used as load-bearing components in buildings such as columns, shear walls, and foundations. They are occasionally used in bridge structures, as well. High-strength materials are feasible for columns, especially on lower floors of a tall building where the loads will be greatest. This is because the higher the material strength, the smaller the section size required to resist the same load. This frees up more usable floor space and requires less material and construction work. High-strength concrete is relatively more brittle than normal concrete, especially when it is not reinforced. However, if it is used in a composite manner with the steel section, its behavior becomes more ductile depending on the steel contribution ratio. The cost to the strength ratio of high-strength steel section is found to be lower than that of other materials, and thus high-strength steel sections have better economic benefits compared with the normal-strength steel. It is a well-known fact that slender members made of high-strength steel are prone to buckling failure in which the material strength may not be fully utilized. Therefore, its use may not be effective in reducing the cross-section area of the slender steel members. However, if high-strength steel section is used with concrete, especially with high-strength concrete, the buckling of steel section can be effectively restrained and the steel strength can be fully utilized. This opens new opportunity to provide more options for designers to adopt innovative methods of design involving the use of high-strength steel and concrete materials for structural application in modern construction.

Although codes of practice on steel-concrete composite structures are available in many countries such as the United States, Australia, Japan, and many countries across Europe, they do not provide sufficient guidelines on the use of high-strength construction materials in such applications. The design method described in this book is based on Eurocode 4 (EN 1994-1-1, 2004) for the design of steel-concrete composite members with special considerations for higher strength steel and concrete materials. More than 2000 test data sets collected from the literature on steel-concrete composite members with normal- and high-strength materials have been analyzed to formulate the design guide proposed in this book. The background knowledge leading to the development of such design guide is explained. Clear guidance is also provided to select matching concrete and steel grades for the design of high-strength composite members. Quality control of high-strength concrete materials and weldability of high-strength steel sections are emphasized. Fire protection and fire resistance design method in accordance with EN 1994-1-2 (2005) are provided for the high-strength composite member based on the latest research works done by the fellow researchers.

Finally, good detailing practices are provided for typical joints between steel-concrete composite members and the other structural components.

Some of the design calculations may involve tedious formulas and require iterative steps to find an optimum solution. Opportunity for error is great if such calculations are done by hand. Excel spreadsheets are provided in this book, addressing this need by automating the calculations. Readers are encouraged to test and familiarize themselves with the program before using it for design.

I would like to thank the expert committee members from the Singapore Structural Steel Society and the Building Construction Authority of Singapore for their comments and suggestions in reviewing the design guide BC4 (2021), which inspired us to write this book. The technical assistance provided by Yongnam Engineering and Construction Pte Ltd and JFE Steel Corporation is highly appreciated. They have been very supportive in sharing their expertise and knowledge involving the use of composite members for high-rise building construction in Singapore and Japan.

Finally, a special thank to Li Shan, Xiong Dexin, Wang Yanbo, Huang Zhenyu, and Krishna Padmaja for their hard work and dedication to making every critical project a success. Their comments and contributions to various parts of the book are greatly appreciated.

Richard Liew
Professor, National University of Singapore

March 2021

Reference

BC4 (2021), Design guide for composite steel and concrete members with high strength materials - An Extension of EN1994-1-1 Method to C90/105 Concrete and S550 Steel Section, Building and Construction Authority, 40pp.

Introduction

1

Outline

1.1 Concrete filled steel tubular columns

Concrete filled steel tubular (CFST) column, comprising a hollow steel tube infilled with concrete with or without additional reinforcements or steel section, has been widely used in high-rise building construction. The local buckling of the outer steel tube is delayed or even prevented by the concrete core while the inner concrete core is confined by the steel tube providing enhancement in strength and ductility under high compressive load. The steel tube can serve as permanent formwork for concrete casting and thus it eliminates the need of additional work and leads to fast track construction.

CFST columns have various composite cross-sections as shown in Fig. 1.1. Circular, square and rectangular sections are commonly adopted while polygonal or elliptical sections may also be used for architectural and functional requirements. Conventionally, only plain concrete is filled into the hollow steel sections. Nowadays, the concrete core may be reinforced by fibers or steel bars to enhance ductility and fire resistance of the column. For convenience, the reinforcements can be replaced by an internal steel tube which can provide higher confinement to the concrete core. Other steel sections, such as solid steel section or H-section, can be inserted into the concrete core to further enhance the compression resistance and reduce the column size, but may be obstructed by the shear studs and/or the internal diaphragm plates at the junction of beam to column joints. For columns subjected to high flexural loading, concrete filled double-skin sections with inner tube not filled with concrete can be used to increase the flexural stiffness using less material.

1.2 Concrete encased steel columns

Concrete encased steel (CES) column is another form of composite columns that synergize the merits of steel and concrete materials, which comes in square, rectangular and circular cross-sections, etc. Different from CFST columns, the steel section of fully encased CES columns is completely surrounded by concrete, thus eliminating the need of fire and corrosion protection. The embedded steel section functions as temporary load-bearing member at the construction stage thus expediting

Design of Steel-Concrete Composite Structures Using High Strength Materials. DOI: 10.1016/C2019-0-05474-X

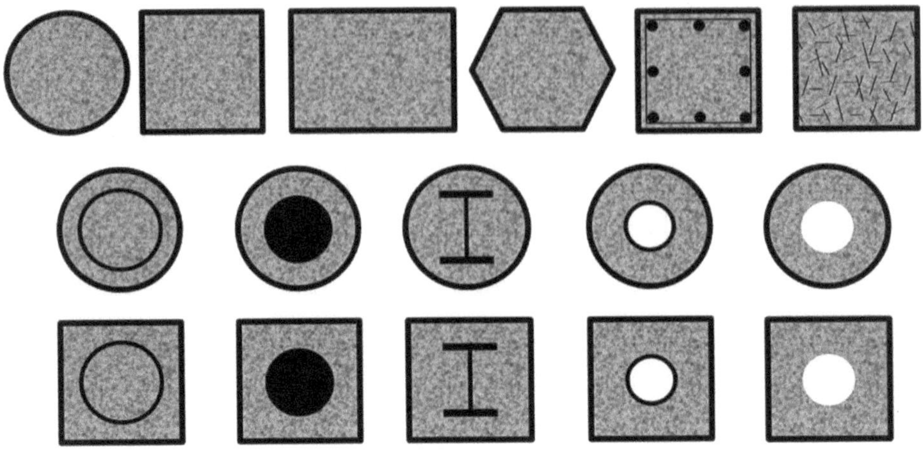

Fig. 1.1 Types of cross-section of CFST columns.

the construction process, and it offers excellent shear resistance and energy dissipation capacity. The embedded steel profile can be H-section, cruciform section or twin section depending on structural requirements which are related to strength, stiffness and ductility demand as shown in Fig. 1.2 (a). As a good practice, every alternate longitudinal reinforcing bar should be laterally supported by adopting diagonal crossties as shown in Fig. 1.2 (b). As for the partially encased columns, the maximum width to thickness ratio of steel flange shall be limited to avoid local buckling.

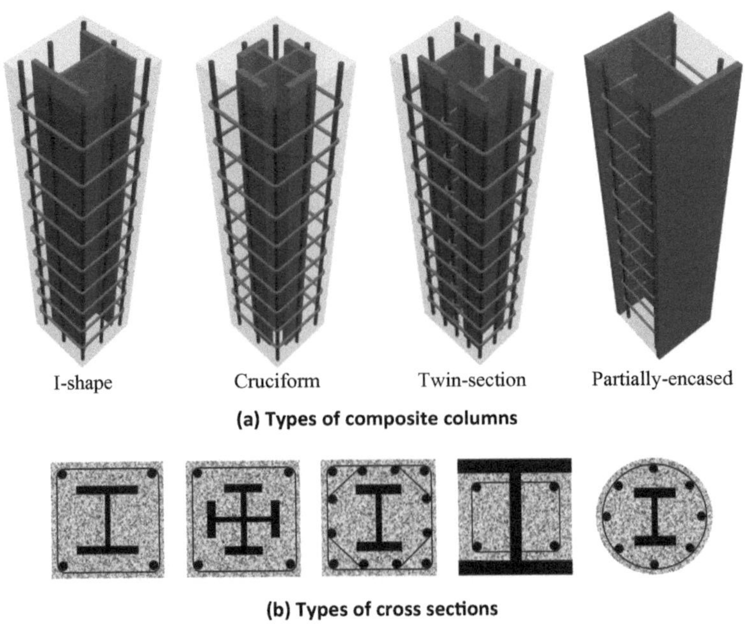

Fig. 1.2 Types of concrete encased steel composite columns.

1.3 Applications of high strength materials

With an aim towards a more sustainable construction, high strength materials may reduce the use of construction materials, thus reducing the use of water, energy and manpower in handling such materials. Demand for and use of high strength materials for high-rise buildings began in the 1970s, primarily in the U.S.A. Nowadays, high strength construction materials are mostly used in Asia region for high rise construction. Table 1.1 summarizes some of the buildings using high strength steel and concrete materials which focused mainly on composite column construction.

Table 1.1 Buildings using high strength materials in composite members.

Project	Country	Year Completed	Building Height (m)/Storey	Concrete (f_{ck}, N/mm^2)	Steel (f_y, N/mm^2)	Used in
Pacific First Centre	U.S.A	1989	185 m	131	350	CFST columns
Two-Union building	U.S.A	1989	226 m	131	350	CFST columns
Gateway Tower	U.S.A	1990	220 m	117.2	350	CFST columns
Lotte World Tower	Korea	2015	555 m	–	570	Outriggers, trusses, and CFST columns
W-Comfort Towers	Japan	2004	178 m	100	–	CFST columns
Obayashi Technical Research Institute	Japan	2010	Multi-storey	160	700	CFST columns
Otemachi Tower	Japan	2014	200 m	150	780	CFST columns
Abeno Harukas	Japan	2014	300 m	150	590	CFST columns
Taipei 101	Taiwan	2004	508 m	70	510	CFST columns
Guangzhou West Tower	China	2010	432 m	90	345	CFST bracings
Goldin 117 Tower	China	2015	597 m	70	390	CFST columns
Robinson Tower	Singapore	2019	17 storeys	60	S355	CFST Columns
Integrated development at 88 Market Street	Singapore	2021	51 storeys	65	S355	CFST Columns
New Afro-Asia i-Mark Building,	Singapore	2020	19 storeys	60	S355	CFST Columns

(*Continued*)

Table 1.1 (*Cont'd*).

Project	Country	Year Completed	Building Height (m)/Storey	Concrete (f_{ck}, N/mm^2)	Steel (f_y, N/mm^2)	Used in
Outram Community Hospital	Singapore	2019	24 Storeys + 4 basements	100	355	CES columns
National Cancer Centre	Singapore	2020	24 storeys + 4 basements	70	460	CES columns

(a) PETRONAS Tower Grade 80 concrete

(b) Sail, Singapore, Grade 80 concrete

(c) World Financial Centre Shanghai, Grade 450 steel

(d) Sky Tree Tokyo, Grade 700 steel

Fig. 1.3 High-rise construction utilizing high strength steel and concrete.

Fig. 1.3(a) shows the PETRONAS Tower in Kuala Lumpur, Malaysia, which is an 88-storey building utilizing Grade 80 high strength concrete for columns with outer diameter up to 2.4 m. The Sail at the Marina Bay Singapore shown in Fig. 1.3(b) is a 70-storey residential building with a height of 245 m, also utilizing Grade

Fig. 1.4 Concrete filled tubes for high-rise construction.

80 high strength concrete with column size about 2.0 m diameter. The Hong Kong International Commence Centre with 110 Stories and 480 m height was constructed using Grade 90 concrete. WFC Shanghai, as shown in Fig. 1.3(c), utilized Grade 450 steel plate of thickness up to 100 mm for the composite columns. Part of the structure of the Tokyo Sky Tree (Fig. 1.3(d)) in Japan was constructed using Grade 700 steel tubes. These are the strong evidences of using high strength steel and concrete materials in modern high-rise construction.

Fig. 1.4 shows the construction of 50-storey high-rise buildings utilizing concrete filled tubular columns in which the largest column diameter is about 1.5 m. If an ultra-high strength concrete C190 is used, the size of such columns can be reduced approximately by half.

Recent breakthrough in application of high strength steel and concrete was seen in the construction of Techno Station in Tokyo Japan, as shown in Fig. 1.5. The building, which utilized concrete filled tubes with 780 MPa high strength steel and 160 MPa

Fig. 1.5 Techno Station, Tokyo, Japan utilizing Grade 160 concrete and Grade 780 (Endo, 2011).

Fig. 1.6 The Otemachi Tower, Tokyo, Japan (Syuichi et al. 2013).

ultra-high strength concrete, was able to reduce the column dimension from 800 mm (based on normal strength materials) to 500 mm. The design was able to generate large workspace thus unlocking the valuable space for commercial uses.

Fig. 1.6 shows a recently completed building using CFST columns in Japan which consists of office, hotel and retailer space. The building of 187 m in height is considered to be a high-rise building in Japan, which varies in span length at the fourth floor and the 32nd floor. To overcome the challenge of the span changing floors, the CFST columns and mega trusses have been employed. The CFST columns at lower stories comprise of steel tubes of 780 N/mm^2 in tensile strength and concrete of 150 N/mm^2 in compressive strength.

In Japan, many of the buildings were constructed using steel box columns infilled with concrete such as the one shown in Fig. 1.7. The load ratio (axial load/ load carrying capacity of the composite column) of these columns is relatively low in the range of 0.2–0.3 since they are designed primary for earthquake loads. As a result of this, fire protection is normally not required at the external surface of the box columns.

Fig. 1.8 shows the Robinson Tower which comprises a 17-storey 'crystalline' like office tower hovering above an elevated seven-storey retail podium, separated by a striking roof garden. The mega columns at the podium supporting the storey loads above are made of rectangular steel box column infilled with C60/70 concrete.

Concrete Encased Steel (CES) columns have been widely used in the construction of high-rise buildings. They offer advantages in terms of load-carrying capacity, ductility

Fig. 1.7 Concrete filled tubular columns in multi-storey buildings without external fire protection.

Fig. 1.8 Robinson tower Singapore.

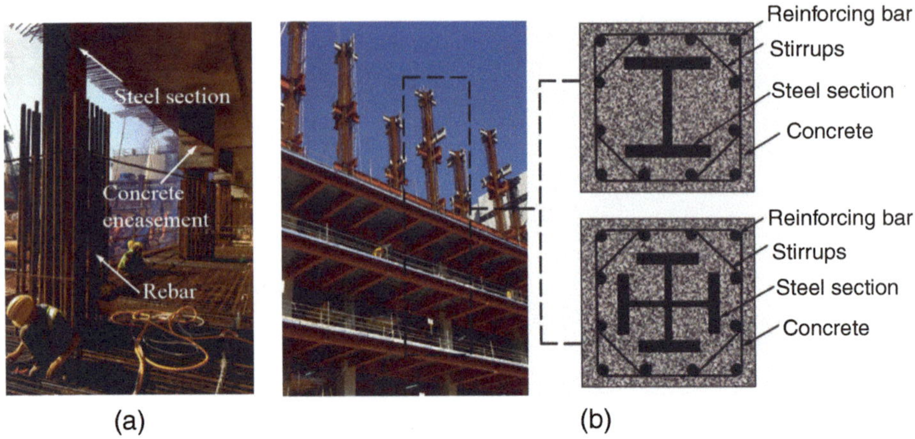

Fig. 1.9 Concrete Encased Steel composite columns used in high-rise buildings: (a) Basement construction; (b) Superstructure construction. (Lai et al., 2020a).

and fire resistance as compared with conventional Reinforced Concrete (RC) columns. In top-down constructions, steel king posts, which are pre-driven to the ground, are often made of steel sections with sufficient capacity to resist the axial force transmitted from superstructures during the construction. After excavation to expose the steel king post, reinforcement cage is installed and finally the steel column is encased by reinforced concrete to form concrete encased steel composite column, in which no additional fire protection is needed. To differentiate CES columns from RC columns and bare steel columns, some limitations on structural steel are stipulated, which vary from code to code. As specified in AISC 360–16, the cross-section area of steel core shall comprise at least 1% of total composite section, while EN 1994–1–1 and AS/NZS 2327 use the steel contribution ratio for judgment, which considers not only cross-section area but also the yield strength of steel section. In both codes, the steel contribution ratio is limited to 0.2 to 0.9, otherwise it shall be regarded as RC columns or steel columns. According to Chinese Code JGJ 138–2016, the steel area ratio shall be within 4% to 15%. If exceeding 15%, more reinforcing bars and hoop bars shall be provided.

A general view of CES columns is shown in Fig. 1.9. CES columns have been successfully used in Singapore in both basement construction and the superstructures above the ground. To provide sufficient confinement effect to the core concrete and ensure every longitudinal bar is laterally supported, multiple stirrups are commonly adopted as given in Fig. 1.9. The shape of steel section also varies from H-shape to cruciform shape, depending on the specific structural requirement.

Fig. 1.10 shows a recently completed (2019) hospital in Singapore constructed with high strength CES columns, which provides 550 hospital beds. This is a 24-storey mix-used building with 4 basements occupying a total floor area of 146,000 m^2. The basement construction employed CES mega columns with C100 concrete and S355 steel, and normal concrete grade was used in the upper storeys. Polypropylene (PP) Fiber of 2 kg/m^3 was added to the high strength concrete mix to prevent concrete spalling in fire situation.

Fig. 1.10 Outram Community Hospital, Singapore, utilizing Grade C100 concrete and S355 steel section.

As the first building utilizing high strength CES columns in Singapore, the construction of Outram Community Hospital was started in 2015 and completed in 2019.

Long-term durability and ductility of high strength materials are two important factors for consideration for their adoption in high-rise construction. The fact that high strength steel and concrete materials have been used in countries which have seismic activities such as U.S.A, Japan, Korea and China, indicates that the ductility issues of such materials could be resolved by research and development of new materials subjecting them to cyclic tests and advanced finite element analyses, and most importantly though stringent control of material quality at the factory and the site. As material research and manufacturing technology improve, the use of higher strength concrete and steel materials will continue to increase as wider applications are being sought in the construction of modern cities.

1.4 Construction method

The use of high strength steel has significant advantages for tall building construction, and the application is beneficial for the construction of mega columns, out-rigger and belt truss system, transfer girders/trusses, king posts for basement and top-down construction as shown in Fig. 1.11.

Concrete encased steel composite column is a preferable option for top–down construction, which is commonly adopted for deep excavation projects including metro station and tunnel construction. Top–down construction is an efficient method to shorten construction time as it enables the superstructure and substructures to be built simultaneously.

The underground space is constructed from top to bottom along with soil excavation. As demonstrated in Fig. 1.11, the diaphragm wall is firstly constructed to isolate the worksite and it also functions as temporary soil retention system in the excavation stage. Bore piles are then drilled deep into the foundation, followed by rebar insertion, concreting work and the embedment of steel king posts. Fig. 1.12 shows the top-down

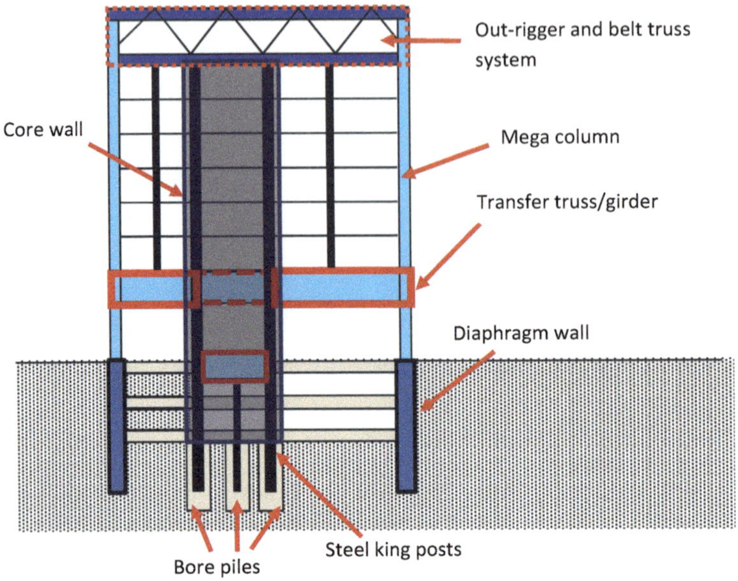

Fig. 1.11 Use of high strength steel in tall building construction.

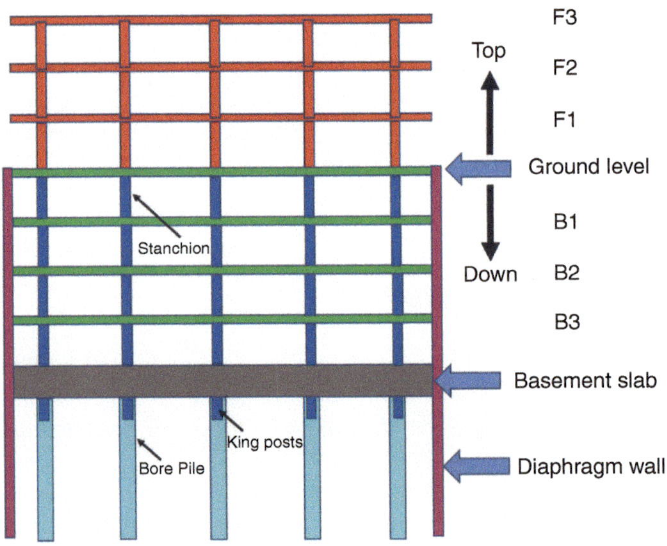

Fig. 1.12 Steel king posts to be encased with concrete to form composite columns for top–down construction.

construction sequence. After the construction of bore piles with plunged in steel king posts, the ground slab is casted and excavation work proceeds downward to form the first level of basement (B1). Same procedures repeat for the excavation and slab casting in the subsequence basement level (B2, B3 etc.). The basement slabs provide lateral bracing to the perimeter diaphragm wall.

Fig. 1.13 Top–down construction: (a) Soil excavation; (b) Reinforcement work of basement slab.

Two site photos showing soil excavation and slab reinforcement work are given in Fig. 1.13. The steel king posts sustaining construction load can be either a single H-section or two H-sections being tied together. After the reinforcement cage installation and concrete casting, composite CES column is formed to resist permanent and imposed loads.

1.5 Design guide

Current design codes for steel-concrete composite columns may be only applicable for normal strength concrete and steel as shown in Table 1.2. For example, Eurocode 4 (EN 1994–1–1, 2004) applies to composite columns with normal weight concrete of strength classes C20/25 to C50/60 and steel grades S235 to S460, American code AISC 360–16 only applies to composite columns with normal weight concrete cylinder strength from 21 MPa to 69 MPa and steel yield strength up to 525 MPa (AISC, 2016), and Chinese code GB 50936 only applies to composite columns with normal weight concrete cylinder strength from 25 MPa to 67 MPa and steel yield strength from 235 MPa to 420 MPa (GB 50936, 2014). Although the Japanese's code allows the use of high strength concrete with compression strength up to 90 N/mm^2 (AIJ, 1997), the steel is limited to normal strength.

Table 1.2 Limitation on characteristic strength (N/mm^2) in modern design codes.

Codes	Steel yield strength (N/mm^2)	Concrete cylinder strength, (N/mm^2)
EN 1994–1–1	235 ~ 460	20 ~ 50
AISC 360–16	≤525	21 ~ 69
GB 50936	235 ~ 420	25 ~ 67
Architectural Institute of Japan (AIJ)	235 ~ 440	18 ~ 90
EN 1992–1–1	–	12 ~ 90
EN 1993–1–1, EN 1993–1–12	235 ~ 700	–

Table 1.2 also shows that the Eurocode 2 (EN 1992–1–1, 2004) for concrete structural design allows Grade 90 concrete and Eurocode 3 (EN 1993–1–1, 2005; EN 1993–1–12, 2007) for steel structural design allows S700 steel, but the Eurocode 4 (EN 1994–1–1, 2004) for steel-concrete composite structural design restricts the concrete strength up to Grade 50 and steel strength up to S460. It is well known that the composite structural members generally exhibit better ductility and higher buckling resistance compared with individual steel or reinforced concrete members. However, Eurocode 4 (design of composite steel and concrete structures) gives narrower range of material strength for steel and concrete compared with Eurocode 2 (design of concrete structures) and Eurocode 3 (design of steel structures). In view of the recent research done on CFST columns and CES columns, the material strength range in Eurocode 4 can be further extended to match those in Eurocodes 2 and 3 in order to fully realize the beneficial effects of composite construction.

A series of design guides on composite structures have been produced by the American, Chinese, European, and Japanese codes. However, they do not provide guidance on the design of CFST and CES members using high strength concrete and high strength steel. This book fills the gap by allowing the design of composite columns with high strength concrete mm^2 and yield strength steel (see Chapter 4). The design is based on Eurocode 4 (EN 1994–1–1, 2004; EN 1994–1–2, 2005) for the design of composite columns with special considerations for the high strength concrete and high strength steel. The design method has been proposed based on the latest research done in the structural laboratory at the National University of Singapore, and the proposed method is benchmarked against the test data collected worldwide as shown in Chapter 3 and numerical results performed using advanced analysis method as reported in Chapter 5.

Special considerations for fire resistance design, fabrication of high strength steel sections, and preparation of high strength concrete are emphasized and guidance are provided in Chapters 6 and 7. Chapter 8 of the book provides some drawings and detailing for various beams to column joints, column spice joints and column base connections involving concrete composite members.

The design calculations involving long formulas and many iterative steps might be too complex that few engineers ever attempt to solve it by hand. Practical design problems can take days or weeks to find an optimum by hand and the opportunity for error is great. Appendices A to C provide the design flowchart, worked examples, and excel spreadsheets to facilitate complex calculations. Those problems in the work examples are solved in seconds with an Excel program and are errorless once the program is tested and proven. Readers are encouraged to test and familiarize with the program before using it for design.

Materials

2

Outline

2.1 Concrete

American Concrete Institute (ACI, 2010) defines high strength concrete as concrete with a cylinder compressive strength greater than 50N/mm^2. The definition of high strength concrete is implicitly reflected in EN 1992–1–2 (2004) for fire resistant design of concrete structures. For high strength concrete, special care is required for production and testing and at which special structural design requirements may be needed. As technology progresses and the use of concrete with even higher compressive strength evolves, the definition of high strength concrete is likely to be revised.

High strength concrete is made possible by reducing porosity, inhomogeneity, and micro-cracks in the hydrated cement paste and in the interfacial transition zone between cement paste and aggregates. The utilization of fine pozzolanic materials in high strength concrete leads to a reduction of the size of the crystalline compounds, particularly, calcium hydroxide. Consequently, there is a reduction of the thickness of the interfacial transition zone. The densification of the interfacial transition zone allows for efficient load transfer between the cement mortar and the coarse aggregate, contributing to the strength of the concrete. For high strength concrete with extremely dense matrix, a weak aggregate may become the weak link for the concrete strength.

It is important to note that high strength concrete (HSC) and high performance concrete (HPC) are not synonymous. HSC exhibits higher strength and secant modulus whereas HPC has their ingredients and mix proportions specifically prepared as to possess properties for the expected use of the structure such as higher strength, better ductility and lower permeability.

The concrete strength classes, as shown in Table 2.1 and Table 2.2, for normal strength concrete (NSC) and HSC can be used for the design of CFST columns and CES columns. Concrete class higher than C90/105 can only be used with further investigation and advices from specialists using performance-based design approach.

Design of Steel-Concrete Composite Structures Using High Strength Materials. DOI: 10.1016/C2019-0-05474-X

Table 2.1 Strength classes of normal strength concrete.

Strength class	**C12/15**	**C16/20**	**C20/25**	**C25/30**	**C30/37**	**C35/45**	**C40/50**	**C45/55**	**C50/60**
Cylinder strength (f_{ck}, **N/mm²**)	12	16	20	25	30	35	40	45	50
Cube strength ($f_{ck,cube}$, **N/mm²**)	15	20	25	30	37	45	50	55	60
Modulus of elasticity (E_{cm},**GPa**)	27	29	30	31	33	34	35	36	37

Table 2.2 Strength classes of high strength concrete.

Strength class	**C55/67**	**C60/75**	**C70/85**	**C80/95**	**C90/105**
Cylinder strength (f_{ck}, **N/mm²**)	55	60	70	80	90
Cube strength ($f_{ck,cube}$, **N/mm²**)	67	75	85	95	105
Modulus of elasticity (E_{cm}, **GPa**)	38	39	41	42	44

This design method is also applicable to lightweight aggregate concrete which exhibits a density of not more than 2200 kg/m^3 or contains a proportion of artificial or natural lightweight aggregates having a particle density of less than 2000 kg/m^3. Where the lightweight aggregate concrete is used, the modulus of elasticity should be determined based on E_{cm} of NSC or HSC with the same value of f_{ck}.

$$E_{lcm} = E_{cm}\left(\rho_c / 2200\right)^2 \tag{2.1}$$

where

ρ_c is the density of lightweight aggregate concrete, kg/m^3

E_{cm} , E_{lcm} are the modulus of elasticity of normal weight concrete and lightweight concrete, respectively

Concrete with the cylinder compressive strength higher than 90N/mm^2 is referred herein as ultra-high strength concrete (UHSC).

2.2 Structural steel

In general, the steel with yield strength lower and equal to 460N/mm^2 is defined as mild steel, whereas those with yield strength higher than 460N/mm^2 are defined as high tensile steel.

Two types of high tensile steel should be distinguished which are mostly used in building and offshore structures. One type is called Quench and tempered (QT) steel which is manufactured based on quenching and tempering processes after rolling. Quenching imparts a high degree of hardness, while leading to a high strength. However, the quenched steel is too brittle to be used. It is returned to the furnace for tempering which will reduce the hardness and then achieve better ductility. The technical delivery conditions for flat products of QT steels are documented by EN 10025–6 (2004). The abbreviation form of delivery condition is Q.

Another type of high tensile steel is TMCP steel which is manufactured based on thermo-mechanically controlled process during rolling, in which case the rolling process is accompanied by a quenching process. The technical delivery conditions of TMCP plates are included in EN 10149–2 (1996). The abbreviation form of delivery condition is M.

The strength classes as shown in Table 2.3 for mild steel and high tensile steel can be used for the design of CFST columns with hot-rolled, cold-formed or welded steel sections. The modulus of elasticity of steel is taken as 210 GPa. For the structural steel materials to be used in buildings, they should be in compliance with the BC1:2012: Design Guide on Use of Alternative Structural Steel to Eurocode 3 (BC1, 2012).

The limiting values of ratio f_u/f_y, elongation at failure and ultimate strain ε_u are recommended in Table 2.4 for mild steel and high tensile steel.

Table 2.3 Strength classes of mild steel and high tensile steel.

Grade	**Nominal values of yield strength f_y (N/mm^2) with thickness (mm) less than or equal to**					
	16	40	63	80	100	150
S235	235	225	215	215	215	195
S275	275	265	255	245	235	225
S355	355	345	335	325	315	295
S420	420	400	390	370	360	340
S460	460	440	430	410	400	380
S500	500	500	480	480	480	440
S550	550	550	530	530	530	490
S620	620	620	580	580	580	560
S690	690	690	650	650	650	630

Table 2.4 Limitations on ductility, elongation at failure and ultimate strain for mild steel and high tensile steel.

Steel	**Ratio f_u/f_y**	**Elongation at failure**	**Ultimate strain**
Mild steel	≥1.10	15%	$\varepsilon_u \geq 15\varepsilon_y$ ($\varepsilon_y = f_y/E$)
High tensile steel	≥1.05	10%	$\varepsilon_u \geq 15\varepsilon_y$

Table 2.5 Maximum permissible plate thickness (mm) for mild steel.

Steel grade	Sub-grade	Charpy Energy		Stress level		
		At T (°C)	J_{min}	$\sigma_{Ed} = 0.75f_y(t)$	$\sigma_{Ed} = 0.50f_y(t)$	$\sigma_{Ed} = 0.25f_y(t)$
S235	JR	20	27	60	90	135
	J0	0	27	90	125	170
	J2	-20	27	125	170	200
S275	JR	20	27	55	80	125
	J0	0	27	75	115	165
	J2	-20	27	110	155	200
	M, N	-20	40	135	180	200
	ML, NL	-50	27	185	200	230
S355	JR	20	27	40	65	110
	J0	0	27	60	95	150
	J2	-20	27	90	135	200
	K2, M, N	-20	40	110	155	200
	ML, NL	-50	27	155	200	210
S420	M, N	-20	40	95	140	200
	ML, NL	-50	27	135	190	200
S460	Q	-20	30	70	110	175
	M, N	-20	40	90	130	200
	QL	-40	30	105	155	200
	ML, NL	-50	27	125	180	200
	QL1	-60	30	150	200	215

Notes: Linear interpolation can be used in applying this table.
Stress level σEd should be calculated as for serviceability limit state taking into account all combinations of permanent and variable actions as defined in EN 1991.

The maximum permissible plate thickness for mild steel and high tensile steel can be determined in accordance with EN 1993–1–10 and EN 1993–1–12, respectively. The maximum thickness of steel plate could be determined based on Table 2.5 and Table 2.6 where a reference temperature of 10 °C is used.

2.3 Reinforcing steel

The yield strength of reinforcing steel is limited to the range of 400 N/mm^2 to 600 N/mm^2 as conforming to EN 1992–1–1 (2004). The plain round bar, Grade 250, is not recommended because it does not offer any advantages over the ribbed reinforcement and seldom used in practice.

The use of reinforcing steel should be in accordance with BS EN 10080 (2005) which however gives no actual specification for strength class. Where it is the case, BS 4449 (2005) may be referred to. The strength class provided in BS 4449 (2005) is shown in Table 2.7 with an elastic modulus of 210 GPa.

It is noted that the characteristic yield strength of reinforcing steel has been increased from the conventional Grade 460 to Grade 500. Nevertheless, Grade 460 reinforcing steel is still allowed in present guideline in accordance with BS 4449 (1997).

Table 2.6 Maximum permissible plate thickness (mm) for high tensile steel.

Steel grade	Sub-grade	Charpy Energy		Stress level		
		At T (°C)	J_{min}	$\sigma_{Ed} = 0.75f_y(t)$	$\sigma_{Ed} = 0.50f_y(t)$	$\sigma_{Ed} = 0.25f_y(t)$
S500	Q	0	40	55	85	145
	Q	−20	30	65	105	170
	QL	−20	40	80	125	195
	QL	−40	30	100	145	200
	QL1	−40	40	120	170	200
	QL1	-60	30	140	200	205
S550	Q	0	40	50	80	140
	Q	−20	30	60	95	160
	QL	−20	40	75	115	185
	QL	−40	30	90	135	200
	QL1	−40	40	110	160	200
	QL1	-60	30	130	185	200
S620	Q	0	40	45	70	130
	Q	−20	30	55	85	150
	QL	−20	40	65	105	175
	QL	−40	30	80	125	200
	QL1	−40	40	100	145	200
	QL1	-60	30	120	170	200
S690	Q	0	40	40	65	120
	Q	−20	30	50	80	140
	QL	−20	40	60	95	165
	QL	−40	30	75	115	190
	QL1	−40	40	90	135	200
	QL1	-60	30	110	160	200

See Table 2.5 for the explanatory notes.

Table 2.7 Strength class of reinforcing steel, BS 4449 (2005).

Class	Characteristic yield strength (f_{yk},MPa)	Ultimate/yield strength ratio	Ultimate elongation
B500A	500	1.05	2.5%
B500B	500	1.08	5.0%
B500C	500	≥1.15, <1.35	7.5%

2.4 Shear connector

The mechanical characteristics and nominal dimensions of shear studs may be referred to BS EN ISO 13918 (2008) and BS EN ISO 898–1 (1999). Weldability and welding examination of shear studs should be checked in accordance with BS EN ISO 14555 (2006).

High strength concrete provides higher confinement for shear studs locally. Thus, the design shear strength of stud, P_{Rd}, as given in EN 1994–1–1 (2004), is deemed to be appropriate to high strength concrete (An and Cederwall, 1996; Yan and Liew, 2013). The design shear resistance of a headed stud may be calculated as:

$$P_{Rd} = \min\left(\frac{0.8 f_{us} \pi d_s^2}{4\gamma_v}, \frac{0.29\alpha_s d_s^2 \sqrt{f_{ck} E_{cm}}}{\gamma_v}\right) \tag{2.2}$$

where

d_s is the diameter of the shank of the shear stud, 16 mm ≤ d_s ≤ 25 mm

f_{us} is the ultimate strength of the shear stud, ≤ 500 N/mm^2

h_{sc} Is the overall height of the shear stud

γ_v is the partial factor which is 1.25 for the ultimate limit state

$$\alpha_s = 0.2\left(\frac{h_{sc}}{d_s} + 1\right) \text{for } 3 \le h_{sc} \le 4$$

$$1.0 \text{ for } h_{sc} > 4$$

f_{ck} is the characteristic cylinder compressive strength of concrete

E_{cm} is the elastic modulus of concrete

When high strength concrete is used, the design resistance of the connector will be governed by the design shear resistance of the stud rather than by the concrete bearing resistance.

Shear connectors other than the stud type, such as weld beads, welded reinforcements, welded shear keys, etc., are also allowed provided they can perform in accordance with the product manufacturer's recommendations or when specialist's advice is consulted.

2.5 Bolts

The characteristic values of yield strength f_{yb} and ultimate tensile strength f_{ub} for bolt classes 4.6, 4.8, 5.6, 5.8, 6.8, 8.8 and 10.9 are given in Table 2.8.

Bolts from class 4.6 up to and including class 10.9 can be used as non-preloaded connections, whereas for preloaded connections, class 8.8 and 10.9 should be used.

For holding-down bolts used in foundation situations, their characteristics with threads from M16 up to and including M64 of product grade C are provided in BS 7419 (2012). Generally, Class 4.6 and class 8.8 as shown in Table 2.8 are used for the holding-down

Table 2.8 Characteristic values of yield strength and ultimate tensile strength for bolts.

Bolt Class	**4.6**	**4.8**	**5.6**	**5.8**	**6.8**	**8.8**	**10.9**	
	f_{yb}(N/mm^2)	240	320	300	400	480	640	900
	f_{ub}(N/mm^2)	400	400	500	500	600	800	1000

bolts. They shall be fabricated from hot-rolled steel conforming to EN 10025–2 (2004), EN 10025–3 (2004) and EN 10,025–4 (2004). Reinforcing steels may be used as the holding-down bolts in which case they shall be in accordance with BS EN 10,080 (2005) and the steel grade shall be specified. The reinforcing steel should not be used as holding-down bolts with yield strength higher than 300 N/mm^2 (EN 1993–1–8, 2005). The design anchorage resistance of the holding-down bolts against pull-out failure, concrete-cone failure, splitting failure, blow-out failure, etc., can be determined in accordance with DDCEN/TS 1992–4–1 (2009) and DDCEN/TS 1992–4–2 (2009).

Test database

3

Outline

3.1 Test database on CFST columns

3.1.1 General

This chapter expands the database of Goode (2008) to include 2033 test results on CFST columns based on the research work done in NUS (Liew et al., 2013). The new test data also includes recent research on steel tubes infilled with ultra-high strength concrete with cylinder compressive strength greater than 90 N/mm^2. The test results are used for comparison with the resistance predicted by EC4 (EN 1994–1–1, 2004) to establish the design guide of using high strength materials for CFST columns.

In this database, test specimens involving short and long CFSTs subjected to compression, uniaxial bending, and bi-axial bending, are categorized for comparison with EC4 predictions. Tests on encased columns, columns with stainless steel and aluminum steel sections are excluded. Results from tests involving preload effect, sustained loading for creep and shrinkage studies and dynamic loadings are not included. In addition, CFST columns with Class 4 slender sections, in which the d/t ratio exceeds the class 4 limit stipulated in EN 1994–1–1, are also excluded although they were included originally in Goode's database.

In this database, the concrete compressive cylinder strength is in the range of 8.5 N/mm^2 to 243 N/mm^2, and the steel yield strength ranges from 178 N/mm^2 to 853 N/mm^2. The ratio of column height over section smaller dimension is in the range of 0.67 to 60, and the relative slenderness $\bar{\lambda}$ ranges from 0.02 to 1.30 which is within the limit of EC4.

The columns, both circular and rectangular, are divided into three groups: axially loaded cross sections, axially loaded columns, and beam-columns in which members are subject to compression and bending. Axially loaded cross section represents short

Design of Steel-Concrete Composite Structures Using High Strength Materials. DOI: 10.1016/C2019-0-05474-X

columns in which their resistances are not affected by the member length and thus the buckling reduction factor χ equal to 1.0. For those with $\chi < 1.0$, they are grouped into axially loaded columns and beam-columns. The bending moment can be directly induced by the compression action, for example where the bending moment results from the eccentricity of the compression force.

3.2 Influence of concrete strength

The average test/prediction ratios against concrete cylinder strength are shown in Table 3.3. The ratios are categorized into three groups based on concrete strength. By studying the values which are not bracketed, the average test/EC4 value for each type of CFST specimens is greater than unity generally indicating a conservative average prediction by EC4. Table 3.3 also provides the percentage of conservative prediction (i.e., test/EC4 prediction ≥ 1.0) which reflects the reliability of the prediction by EC4. For high strength concrete (50 N/mm^2 < f_{ck} ≤ 90 N/mm^2), the percentages of all circular columns and the axially loaded rectangular cross sections are lower than those of their counterparts with normal strength concrete (f_{ck} ≤ 50 N/mm^2). For ultra-high strength concrete (f_{ck} > 90 N/mm^2), the percentages of all columns are lower than those of their counterparts with normal strength concrete. This reflects the increasing complexity and severity as the concrete strength increases.

For high strength concrete with f_{ck} > 50 N/mm^2 the effective compressive strength of concrete in accordance with EC2 (EN 1992–1–1, 2004) should be used. The effective strength is determined by multiplying the characteristic strength by a reduction factor η as given in Eq. (3.1). For ultra-high strength concrete with f_{ck} > 90 N/mm^2, the reduction factor of $\eta = 0.8$ should apply and the increase of concrete strength due to confinement effect from steel tube should be ignored.

$$\eta \begin{cases} 1.0-(f_{ck}-50)/200 & 50\ \text{N/mm}^2 < f_{ck} \le 90\ \text{N/mm}^2 \\ 0.8 & f_{ck} > 90\ \text{N/mm}^2 \end{cases} \tag{3.1}$$

With the introduction of reduction factor η, the effective compressive strengths are given in Table 3.1 for various high strength concrete classes. Accordingly, the secant modulus for high and ultra-high strength concrete should be modified based on the effective strength as in Eq. (3.2). The modified secant moduli are given in Table 3.2 for various high strength concrete classes.

Table 3.1 Effective compressive strengths of high strength concrete.

Strength classes	C55/67	C60/75	C70/85	C80/95	C90/105
Effective compressive strength (N/mm^2)	53.6	57	63	68	72
Reduction ratio	2.5%	5.0%	10.0%	15.0%	20.0%

Table 3.2 Secant modulus of high strength concrete.

Strength classes	C55/67	C60/75	C70/85	C80/95	C90/105
Modified secant modulus (GPa)	38.0	38.6	39.6	40.4	41.1
Reduction ratio	0.7%	1.3%	2.8%	4.3%	5.9%

$$E_{cm} = 22\left[(\eta \cdot f_{ck} + 8)/10\right]^{0.3} \tag{3.2}$$

The predictions of CFST members using EC4 based on the effective strength and modified secant modulus are given in Table 3.3 where values are in the brackets [ ] and < >. It is found that the percentages of all columns with high strength and ultra-high strength concretes are comparable with those of their counterparts with normal strength concrete. The comparable reliability with normal strength concrete has been achieved for high strength concrete and ultra-high strength concrete with the introduction of concrete strength reduction factor η in Eq. (3.1) and neglect of confinement effect for ultra-high strength concrete. Hence, it is recommended that the EC4 limitation on concrete strength could be extended to C90/105 concrete with the reduction factor. However it is not recommended for the ultra-high strength concrete owing to lesser test data. In case where the ultra-high strength concrete is used in the CFST columns, specialist advice should be sought.

Table 3.3 gives the percentages for all test data in terms of concrete strength. Overall, the percentage decreases with increasing concrete strength (refer to values not in the brackets). When the reduction factor η and neglect of confinement are introduced, the percentages of high strength concrete and ultra-high strength concrete are larger than those of their counterparts with normal strength concrete. The design values (refer to values in ( ), { } and ||) are also provided in Table 3.3 with the introduction of partial factors of 1.5 and 1.0 for concrete and steel, respectively. When the codes specified design values are compared with the test results, the percentages of under-prediction (i.e., test/EC4 prediction <1.0) are less than 3%.

The Test/EC4 ratios are also plotted in Fig. 3.1, with the effective strength and modified modulus of elasticity applied for concrete with strength higher than 50 N/mm^2. In order to further extend the EC4 scope to include the ultra-high strength concrete, more test data should be provided for concrete with strength higher than 100 N/mm^2.

3.3 Influence of steel strength

The average test/prediction ratios against steel yield strength are shown in Table 3.4. All the values are based on the reduction factor η in Eq. (3.1) and the neglect of confinement for ultra-high strength concrete. The ratios are categorized into three groups based on steel strength. It is observed that the average ratio of test/EC4 prediction for each type of column with mild steel ($f_y \leq 460$ N/mm^2) is greater than

Table 3.3 Influence of concrete strength on test/EC4 prediction ratios for CFST members.

Type of column		Compressive cylinder strength of concrete		
		≤50 N/mm²	≤90 N/mm²	>90 N/mm²
Axially loaded circular cross section	Nos.	295	130	44
	Test/EC4 ≥ 1	66.8% (99.3%)	59.2% [66.9%]{97.7%}	47.7% < 100% > |100%|
	Av.	1.068 (1.355)	1.023 [1.062]{1.383}	1.016 < 1.298 > |1.646|
	St. Dev.	0.136 (0.169)	0.111 [0.132]{0.190}	0.104 < 0.139 > |0.153|
Axially loaded circular column	Nos.	383	60	22
	Test/EC4 ≥ 1	85.9% (97.4%)	68.3% [83.3%]{98.3}	81.8% < 100% > |100%|
	Av.	1.186 (1.388)	1.039 [1.075]{1.339}	1.085 < 1.195 > |1.512|
	St. Dev.	0.246 (0.267)	0.110 [0.121]{0.162}	0.095 < 0.093> |0.157|
Circular beam-column	Nos.	240	66	46
	Test/EC4 ≥ 1	82.1% (98.8%)	71.2% [81.8%]{98.5%}	69.6% < 78.3% > |91.3%|
	Av.	1.192 (1.352)	1.086 [1.136]{1.356}	1.008 < 1.121 > |1.378|
	St. Dev.	0.217 (0.237)	0.182 [0.189]{0.216}	0.172 < 0.205 > |0.266|
Axially loaded rectangular cross section	Nos.	282	63	39
	Test/EC4 ≥ 1	80.1% (99.6%)	68.3% [90.5%]{96.8%}	56.4% <89.7% > |100%|
	Av.	1.122 (1.287)	1.068 [1.118]{1.330}	1.032 < 1.136 > |1.321|
	St. Dev.	0.150 (0.196)	0.123 [0.117]{0.168}	0.093 < 0.099 > |0.132|
Axially loaded rectangular column	Nos.	101	40	12
	Test/EC4 ≥ 1	62.4% (94.1%)	70.0% [77.5%]{95.0%}	58.3% < 91.7% > |100%|
	Av.	1.059 (1.220)	1.057 [1.099]{1.321}	1.095 < 1.212 > |1.458|
	St. Dev.	0.140 (0.172)	0.134 [0.140]{0.177}	0.206 < 0.193> |0.233|
Rectangular beam-column	Nos.	160	23	27
	Test/EC4 ≥ 1	73.1% (98.1%)	87.0% [87.0%]{100%}	70.4% < 85.2% > |100%|
	Av.	1.107 (1.338)	1.099 [1.128]{1.461}	1.044 < 1.089 > |1.314|
	St. Dev.	0.279 (0.341)	0.112 [0.102]{0.148}	0.115 < 0.117 > {0.124}
All test data	Nos.	1461	382	190
	Test/EC4 ≥ 1	77.3% (98.3%)	67.0% [78.3%]{97.6%}	62.6% < 90.0% > |97.9%|
	Av.	1.133 (1.339)	1.052 [1.094]{1.361}	1.034 < 1.175 > |1.440|
	St. Dev.	0.210 (0.240)	0.132 [0.141]{0.186}	0.132 < 0.165 > |0.224|

Notes: 1) For the value1, (value2), [value3], {value4}, <value5> and |value6| in the table, value1 is based on the characteristic strengths of steel and concrete; (value2) is based on design strengths; [value3] is based on characteristic strengths with reduction factor η for concrete; {value4} is based on design strengths with reduction factor η for concrete. <value5> is based on characteristic strengths with reduction factor η and neglect of confinement for concrete; |value6| is based on design strengths with reduction factor η and neglect of confinement for concrete.
2) The design partial factor is 1.5 and 1.0 for concrete and steel, respectively.
3) This table does NOT include test specimens with class 4 section as in EC4.
4) *Av.* = Average value; St.Dev. = Standard Deviation.

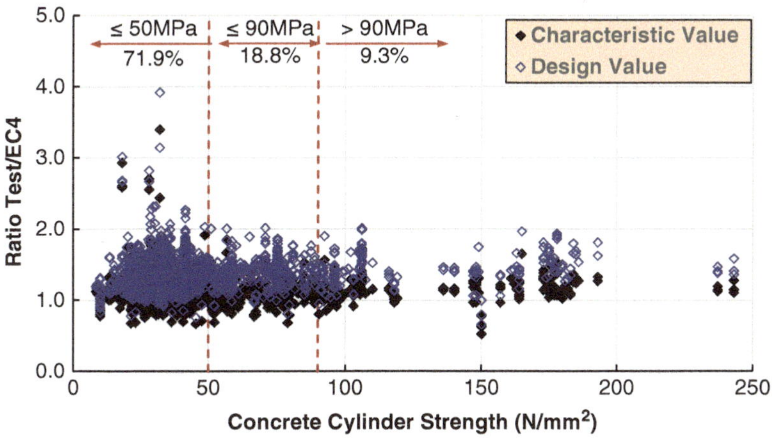

Fig. 3.1 Comparison of test/EC4 prediction ratio against concrete strength for CFST members.

unity. However, this is not true for columns with high tensile steel (f_y > 460 N/mm^2). This might be due to the lack of test data. Thus for the use of high tensile steel, the reliability needs to be further investigated.

Table 3.4 also gives the percent of test/EC4 prediction >1.0 and the average prediction all test data in terms of steel yield strength. The ratios between test and design prediction are also provided. The average test/EC4 prediction ratio for CFST with high tensile steel is higher than those with mild steel, but the standard deviation is higher indicating that wide scattering of results are observed for test specimens involving the use of high tensile steel. When the codes specified design values are compared with the test results, the percentage of under-prediction (i.e., test/EC4 prediction <1.0) is less than 5%, although the average ratio is higher than those with mild steel.

The Test/EC4 ratios are also plotted in Fig. 3.2. More than 90% test data are from CFST columns with mild steels. Test data is insufficient to establish the validity of using the high tensile steels according to EC4. The following section provides additional guideline to limit the use of high tensile steel by selecting matching grades of steel and concrete materials for composite construction.

3.4 Test database on CES columns

3.4.1 General

This design guide compiles a database of CES columns including the experimental research reported in published literatures and the recent experimental work done by the authors (**Lai et al., 2019a, b, c; Lai and Liew, 2020a, b**). The test results are used for comparison with the resistance predicted by EC4 (EN

Table 3.4 Influence of steel strength on test/EC4 prediction ratios for CFST members.

Types of column		Yield strength of steel		
		≤460 N/mm²	≤550 N/mm²	>550 N/mm²
Axially loaded circular cross section	Nos.	450	5	14
	Test/EC4 ≥ 1	71.6% (99.6%)	40.0% (40.0%)	28.6% (100%)
	Av.	1.093 (1.399)	0.922 (1.133)	0.975 (1.180)
	St. Dev.	0.150 (0.189)	0.200 (0.215)	0.068 (0.081)
Axially loaded circular column	Nos.	414	38	13
	Test/EC4 ≥ 1	85.7% (97.6%)	89.5% (100%)	92.3% (92.3%)
	Av.	1.152 (1.378)	1.399 (1.532)	1.160 (1.270)
	St. Dev.	0.167 (0.210)	0.544 (0.526)	0.112 (0.122)
Circular beam-column	Nos.	346	6	–
	Test/EC4 ≥ 1	82.4% (98.0%)	33.3% (83.3%)	–
	Av.	1.175 (1.356)	1.032 (1.367)	–
	St. Dev.	0.211 (0.236)	0.213 (0.268)	–
Axially loaded rectangular cross section	Nos.	308	21	55
	Test/EC4 ≥ 1	84.7% (99.0%)	100% (100%)	65.5% (100%)
	Av.	1.135 (1.324)	1.132 (1.310)	1.048 (1.146)
	St. Dev.	0.147 (0.189)	0.071 (0.096)	0.089 (0.117)
Axially loaded rectangular column	Nos.	145	8	–
	Test/EC4 ≥ 1	67.6% (95.2%)	87.5% (87.5%)	–
	Av.	1.078 (1.262)	1.152 (1.328)	–
	St. Dev.	0.141 (0.181)	0.267 (0.331)	–
Rectangular beam-column	Nos.	187	8	15
	Test/EC4 ≥ 1	73.8% (98.4%)	87.5% (100%)	100% (100%)
	Av.	1.099 (1.338)	1.061 (1.385)	1.221 (1.459)
	St. Dev.	0.255 (0.319)	0.071 (0.119)	0.219 (0.168)
All test data	Nos.	1850	86	97
	Test/EC4 ≥ 1	78.9% (98.3%)	84.9% (94.2%)	69.1% (99.0%)
	Av.	1.128 (1.357)	1.226 (1.410)	1.079 (1.216)
	St. Dev.	0.181 (0.222)	0.409 (0.395)	0.141 (0.165)

Notes: 1) For the value1, (value2) in the table, value1 is based on characteristic strengths of steel and concrete; (value2) is based on design strengths. For concrete with $f_{ck} > 50$ N/mm², the reduction factor η is considered for the concrete compressive strength and the secant modulus of concrete is modified accordingly. For concrete with $f_{ck} > 90$N/mm², confinement effect is ignored.
2) The design partial factor is 1.5 and 1.0 for concrete and steel, respectively.
3) This table does NOT include test specimens with class 4 section as in EC4.
4) *Av.* = Average value; St.Dev. = Standard Deviation.

1994–1–1, 2004) to establish the design guide of using high strength materials for CES columns.

The database comprises 178 test data, which involves short and long CES columns subjected to concentric compression and eccentric compression. Tests on partially encased columns and columns with fiber-reinforced concrete are excluded. Results from tests involving preload effect, sustained loading for creep and shrinkage studies are not included.

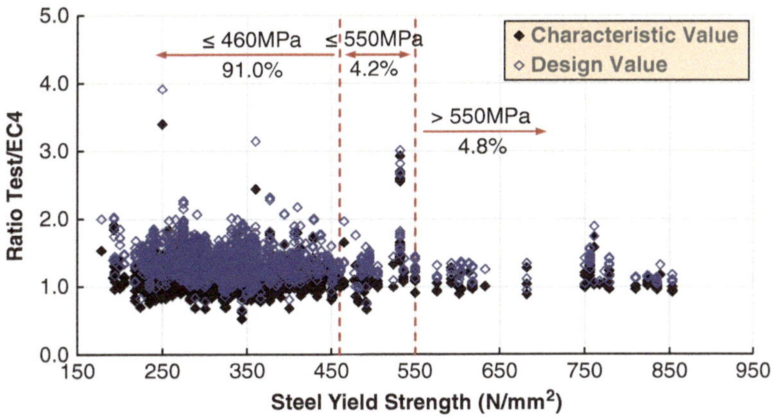

Fig. 3.2 Comparison of test/EC4 prediction ratio against steel strength for CFST members.

In this database, the concrete compressive cylinder strength is in the range of 18 N/mm^2 to 136.6 N/mm^2, and the steel yield strength ranges from 240 N/mm^2 to 913 N/mm^2. The columns are divided into three groups: axially loaded cross sections, axially loaded columns, and beam-columns in which members are subject to compression and bending. Axially loaded cross section represents short columns in which their resistances are not affected by the member length and thus the buckling reduction factor χ equal to 1.0. For those with $\chi < 1.0$, they are grouped into axially loaded columns and beam-columns. The bending moment can be induced by transverse loading applied to columns or directly induced by the compression action, such as the bending moment resulted from the eccentricity of the compression force.

For the CES beam-columns, the methodology proposed by Lai et al. (2019d) is adopted to compare the test results with EC4 prediction. As illustrated in Fig. 3.3, the EC4-predicted result is determined by drawing a straight line from the origin that intersects with the EC4- N-M curve. This is compared with the line from the origin to the test result. The lengths of straight line ending at test result (R_{test}) and ending at the predicted result (R_{pre}) are used for comparison.

3.5 Influence of concrete strength

The average test/prediction ratios against concrete cylinder strength are shown in Table 3.5. The ratios are categorized into three groups based on concrete strength. By studying the values which are not bracketed, the average test/EC4 value for each type of CES specimens greater than unity generally indicates a conservative average prediction by EC4. Table 3.5 also provides the percentage of conservative prediction (i.e., test/EC4 prediction ≥ 1.0) which reflects the reliability of the prediction by EC4. For ultra-high strength concrete ($f_{ck} > 90$ N/mm^2), the percentages

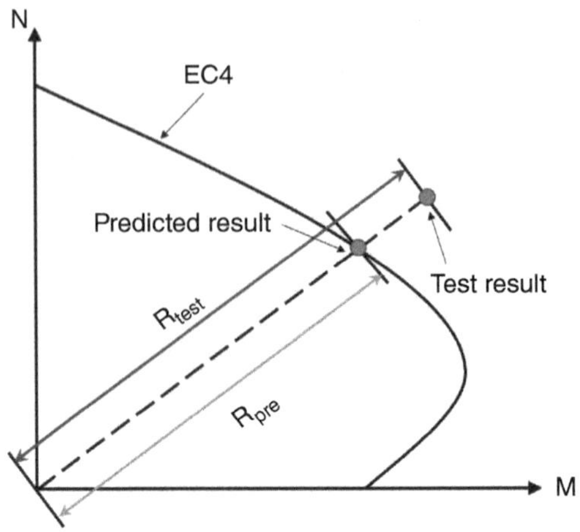

Fig. 3.3 Comparison methodology for CES beam-columns.

Table 3.5 Influence of concrete strength on test/EC4 prediction ratios for CES members.

Type of column		**Compressive cylinder strength of concrete** ≤50 N/mm²	≤90 N/mm²	>90 N/mm²
Axially loaded cross section	Nos.	52	3	26
	Test/EC4 ≥ 1	94.2% (94.2%)	33.3% (66.7%)	42.3% (76.9%)
	Av.	1.127 (1.131)	0.935 (0.982)	0.947 (1.113)
	St. Dev.	0.071(0.074)	0.081 (0.108)	0.097 (0.116)
Axially loaded column	Nos.	15	2	3
	Test/EC4 ≥ 1	93.3% (93.3%)	100% (100%)	100% (100%)
	Av.	1.415 (1.415)	1.785 (1.785)	1.264 (1.382)
	St. Dev.	0.261 (0.261)	0.088 (0.088)	0.087 (0.110)
beam-column	Nos.	63	2	12
	Test/EC4 ≥ 1	84.1% (84.1%)	100% (100%)	16.7% (83.3%)
	Av.	1.194 (1.194)	1.229 (1.324)	0.844 (1.082)
	St. Dev.	0.155 (0.155)	0.009 (0.004)	0.111 (0.096)
All test data	Nos.	130	7	41
	Test/EC4 ≥ 1	89.2% (89.2%)	71.4% (85.7%)	39.0% (80.5%)
	Av.	1.193 (1.194)	1.262 (1.282)	0.940 (1.123)
	St. Dev.	0.170 (0.170)	0.360 (0.345)	0.143 (0.133)

Notes: 1) For the value1, (value2) in the Table, value1 is based on the characteristic strengths of steel and concrete; (value2) is based on characteristic strengths with reduction factor η for concrete (see Eq. (3.1)) and effective yield strength of steel section if material compatibility is not satisfied (see Section 3.3).
2) *Av.* = Average value; St.Dev. = Standard Deviation.

of all columns are lower than those of their counterparts with normal strength concrete. This reflects the increasing complexity and severity as the concrete strength increases.

The calculations of effective compressive strength and secant modulus of high strength concrete introduced in Section 3.1.2 also apply to CES members. The predictions of CES members using EC4 based on the effective strength and modified secant modulus are given in Table 3.5, where values are in the brackets (). It is found that the percentages of all columns with high strength and ultra-high strength concretes are comparable with those of their counterparts with normal strength concrete. The comparable reliability with normal strength concrete has been achieved for high strength concrete and ultra-high strength concrete with the introduction of concrete strength reduction factor, η, in Eq. (3.1). Hence, it is recommended that the EC4 limitation on concrete strength could be extended to C90/105 concrete with the reduction factor. However it is not recommended for the ultra-high strength concrete owing to lesser test data. In case where the ultra-high strength concrete is used in the CES columns, advanced calculation method with proper calibration against test data may be sought in **Lai & Liew (2020 c; 2021).**

The Test/EC4 ratios are also plotted in Fig. 3.4, with the effective strength and modified modulus of elasticity applied for concrete with strength higher than 50 N/mm^2. Although there seems to be a scarcity of experimental research on high strength concrete (50 N/mm^2 < f_{ck} ≤ 90 N/mm^2), the conservative estimation for normal strength concrete (f_{ck} ≤ 50 N/mm^2) and ultra-high strength concrete (f_{ck} > 90 N/mm^2) implies the comparable reliability for high strength concrete. In order to further extend the EC4 scope to include the ultra-high strength concrete, more test data should be provided for concrete with strength higher than 100 N/mm^2.

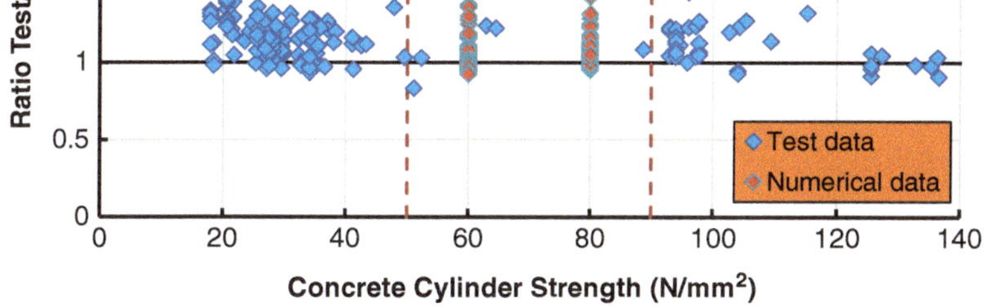

Fig. 3.4 Comparison of test/EC4 prediction ratio against concrete strength for CES members.

Table 3.6 Parameter range in numerical study.

Parameter	Range of parameters
Concrete compressive strength (MPa)	60, 80
Steel section yield strength (MPa)	355, 550
Load eccentricity ratio (e/D); e = load eccentricity; D = cross section dimension	0.1, 0.2, 0.3, 0.4, 0.5, 0.6
Column length L (mm)	2000(L/D = 6.67), 3000(L/D = 10), 4000(L/D = 13.33), 5000(L/D = 16.67), 6000(L/D = 20)
Cross Section type	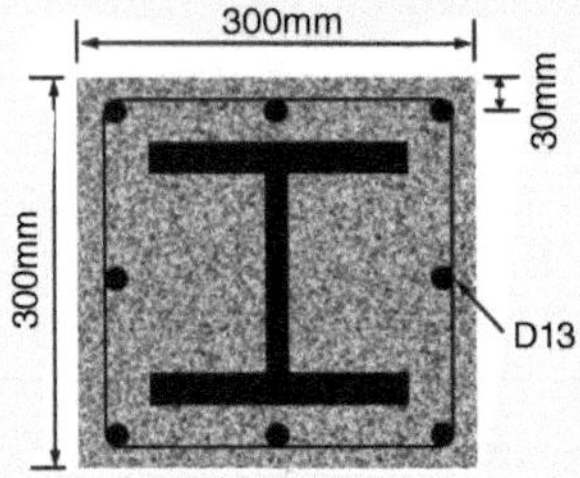 H-200 × 200 × 18 × 10 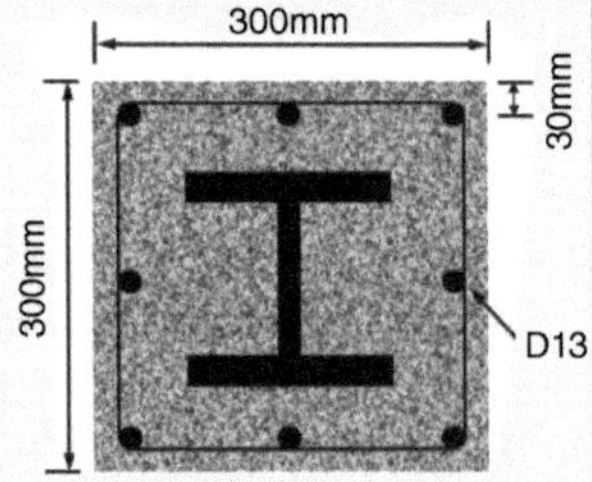 H-150 × 150 × 18 × 10

The test database shown in Fig. 3.4 demonstrates a gap for CES columns with concrete grade spanning C50 to C90 and steel grade within S460 to S550. To fill this gap, additional numerical data is generated using the fiber analysis to enrich the current test data. The accuracy of the fiber analysis method has been validated against the test results. The details of the numerical method are described in Chapter 5, Section 5.4.

Additional numerical analyses are performed based on the parameters described in Table 3.6. The stress-strain relations of concrete, steel section and reinforcing bars used in numerical study are given in detail in **Lai & Liew (2020a).** These numerical data are included in Fig. 3.3 for comparison. It can be observed that the numerical data/prediction ratio is near to 1.0 or higher indicating that the proposed design method is safe to be used for concrete encased steel composite columns with concrete class up to C90/105.

3.6 Influence of steel strength

The average test/prediction ratios against steel yield strength are shown in Table 3.7. All the values are based on the reduction factor η in Eq. (3.1). The ratios are categorized into three groups based on steel strength. It is observed that the average ratio of test/EC4 prediction for each type of column with mild steel ($f_y \leq 460$ N/mm^2)

Table 3.7 Influence of steel strength on test/EC4 prediction ratios for CES members.

Types of column		**Yield strength of steel** ≤460 N/mm²	≤550 N/mm²	>550 N/mm²
Axially loaded cross section	Nos.	70	2	9
	Test/EC4 ≥ 1	84.3% (90.0%)	100% (100%)	0% (66.7%)
	Av.	1.084(1.129)	1.154 (1.240)	0.868 (1.020)
	St. Dev.	0.102 (0.091)	0.025 (0.026)	0.042(0.058)
Axially loaded column	Nos.	20	–	–
	Test/EC4 ≥ 1	95% (95%)	–	–
	Av.	1.429 (1.447)	–	–
	St. Dev.	0.264 (0.258)	–	–
Beam-column	Nos.	65	–	12
	Test/EC4 ≥ 1	84.6% (84.6%)	–	16.7% (83.3%)
	Av.	1.195 (1.195)	–	0.843 (1.082)
	St. Dev.	0.153 (0.153)	–	0.111 (0.096)
All test data	Nos.	155	2	21
	Test/EC4 ≥ 1	85.8% (88.4%)	100% (100%)	9.52% (76.2%)
	Av.	1.175 (1.198)	1.154 (1.240)	0.854 (1.055)
	St. Dev.	0.189 (0.180)	0.025 (0.026)	0.089 (0.087)

Notes: 1) For the value1, (value2) in the Table, value1 is based on the characteristic strengths of steel and concrete; (value2) is based on characteristic strengths with reduction factor η for concrete (see Eq. (3.1)) and effective yield strength of steel section if material compatibility is not satisfied (see Section 3.3).
2) *Av.* = Average value; St.Dev. = Standard Deviation.

is greater than unity. However, for columns with steel yield strength greater than 550 N/mm^2, the average ratio of test/EC4 prediction are less than unity, leading to unsafe design.

Table 3.7 also gives the percent of test/EC4 prediction > 1.0 and the average prediction all test data in terms of steel yield strength. The average test/EC4 prediction ratio for CES columns with high tensile steel is lower than those with mild steel, indicating the yield strength may not be fully utilized when high tensile steel is used, which will be elaborated in Section 3.3.

The Test/EC4 ratios are also plotted in Fig. 3.5. About 90% test data are from CES columns with mild steels. Test data is insufficient to establish the validity of using the high tensile steels according to EC4. To fill the gap of physical tests on CES columns with steel yield strength between 460 MPa to 550 MPa, numerical data was supplemented in Fig. 3.5 using fiber section analysis. The parameters selected in numerical simulation was described in Table 3.6. It can be observed that the numerical data/prediction ratio is near to 1.0 or higher indicating that the proposed design method is safe to be used for concrete encased steel composite columns with steel grade up to S550. In addition, Section 3.3 provides additional guideline to limit the use of high tensile steel by selecting matching grades of steel and concrete materials for composite column construction.

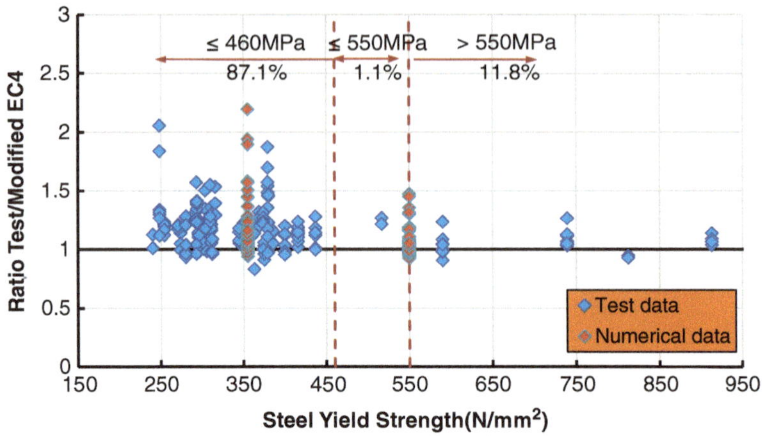

Fig. 3.5 Comparison of test/EC4 prediction ratio against steel strength for CES members.

3.7 Material compatibility between steel grade and concrete class

For high strength concrete filled steel tubular columns and concrete encased steel columns subjected to compression, it is necessary to ensure that yielding of the steel section occurs before the concrete core reaches its maximum stress. Otherwise, the full plastic resistance of the composite section cannot be achieved due to brittle failure of high strength concrete after reaching the maximum stress. Hence, the selections of steel grade and concrete class have to ensure that the yield strain of steel is smaller than the compressive strain of concrete at the peak stress. The yield strain of steel and the strain of concrete at peak stress may be calculated in accordance with EN 1992–1–1 (2004) and EN 1993–1–1 (2005) as:

$$\text{Steel yield strain}: \varepsilon_y = f_y / E_a \tag{3.3}$$

$$\text{Concrete strain at peak stress}\ (\text{‰}): \varepsilon_{c1} = 0.7 f_{cm}^{0.31} < 2.8 \tag{3.4}$$

where $f_{cm} = f_{ck} + 8$ is the mean compressive strength of concrete at 28 days, in N/mm^2. $E_a = 210$ GPa is the elastic modulus of steel tube.

The yield strain of steel and the strain of concrete at peak stress are respectively shown in Table 3.8 and Table 3.9. It is noted that the calculation for the strain of concrete at peak stress ignores the confinement effect from the steel tubes for CFST columns. If the increase of strain by confinement is taken into account, higher steel grade could be used.

Table 3.10 gives recommendation on the matching grades of steel and concrete suitable for use in CFST and CES columns such that $\varepsilon_y < \varepsilon_{cl}$. It is recommended that the steel tubular sections and encased steel sections up to Grade S550 may

Table 3.8 Yield strain of steel.

Steel Grade	f_y	ε_y (‰)
S235	235	1.12
S275	275	1.31
S355	355	1.69
S420	420	2.00
S460	460	2.19
S500	500	2.38
S550	550	2.62
S620	620	2.95
S690	690	3.29

Table 3.9 Strain of concrete at peak stress.

Concrete Class	f_{ck}	f_{cm}	ε_{c1} (‰)
C12/15	12	20	1.77
C16/20	16	24	1.87
C20/25	20	28	1.97
C25/30	25	33	2.07
C30/37	30	38	2.16
C35/45	35	43	2.25
C40/50	40	48	2.32
C45/55	45	53	2.40
C50/60	50	58	2.46
C55/67	55	63	2.53
C60/75	60	68	2.59
C70/85	70	78	2.70
C80/95	80	88	2.80
C90/105	90	98	2.90

be used with concrete grade up to C90/105, although test evidence by Liew et al. (2013) shows that the strain of concrete at peak stress of CFST is very much higher than the concrete without any confinement. In case where higher strength concrete and high strength steel not shown in Table 3.10 need to be used in CFST and CES columns, advanced calculation method such as those reported in Chapter 5 may be used with special attention to validate the accuracy of the predicted results.

Alternatively, the effective yield strength of steel can be determined according to the concrete characteristic strength with strength class up to C90/105 using the following expression:

$$f_{y,eff} = 0.7E_a\left(f_{ck} + 8\right)^{0.31} \tag{3.5}$$

Table 3.10 Compatibility between steel and concrete in CFST and CES columns.

	S235	S275	S355	S420	S460	S500	S550	S620	S690
C12/15	√	√	√	×	×	×	×	×	×
C16/20	√	√	√	×	×	×	×	×	×
C20/25	√	√	√	×	×	×	×	×	×
C25/30	√	√	√	√	×	×	×	×	×
C30/37	√	√	√	√	×	×	×	×	×
C35/45	√	√	√	√	√	×	×	×	×
C40/50	√	√	√	√	√	×	×	×	×
C45/55	√	√	√	√	√	√	×	×	×
C50/60	√	√	√	√	√	√	×	×	×
C55/67	√	√	√	√	√	√	×	×	×
C60/75	√	√	√	√	√	√	×	×	×
C70/85	√	√	√	√	√	√	√	×	×
C80/95	√	√	√	√	√	√	√	×	×
C90/105	√	√	√	√	√	√	√	×	×

Notes: "√" indicates compatible materials and "×" is not recommended.

where

$fy_{,eff}$ is the effective yield strength of steel

f_{ck} is the characteristic cylinder compressive strength of concrete

E_a is the modulus of Elasticity of steel

In case where the steel yield strength of CES specimens in the database exceeds the effective yield strength calculated using Eq. (3.5), the predicted results plotted in Fig. 3.4 and Fig. 3.5 are based on the effective yield strength instead of the steel yield strength.

Design of steel-concrete composite columns considering high strength materials

4

Outline

4.1 General

This design guide applies to CFST and CES columns with steel grade up to S550 and concrete class up to C90/C105. The partial factors for concrete, steel and reinforcement are shown in Table 4.1 (EN 1992–1–1, 2004; EN 1993–1–1, 2005; EN 1993–1–8, 2005; EN 1994–1–1, 2004).

The scope of the present method is limited to the types of CFST and CES sections shown in Fig. 4.1. It is not applicable to laced or battened CFST columns which consist of two or more discontinuously connected sections. Where it is the case, the two sections must be braced to ensure that shear deformation is negligible and does not affect the overall stability of the composite column.

The ratio of the depth to the width of the composite cross-section should be within the limits of 0.2 and 5.0. The longitudinal reinforcement that may be used in calculation should not exceed 6% of the concrete area. Longitudinal reinforcement in concrete filled tube may not be necessary if it is not required for fire resistance.

Design of Steel-Concrete Composite Structures Using High Strength Materials. DOI: 10.1016/C2019-0-05474-X

Table 4.1 Partial factors of materials.

Concrete	Steel	Reinforcement	Shear Connector
$\gamma_c = 1.5$	$\gamma_a = 1.0$	$\gamma_s = 1.15$	$\gamma_v = 1.25$

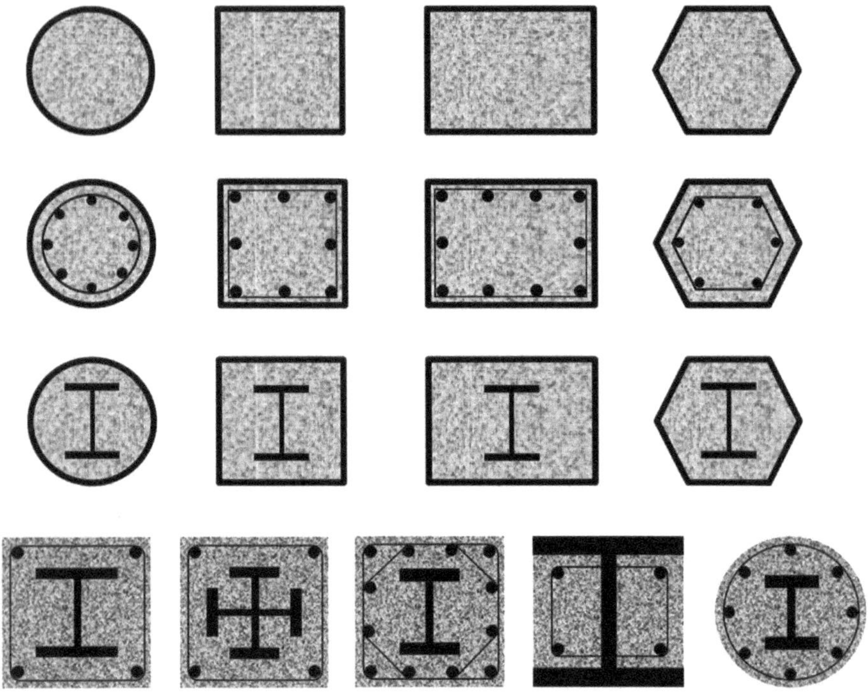

Fig. 4.1 Types of double symmetric CFST and CES sections.

The steel contribution ratio δ is defined as:

$$\delta = \frac{A_a f_{yd} + A_e f_{ed}}{N_{pl,Rd}} \tag{4.1}$$

A_a, A_e are the cross-sectional area of steel tube and encased steel section respectively

f_{yd}, f_{ed} are the design strength of steel tube and encased steel section respectively

$N_{pl,Rd}$ is the axial resistance of composite section defined by Eq. (4.3)

The steel contribution ratio δ should fulfill the following condition:

$$0.2 \leq \delta \leq 0.9 \tag{4.2}$$

The composite columns limited in the proposed design guide should be checked for:

- Resistance to local bucking in accordance with Section 4.2
- Resistance of cross-section in accordance with Section 4.3
- Resistance of member in accordance with Section 4.4
- Resistance of longitudinal shear in accordance with Section 4.5
- Introduction of loads in accordance with Section 4.6

The general method as given in EN 1994–1–1 (2004) may also be extended to composite columns with high strength materials. The general method could be implemented by means of advanced finite element analysis. The details of which are not provided in this design guide.

4.2 Local buckling

The effects of local buckling may be neglected for a steel section fully encased in concrete of CFST columns or CES columns. For outer tubes and partially encased steel sections, the maximum values of Table 4.2 should not be exceeded.

The local buckling of outer tube could be prevented by increasing the plate thickness. Alternatively, stiffener plates to be welded on the steel plate in the longitudinally direction of the column, as shown in Fig. 4.2, may be used to reduce the effective width of the steel plates.

Table 4.2 Maximum values (*d*/*t*), (h/t) for local buckling.

Concrete filled Cross-section	Maximum value in expressions	Maximum values according to steel grades						
		S275	S355	S460	S500	S550	S620	S690
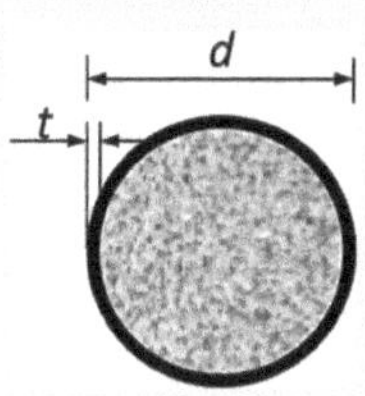	$(d/t)_{max} = 90\frac{235}{f_y}$	77	60	46	42	38	34	30
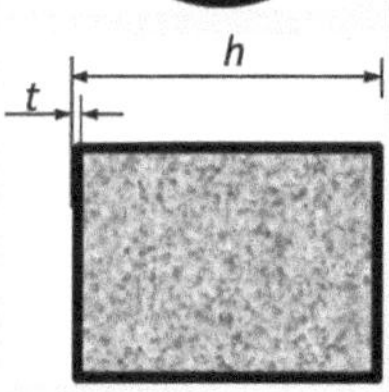	$(h/t)_{max} = 52\sqrt{\frac{235}{f_y}}$	48	42	37	35	34	32	30
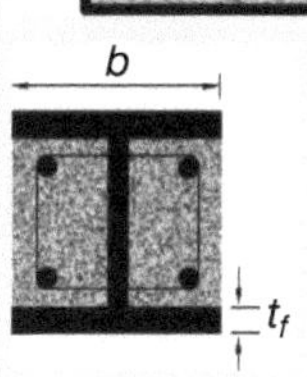	$(b/t_f)_{max} = 44\sqrt{\frac{235}{f_y}}$	41	36	31	30	29	27	26

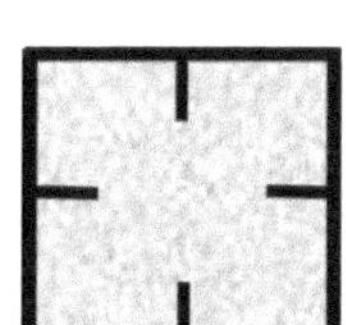

Fig. 4.2 Prevention of local buckling of outer tube of CFST columns using plate stiffeners.

4.3 Resistance of cross sections

4.3.1 Resistance to compression

The plastic design resistance to compression of a composite column cross-section $N_{pl,Rd}$ is calculated by adding the plastic resistances of its components:

$$\text{For CFST section: } N_{pl,Rd} = A_a f_{yd} + A_c f_{cd} + A_s f_{sd} + A_e f_{ed} \tag{4.3a}$$

$$\text{For CES section: } N_{pl,Rd} = 0.85 A_c f_{cd} + A_s f_{sd} + A_e f_{ed} \tag{4.3b}$$

where

A_a, A_c, A_s, A_e are the cross-sectional area of steel tube, concrete, reinforcements and encased steel section respectively

f_{yd}, f_{cd}, f_{sd}, f_{ed} are the design strength of steel tube, concrete, reinforcements and encased steel section respectively

For circular CFST columns, the increase of concrete strength due to confinement effect may be allowed provided that the relative slenderness $\bar{\lambda}$ of the member does not exceed 0.5 and $e/D < 0.1$, where e is the eccentricity of loading given by M_{Ed}/N_{Ed} and D is the external dimension of the column section. The plastic resistance to axial compression considering the confinement effect may then be calculated as:

$$N_{pl,Rd} = \eta_a A_a f_{yd} + A_c f_{cd}\left(1 + \eta_c \frac{t}{d}\frac{f_y}{f_{ck}}\right) + A_s f_{sd} + A_e f_{ed} \tag{4.4}$$

where

f_y,f_{ck} are the characteristic strength of steel tube and concrete, respectively

t is the thickness of the steel tube

$$\eta a = \eta_{a0} + (1 - \eta_{a0})(10e/D) \text{ for } e/D \le 0.1$$

$$1.0 \text{ for } e/D > 0.1$$

$$\eta_{a0} = 0.25(3 + 2\bar{\lambda}) \text{ but } \le 1.0$$

$$\eta_c = \eta_{c0}(1 - 10e/d) \text{ for } e/D \le 0.1$$

$$1.0 \text{ for } e/D > 0.1$$

$$\eta_{c0} = 4.9 - 18.5\bar{\lambda} + 17\bar{\lambda}^2 \text{ but } \ge 0$$

For high strength concrete (f_{ck} > 50N/mm^2), the compressive strength f_{ck} or f_{cd} should be reduced by the reduction factor of η given in Eq. (3.1).

4.3.2 Resistance to shear forces

The shear forces $V_{a,} E_d$, $V_{e,} E_d$, $V_{c,} E_d$ acting on the outer steel tube, internal steel section and concrete section, respectively, can be calculated as:

$$V_{a,ED} = V_{Ed} \frac{M_{pl,a,Rd}}{M_{pl,Rd}} \tag{4.5}$$

$$V_{e,Ed} = V_{Ed} \frac{M_{pl,e,Rd}}{M_{pl,Rd}} \tag{4.6}$$

$$V_{c,Ed} = V_{Ed} - V_{a,Ed} - V_{e,Ed} \tag{4.7}$$

where

$M_{pl,a,Rd}$ is the plastic moment resistance of the steel tube

$M_{pl,e,Rd}$ is the plastic moment resistance of the inner steel section

V_{Ed} is the design shear force

For simplification, V_{Ed} may be assumed to act on the steel tube and inner steel section only. Thus, the shear forces $V_{a,Ed}$, $V_{e,Ed}$ can be calculated as:

$$V_{a,Ed} = V_{Ed} \frac{M_{pl,a,Rd}}{M_{pl,a,Rd} + M_{pl,e,Rd}} \tag{4.8}$$

$$V_{e,Ed} = V_{Ed} \frac{M_{pl,e,Rd}}{M_{pl,a,Rd} + M_{pl,e,Rd}} \tag{4.9}$$

In case where the shear force $V_{a,Ed}$ on the steel section exceeds 50% of its design shear resistance $V_{pl,a,Rd}$ of the steel tube, or the shear force on the inner steel section $V_{e,Ed}$ exceeds 50% of its design shear resistance, $V_{pl,e,Rd}$, the influence of transverse shear forces on the resistance to bending and normal force should be considered when determining the interaction curve. The consideration should be taken into account by a reduced design steel strength $(1 - \rho) f_{yd}$ for the steel tube or $(1 - \rho) f_{ed}$ for the encased steel section in their shear areas. Alternatively, the consideration can be taken into account by a reduced web thickness of the shear area as shown in Fig. 4.3.

The reduction factor ρ can be calculated as:

$$\text{For the outer steel tube } \rho = \left(\frac{2V_{a,Ed}}{V_{pl,a,Ed}} - 1 \right)^2 \tag{4.10}$$

$$\text{For the internal steel section } \rho = \left(\frac{2V_{e,Ed}}{V_{pl,e,Ed}} - 1 \right)^2 \tag{4.11}$$

The design shear resistance $V_{pl,a,Rd}$ and $V_{pl,e,Rd}$ can be calculated as:

$$\text{For the outer steel tube } V_{pl,a,Rd} = \frac{f_{yd} A_{V,a}}{\sqrt{3}} \tag{4.12}$$

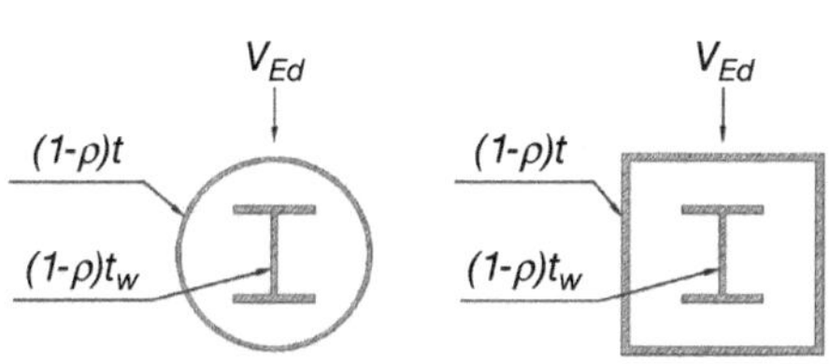

Fig. 4.3 Reduced web thickness of shear area.

$$\text{For the internal steel section } V_{pl,e,Rd} = \frac{f_{ed} A_{V,e}}{\sqrt{3}} \tag{4.13}$$

where

$A_{V,a}$, $A_{V,e}$ are the shear areas of the steel tube and the internal steel section which are determined in accordance with EN 1993–1–1 (2005)

4.3.3 Resistance to combined compression and bending

The resistance of a cross-section to combined compression and moments may be calculated based on interaction curve assuming rectangular stress blocks as shown in Fig. 4.4, taking account of the design shear force in accordance with Section 4.3.2. The tensile strength of the concrete may be neglected.

As a simplification, the interaction curve is a polygonal diagram as shown in Fig. 4.5. The plastic stress distributions of a CFST cross section or CES cross section for the points A, B, C and D are also show in Fig. 4.5.

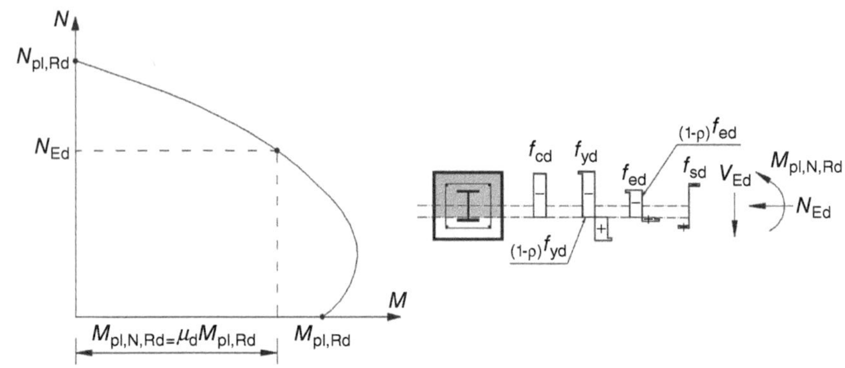

Fig. 4.4 Interaction curve for combined compression and bending.

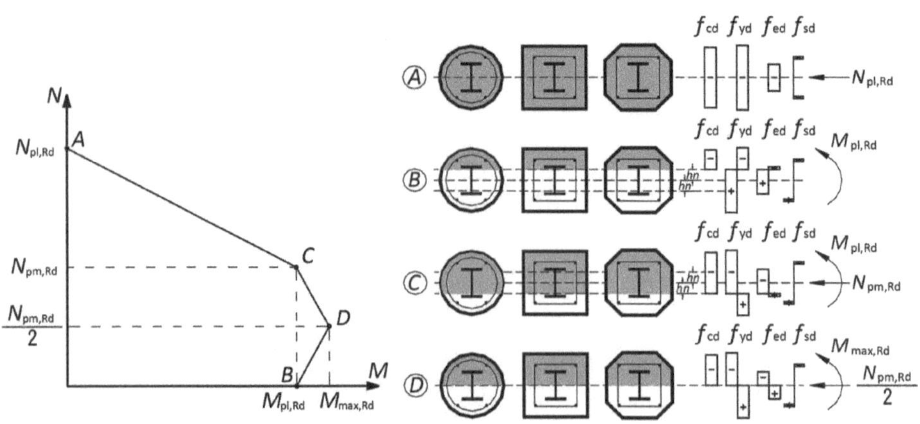

Fig. 4.5 Simplified interaction curve and corresponding stress distributions.

Table 4.3 Section analysis of circular CFST column.

Point	Defining equations
	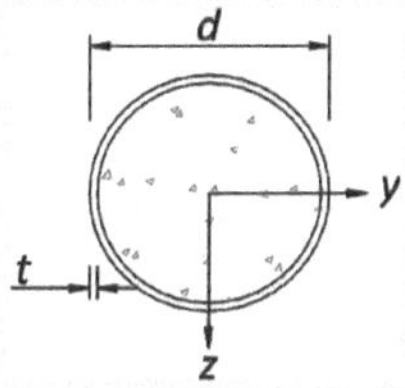
A	$N_{pl,Rd} = \eta_a A_a f_{yd} + A_c f_{cd}\left(1+\eta_c \frac{t}{d}\frac{f_y}{f_{ck}}\right)$
B	$h_n = \frac{A_c f_{cd}}{2df_{cd} + 4t\left(2f_{yd} - f_{cd}\right)}$ $W_{pc} = \frac{(d-2t)^3}{6}$ $W_{pc,n} = (d-2t)h_n^2$ $W_{pa} = \frac{d^3}{6} - W_{pc}$ $W_{pa,n} = dh_n^2 - W_{pc,n}$ $M_{pl,Rd} = \left(W_{pa} - W_{pa,n}\right)f_{yd} + 0.5\left(W_{pc} - W_{pc,n}\right)f_{cd}$
C	$N_{pm,Rd} = A_c f_{cd}\left(1 = \eta_c \frac{t}{d}\frac{f_y}{f_{ck}}\right)$
D	$M_{pl,Rd} = W_{pa} f_{yd} + 0.5 W_{pa} f_{cd}$

Note: For high strength concrete ($f_{ck} > 50\text{N/mm}^2$), the compressive strength f_{ck} or f_{cd} should be reduced by the reduction factor of η given in Eq. (3.1).

The design formulae for the section analysis of circular and rectangular CFST columns can be determined from Table 4.3 and Table 4.4 respectively. The design formulae for CES section can be determined from Table 4.5.

4.4 Resistance of members

4.4.1 Resistance to compression

For a member subject only to axial compression, Clause 6.7.3.5(2) in EN 1994–1–1 enables buckling curves to be used. This is a useful simplification because these curves allow for member initial imperfections. The design effect due to axial compression N_{Ed} in a member should satisfy:

$$\frac{N_{Ed}}{\chi N_{pl,Rd}} \leq 1 \tag{4.14}$$

Table 4.4 Section analysis of rectangular CFST column.

Point	Defining equations
	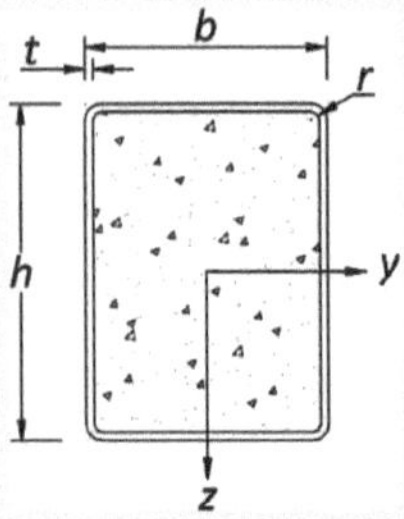
A	$N_{pl,Rd} = A_a f_{yd} + A_c f_{cd}$
B	$h_n = \dfrac{A_c f_{cd}}{2bf_{cd} + 4t\left(2f_{yd} - f_{cd}\right)}$ $W_{pc} = \dfrac{(b-2t)(h-2t)}{4} - \dfrac{2}{3}r^3 - r^2(4-p)\left(\dfrac{h}{2} - t - r\right)$ $W_{pc,n} = (b-2t)h_n^2$ $W_{pa} = \dfrac{bn^2}{4} - \dfrac{2}{3}(r+t)^3 - (r+t)^2(4-p)\left(\dfrac{h}{2} - t - r\right) - W_{pc}$ $W_{pc,n} = bh_n^2 - W_{pc,n}$ $M_{pl,Rd} = \left(W_{pa} - W_{pa,n}\right)f_{yd} + 0.5\left(W_{pc} - W_{pc,n}\right)f_{cd}$
C	$N_{pm,Rd} = A_c f_{cd}$
D	$M_{pl,Rd} = W_{pa} f_{yd} + 0.5_{pc} f_{cd}$

Note: For high strength concrete ($f_{ck} > 50\text{N/mm}^2$), the compressive strength f_{ck} or f_{cd} should be reduced by the reduction factor of η given in Eq. (3.1).
For bending about the weak axis, the dimensions b and h are interchangeable.

For axial compression in members the value of χ for the appropriate non-dimensional slenderness $\bar{\lambda}$ should be determined from the relevant buckling curve according to:

$$\chi = \frac{1}{\Phi + \sqrt{\Phi^2 - \lambda^{-2}}} \text{ but } \chi \leq 1 \tag{4.15}$$

where

$\Phi = 0.5\left[1 + \alpha\left(\bar{\lambda} - 0.2\right) + \bar{\lambda}^2\right]$

α = the imperfection factor

$\bar{\lambda}$ = the relative slenderness for the plane of bending and equal to $\sqrt{N_{pl,Rk} / N_{cr}}$

$N_{pl,}R_k$ = the characteristic value of the plastic resistance to compression $N_{pl,Rd}$ in which the material characteristic strengths rather than the design strength should be used.

N_{cr} = the elastic critical normal force for the relevant buckling mode

Table 4.5 Section analysis of CES column.

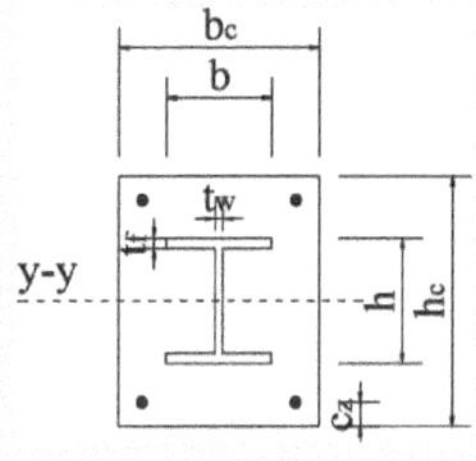

Point	Defining equations
A	$N_A = A_a f_{yd} + A_s f_{sd} + A_c(\alpha_c f_{cd})$
B	a) For h_n below the flange $\left(h_n < \frac{h}{2} - t_f\right)$:
	$h_n = \frac{(A_c + A_s)\alpha_c f_{cd} - 2A_s f_{sd}}{2\left[(b_c - t_w)\alpha_c f_{cd} + 2t_w f_{yd}\right]}; W_{pan} = t_w h_n^2$
	b) For h_n within the flange $\left(\frac{h}{2} - t_f < h_n < \frac{h}{2}\right)$:
	$h_n \frac{(A_c + A_a + A_s - bh)\alpha_c f_{cd} - 2(A_a - bh) f_{yd} - 2A_s f_{sd}}{2\left[(b_c - b)\alpha_c f_{cd} + 2bf_{yd}\right]}$
	$W_{pan} = bh_n^2 - \frac{(b - t_w)(h - 2t_f)^2}{4}$
	c) For h_n above the flange $\left(h_n > \frac{h}{2}\right)$:
	$h_n = \frac{(A_c + A_a + A_s)\alpha_c f_{cd} - 2A_a f_{yd} - 2A_s f_{sd}}{2b_c(\alpha_c f_{cd})}$
	$W_{pan} = W_{pa}$ = plastic section modulus about y-axis
	$W_{psn} = \sum_{i=1}^{n} \lvert A_{s,i} e_{z,i} \rvert; W_{pcn} = bch_n^2 - W_{pan} - W_{psn}$
	$N_B = 0$
	$M_B = M_D - W_{pan} f_{yd} - W_{psn} f_{sd} - \frac{1}{2} W_{pcn}(\alpha_c f_{cd})$
C	$N_C = A_c(\alpha_c f_{cd})$
	$M_C = M_B$

(Continued)

Table 4.5 *(cont'd)*

Point	Defining equations
D	$W_{pa} = bt_f\left(h - t_f\right) + \frac{\left(h - 2t_f\right)^2 t_w}{4}$ $W_{ps} = As\left(\frac{h_c}{2} - c_z\right)$ $W_{pc} = \frac{1}{4}b_c h_c^2 - w_{pa} - w_{ps}$ $N_D = \frac{1}{2}A_c\left(\alpha_c f_{cd}\right)$ $M_D = W_{pa} f_{yd} + W_{ps} f_{sd} + \frac{1}{2}W_{pc}\left(\alpha_c f_{cd}\right)$

Note: For high strength concrete ($f_{ck} > 50$ N/mm^2), the compressive strength f_{ck} or f_{cd} should be reduced by the reduction factor of η given in Eq. (3.1).
For CES column: $\alpha_c = 0.85$

Table 4.6 Imperfection factors for buckling curves.

Buckling curve	a_0	**a**	**b**	**c**	**d**
Imperfection factor	0.13	0.21	0.34	0.49	0.76

The imperfection factor α corresponding to an appropriate buckling curve should be obtained from Table 4.6.

The buckling curves and member imperfections for CFST and CES composite columns can be determined from Table 4.7.

The elastic critical normal force N_{cr} for the relevant buckling mode is determined by:

$$N_{cr} = \frac{\pi^2 (EI)_{eff}}{L_{eff}^2} \tag{4.16}$$

L_{eff} is the buckling length of a composite column for the relevant buckling mode. In the absence of Eurocode guidance, buckling lengths given in BS 5950: Part 1 (2000) are therefore recommended as shown in Table 4.8, where L is the system length of composite column.

The boundary conditions and corresponding buckling lengths are illustrated in Fig. 4.6.

The effective flexural stiffness of a composite column $(EI)_{eff}$ may be calculated as:

$$(EI)_{eff} = E_a I_a + E_s I_s + E_e I_e + 0.6_{cm} I_c \tag{4.17}$$

where

I_a, I_c, I_s, I_e are the second moments of area of the steel tube, the un-cracked concrete, the reinforcements and the encased steel section for the bending plane being considered

Table 4.7 Buckling curves and member imperfections for CFST and CES composite columns.

Cross section	Limits	Axis of buckling	Buckling curve	Member imperfection
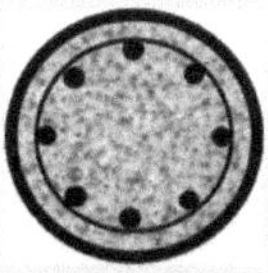	$\rho_s \le 3\%$	any	*a*	*L*/300
	$\rho_s > 3\%$	any	*b*	*L*/200
	–	any	*b*	*L*/200
	–	any	*b*	*L*/200
	–	major minor	*b* *c*	*L*/200 *L*/150
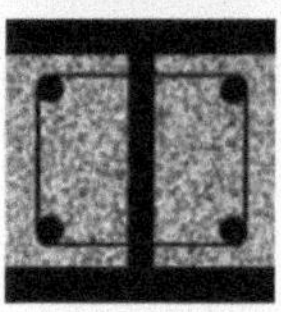	–	major minor	*b* *c*	*L*/200 *L*/150

Note: ρ_s is the area ratio of reinforcements relative to the concrete area.

E_a, E_{cm}, E_s, E_e are the modulus of elasticity of the steel tube, the un-cracked concrete, the reinforcements and the encased steel section.

The influence of long-term effects on the effective flexural stiffness $(EI)_{eff}$ should be accounted for. The modulus of elasticity of concrete E_{cm} should be reduced to the value $E_{c,eff}$ in accordance with the following equation:

$$E_{c,eff}=E_{cm}\frac{1}{1+(N_{G,Ed}/N_{Ed})\varphi_t} \tag{4.18}$$

where

$N_{G,Ed}$ is the part of the normal force that is permanent

φ_t is the creep coefficient

Table 4.8 Buckling lengths for composite columns.

a) Non-sway mode			
End restraint in the plane under consideration by other parts of the structure			L_{eff}
Effectively held in position at both ends	**1)** Effectively restrained in direction at both ends		0.7 *L*
	2) Partially restrained in direction at both ends		0.85 *L*
	3) Restrained in direction at one end		0.85 *L*
	4) Not restrained in direction at either end		1.0 *L*
b) Sway mode			
One end	Other end		L_{eff}
Effectively held in position and restrained in direction	Not held in position	**5)** Effectively restrained indirection	1.2 *L*
		6) Partially restrained indirection	1.5 *L*
		7) Not restrained in direction	2.0 *L*

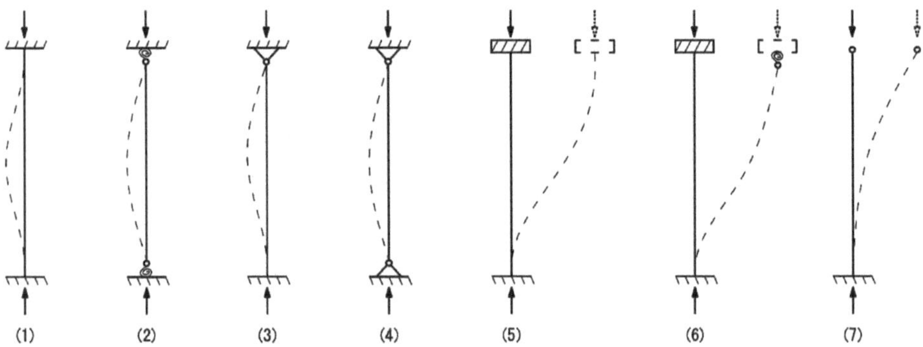

Fig. 4.6 Buckling lengths for composite columns.

It should be noted that for high strength concrete ($f_{ck} > 50$ N/mm^2), a reduced E_{cm} value should be used in accordance with Table 3.2.

BC2 (2008) allows for reduced creep effects with increasing concrete strength. Thus, the creep coefficient φ_t is conservatively taken as that for normal strength concrete when high strength concrete is used. The creep coefficient φ_t may be calculated from (EN 1992–1–1, 2004):

$$\varphi_t = \varphi_0 \beta_c^{(t,t_0)} \tag{4.19}$$

where

$$\varphi_0 = \varphi_{RH} \beta^{(f_{cm})} \beta^{(t_0)}$$

$$\varphi_{RH} = 1 + \frac{1 - RH/100}{0.1\sqrt[3]{h_0}} \text{ for } f_{cm} \leq 35MPa$$

$$\left[1+\frac{1-RH/100}{0.1\sqrt[3]{h_0}}\alpha_1\right]\alpha_2 \text{ for } f_{cm} > 35MPa$$

RH is the relative humidity of the ambient environmental in percent

$$\beta^{f_{(cm)}} = \frac{16.8}{\sqrt{f_{cm}}}$$

f_{cm} is the mean compressive strength of concrete at the age of 28 days and equal to (f_{ck} + 8) N/mm^2

$$\beta^{(t_0)} = \frac{1}{0.1+t_0^{0.2}}$$

$$h_0 = \frac{2A_c}{u}$$

A_c is the cross-sectional area of concrete
u is the perimeter of the concrete section

$$\beta_c^{(t,t_0)} = \left(\frac{t-t_0}{\beta_H+t-t_0}\right)^{0.3}$$

$$= 1.0 \text{ when } t \rightarrow \infty$$

t is the age of concrete in days at the moment considered
t_0 is the age of concrete in days at first loading in days

$$\beta_H = 1.5\left[1+(0.012RH)^{18}\right]h_0 + 250 \leq 1500 \text{ for } f_{cm} \leq 35\ MPa$$

$$1.5\left[1+(0.012RH)^{18}\right]h_0 + 250\alpha_3 \leq 1500\alpha_3 \text{ for } f_{cm} > 35\ MPa$$

$$\alpha_1 = \left[\frac{35}{f_{cm}}\right]^{0.7}$$

$$\alpha_2 = \left[\frac{35}{f_{cm}}\right]^{0.2}$$

$$\alpha_3 = \left[\frac{35}{f_{cm}}\right]^{0.5}$$

In practice, the age of concrete at the moment considered, t, can be conservatively taken as infinity. For the age of concrete on first loading by effects of creep, although EN 1994–1–1 (2004) recommends t_0 = 1 day, it is actually the judgement of designer to determine t_0 since it makes quite a difference whether this age is assumed to be 1 day or 1 month. When t_0 > 100, its effect on creep coefficient is not significant and it is sufficiently accurate to assume t_0 = 100.

For concrete infilled in steel tubes, the relative humidity *RH* can be assumed to be 50% conforming to EN 1992–1–1 (2004) regarding condition of concrete inside the steel tube, while it can be assumed as 80% for concrete encased steel columns to be used in a tropical environment.

4.4.2 Resistance to combined compression and uniaxial bending

The following expression based on the interaction curve determined according to Section 4.3.3 should be satisfied:

$$\frac{M_{Ed}}{M_{pl,N,Rd}} = \frac{M_{Ed}}{\mu_d M_{pl,Rd}} \leq \alpha_M \tag{4.20}$$

where

M_{Ed} Is the greatest of the end moments and the maximum bending moment within the column length, including imperfections and second order effects

$M_{pl,N,Rd}$ Is the plastic bending resistance taking into account the normal force N_{Ed}, given by $\mu_d M_{pl,Rd}$

$M_{pl,Rd}$ Is the plastic bending resistance given by Point B in Section 4.3.3.

$\alpha_M = 0.9$ for *S*235, *S*275, *S*355

0.8 for other steel grades

Within the column length, second-order effects may be allowed for by multiplying the greatest first-order design bending moment by a factor *k* given by:

$$k = \frac{\beta}{1 - N_{Ed} / N_{cr,eff}} \tag{4.21}$$

Second-order effect should be considered for both moments from initial member imperfection and from first-order analysis, as illustrated in Fig. 4.7 (Johnson and Anderson, 2004).

Thus, the design moment, considering second-order effect, is calculated as:

$$M_{Ed} = k_0 N_{Ed} e_0 + k_1 M_{Ed,1} \geq M_{Ed,1} \tag{4.22}$$

where

$M_{Ed,1}$ Is the maximum first-order design moment in column length

e_0 Is the member imperfection, given by Table 4.7

The equivalent moment factor β can be determined as given in Table 4.9.

$N_{cr,eff}$ is the critical normal force for the relevant axis and corresponding to the effective flexural stiffness with the effective length taken as the column length. It can be calculated as:

$$N_{cr,eff} = \frac{\pi^2 (EI)_{eff,II}}{L^2} \tag{4.23}$$

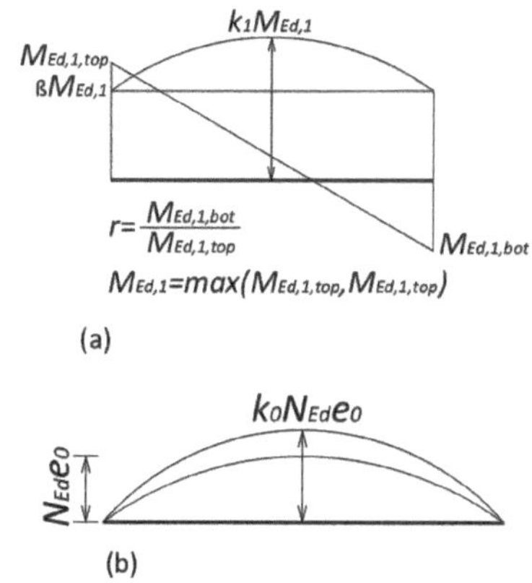

Fig. 4.7 Amplifications for moments from first-order analysis and member imperfection. (a) Moment from first-order analysis (b) Moment from imperfection.

Table 4.9 Equivalent moment factor β.

Moment distribution	**Moment factors β**	**Comment**
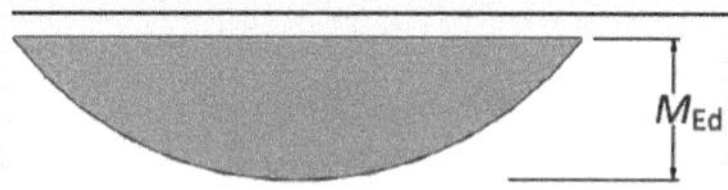 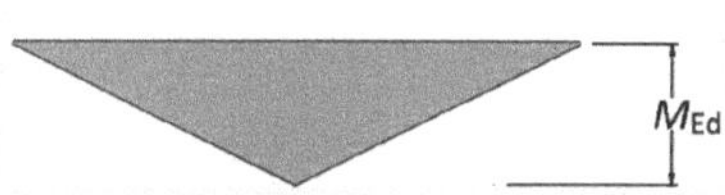	First-order bending moment from member imperfection or lateral load: $\beta = 1.0$	M_{Ed} is the maximum bending moment within the column length ignoring second-order effects
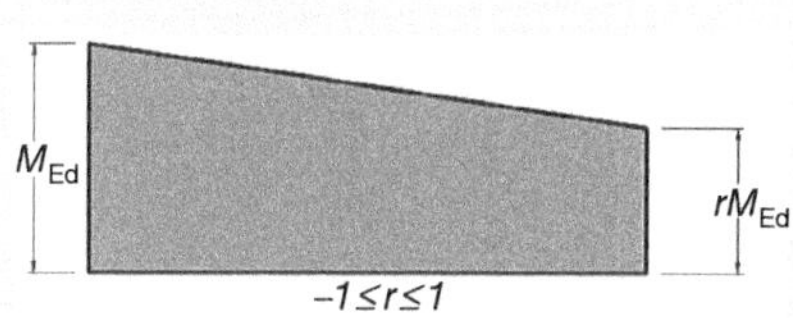	End moments: $\beta = max(0.44,\ 0.66 + 0.44r)$	M_{Ed} and rM_{Ed} are the end moments from first-order or second-order global effects

The design value of effective flexural stiffness $(EI)_{eff,II}$ should be determined from the following expression:

$$(EI)_{eff,II} = 0.9\left(E_a I_a + E_s I_s + E_e I_e + 0.5 E_{c,eff} I_c\right) \tag{4.24}$$

4.4.3 Resistance to combined compression and biaxial bending

For combined compression and biaxial bending, the following conditions should be satisfied for the stability check within the column length and for the check at the column ends:

$$\frac{M_{y,Ed}}{\mu_{dy} M_{pl,y,Rd}} \le \alpha_{M,y} \tag{4.25}$$

$$\frac{M_{z,Ed}}{\mu_{dz} M_{pl,z,Rd}} \le \alpha_{M,z} \tag{4.26}$$

$$\frac{M_{y,Ed}}{\mu_{dy} M_{pl,y,Rd}} + \frac{M_{z,Ed}}{\mu_{dz} M_{pl,z,Rd}} \le 1 \tag{4.27}$$

$M_{y,Ed}$, $M_{z,Ed}$ are the design bending moments around $y-y$ or $z-z$ axis including second-order effects and imperfects

$M_{pl,y,Rd}$, $M_{pl,z,Rd}$ are the plastic bending resistances around $y-y$ or $z-z$ axis

$\alpha_{M,y}$, $\alpha_{M,z}$ = 0.9 for $S235$, $S275$, $S355$

0.8 for other steel grades

The value $\mu_d = \mu_{dy}$ or μ_{dz} as shown Fig. 4.8 in refers to the design plastic resistance moment $M_{pl,Rd}$ for the plane of bending being considered. Values μ_d greater than 1.0 should only be used where the bending moment M_{Ed} depends directly on the compression force N_{Ed}, for example where the moment M_{Ed} results from an eccentricity of the normal force N_{Ed}.

For composite columns and compression members with biaxial bending the values μ_{dy} and μ_{dz} as shown in Fig. 4.8 may be calculated separately for each axis. Imperfections should be considered only in the plane in which failure is expected to occur. If it is not evident which plane is the more critical, checks should be made for both planes. Irrespective of axis, the value μ_d can be interpolated according to Fig. 4.8.

$$N_{Ed} \le N_{pm,Rd}/2: \quad \mu_d = 1 + \frac{2N_{Ed}}{N_{pm,Rd}}\left(\frac{M_{max,Rd}}{M_{pl,Rd}} - 1\right) \tag{4.28}$$

$$N_{pm,Rd}/2 < N_{Ed} \le N_{pm,Rd}: \quad \mu_d = 1 + \frac{2N_{pm,Rd} - N_{Ed}}{N_{pm,Rd}}\left(\frac{M_{max,Rd}}{M_{pl,Rd}} - 1\right) \tag{4.29}$$

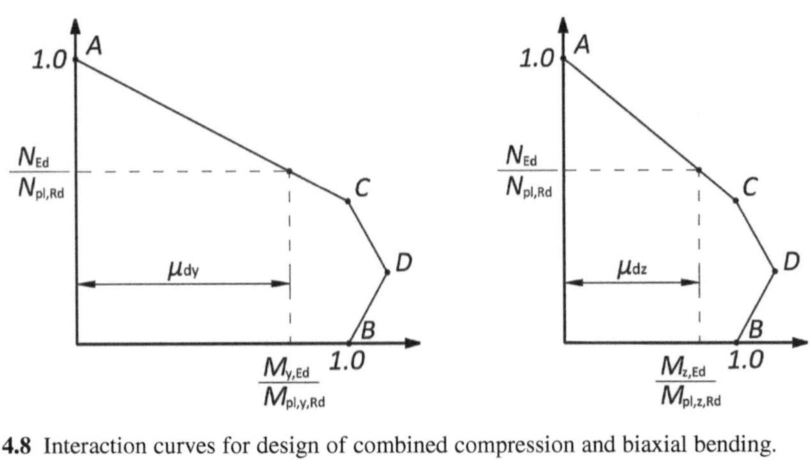

Fig. 4.8 Interaction curves for design of combined compression and biaxial bending.

$$N_{Ed} > N_{pm,Rd}: \quad \mu_d = \frac{N_{pl,Rd} - N_{Ed}}{N_{pl,Rd} - N_{pm,Rd}} \tag{4.30}$$

4.5 Longitudinal shear

The longitudinal shear at the interface between concrete and steel should be checked if it is caused by transverse loads and/or end moments. The design shear strength, τ_{Rd}, depends on surface characteristics of the steel sections, tensile strength, and confinement of the concrete. For high strength concrete, the design shear strength is deemed to be higher than that of normal strength concrete owing to higher tensile strength (Radhika and Baskar, 2012; Roseline and Tensing, 2013). Provided that the surface of the steel section in contact with the concrete is unpainted and free from oil, grease and loose scale or rust, the design shear strength at interface of steel and concrete can be taken as the values in Table 4.10.

The values of design shear strength given in Table 4.10 is for a minimum concrete cover of 40 mm. For greater concrete cover, higher values of τ_{Rd} may be used with an amplification factor, β_c, given as:

$$\beta_c = \min\left[1 + 0.02c_z\left(1 - \frac{40}{c_z}\right), 2.5\right] \tag{4.31}$$

where

c_z is the nominal value of concrete cover in mm

4.6 Load introduction

Shear studs should be provided in the load introduction area if the design shear strength is exceeded at the interface between concrete and steel tube or encased steel section. In absence of a more accurate method, the introduction length should not exceed $2d$ or $L/3$ as shown in Fig. 4.9, where d is the minimum transverse dimension of the CFST column and L is the system length of the column. The design shear strength of stud is given in Section 2.4.

Table 4.10 Design shear strength at the interface between concrete and steel.

Type of cross section	τ_{Rd} (MPa)
Fully concrete encased steel sections	0.30
Circular hollow sections	0.55
Rectangular hollow sections	0.40
Flanges of partially encased section	0.20
Webs of partially encased section	0.00

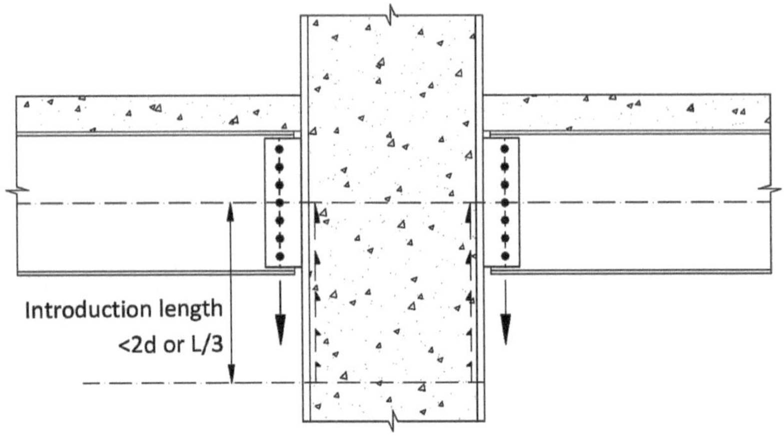

Fig. 4.9 Load introduction area.

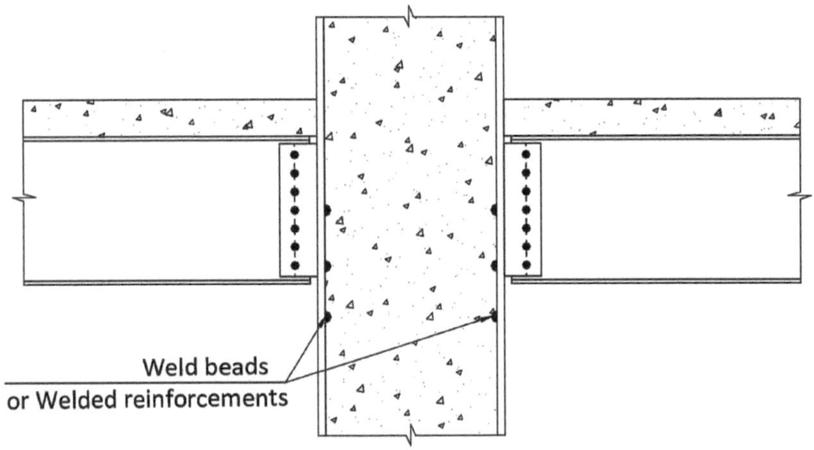

Fig. 4.10 Load introduction to concrete filled steel tubes by weld beads or welded reinforcements.

Alternatively, the shear studs can be replaced by welded reinforcements or weld beads as shown in Fig. 4.10 and Fig. 4.11 (Miyao et al., 1997; Takaki et al., 1997; Takaki et al., 1999). The design shear resistance is calculated as:

$$P_{Rw} = \beta_w N_w A_w f_{cN} \tag{4.32}$$

where

$\beta_w = 1.54 - 0.0143D / t_c$, for $D / t_c > 55$, $\beta_w = 0.7535$

D is the column diameter

t_c is the thickness of column tube

Fig. 4.11 Practical use of weld beads as load introduction to concrete filled steel tubes (Courtesy of JFE Corporation).

N_w is the number of welded reinforcements or weld beads, ≤3
A_w is the projected cross-sectional area of welded reinforcement or weld bead
f_{cN} is the bearing strength of concrete, = $min(A_c/A_w,5)f_{cd}$
f_{cd} is the design strength of concrete
A_c is the cross-sectional area of concrete

4.7 Differential shortening

During construction, differential shortening will occur between columns and core walls due to different material strengths, stress levels, and long-term creep and shrinkage of concrete. The differential shortening may be insignificant and generally ignored for low rise buildings. However, for high-rise buildings it should be taken into account since the cumulative shortening may result in cracks of cladding, uneven floor finishes, plumbing lines and roof waterproof systems. It may also disable the operation of lifts owing to inclined lift shafts. The differential shortening is mainly induced by long-term creep and shrinkage. It depends on axial stress level, age at first loading, and their evolutions in time. Generally, there are three ways to reduce the effects of creep and shrinkage in terms of material preparation, design and construction considerations.

Aggregates play an important role in both creep and shrinkage. Increase of fraction, size and modulus of aggregates would cause a decrease of creep as well as shrinkage. In addition, high strength concrete exhibits less creep than normal strength concrete. In situation of large differential column shortening, high strength concrete is recommended.

More accurate analysis may be used to determine the differential shortening between columns and walls, considering the effects of time dependent creep and shrinkage strains. The time dependent creep and shrinkage strains of concrete can be calculated in accordance with EN 1992–1–1. Generally, sequential construction in the said analyses should be considered since construction rate and method affect the magnitudes of creep and shrinkage strain. When the differential shortening is significant according to the analyses, it can be reduced by adjusting column size, concrete strength and steel contribution ratio, etc. Alternatively, the column length can be corrected based on the calculated differential shortening. The correction may not necessarily be done for each storey since the differential shortening is generally small per floor. In practice, correction for differential shortening is done only after the construction of several storeys.

In terms of construction consideration, the central core walls are generally constructed ahead before the construction of perimeter columns. As a result, the early stage creep and shrinkage are eliminated when the perimeter columns relate to the core walls by floor beams or outriggers. Simple connections allowing for vertical slip can be used for the floor beams and outriggers to relief the internal forces induced by the differential shortening. In case where rigid connections are adopted, the connections can be made simple in the construction stage and become rigid after the creep and shrinkage have sufficiently developed. This means that the internal forces at supports are redistributed to other locations which should be strengthened accordingly.

Fig. 4.12 shows the outrigger truss connecting the internal concrete core wall to a parameter column at the outer edge of the building. The outrigger to core wall connection is a rigid connection, and the outrigger to exterior column connection can be either a rigid or hinged (pinned) connection; the latter is to ensure that exterior column carry only axial force. Movement joint is provided to allow vertical movement between the outrigger truss and the external column during construction so that the permanent and construction loads will not be transferred to the outrigger truss during construction. The gap for the movement joint will be welded after the construction so that the outrigger truss is functioning as part of the lateral load resisting system.

Fig. 4.12 Allowing for column shortening in outrigger truss construction. (a) Outrigger and belt truss system (b) Movement joint to allow vertical shortening of column during construction.

4.8 Summary

Based on calibration with 2033 test data for CFST columns and 178 test data for CES columns, the current EC4 method can be safely extended to the design of concrete filled steel tubular (CFST) columns and concrete encased steel (CES) columns with steel strength up to 550 N/mm^2 and concrete compressive cylinder strength up to 90 N/mm^2, with the following modifications:

- Class 4 steel section should not be allowed to avoid local buckling of steel section. For fully encased steel composite section, such requirement does not apply.
- Matching grades of steel and concrete materials must be observed. Table 3.10 or Eq. (3.5) may be used as a guide to select the grade of steel and class of concrete for the design of CFST columns to avoid the crushing of the core concrete before yielding of steel
- Strength reduction factor, η, (refer to Eq. 3.1) should be applied for high strength concrete with cylinder compressive strength greater than 50 N/mm2. Accordingly, the secant modulus of concrete should be modified based on Eq. (3.2)

Although the proposed design method may be applied to CFST columns and CES columns with ultra-high strength concrete with compressive cylinder strength higher than 90 N/mm^2, more tests are needed to justify the use of the present method. A conservative approach to design the CFST columns with the ultra-high strength concrete is to adopt the concrete strength reduction factor η equal to 0.8 and ignore the concrete confinement effect by the steel tube.

Steel with grades higher than 550 N/mm^2 may also be used, provided a more accurate assessment on the concrete strain at peak stress, considering the tri-axial confinement effect from the steel tube, is carried out.

Design flow charts are given in Appendix A for the design of concrete filled tubular members and concrete encased steel members with an extension of Eurocode 4 Method to C90/105 Concrete and S550 Steel. Work examples are provided in Appendix B to illustrate the design principles highlighted in this chapter. Finally, spreadsheet calculations are provided in Appendix C and readers can download the spreadsheet programs from the website and check the output against the work examples to familiar with the programs.

Behaviour and analysis of high strength composite columns

5

Outline

5.1 General

The previous chapter outlines the simplified design method for high strength composite members subject to compression and bending. This chapter presents background information and advanced calculation methods which include geometric and material nonlinearities to study the fundamental behavior of composite members under different loading scenarios. These methods may be used when the assumptions inherited in the simplified design method are not satisfied, and for composite members with asymmetric sections or non-uniform cross section along its length. The numerical models are calibrated against the experimental results to predict the cross-section capacity and load-displacement response of composite columns. Section 5.2 deals with concrete encased composite columns. A model to capture the premature failure of concrete cover and concrete confinement due to the stirrups and steel section is proposed to predict the load displacement response of concrete encased composite steel members. Section 5.3 investigates the confinement of concrete in concrete filled steel tubular members. The proposed numerical methods provide insights to some of the pertaining issues related to the high strength steel-concrete composite members. It is noteworthy that only static loading is concerned while the dynamic and seismic behavior are excluded pending for future research.

Design of Steel-Concrete Composite Structures Using High Strength Materials. DOI: 10.1016/C2019-0-05474-X

5.2 Concrete encased steel members

5.2.1 Premature cover spalling of high strength concrete

5.2.1.1 Cover spalling phenomenon

Premature cover spalling is a peculiar problem of high strength concrete loaded under high compressive stress due to concentric compression or eccentric compression with very small eccentricity (**Forster et al. 1998; 2001**). This issue was initially discovered in high strength Reinforced Concrete (RC) columns and recently observed in high strength concrete encased steel (CES) composite columns (**Lai et al. 2019 b, c**). As shown in Fig. 5.1, spalling refers to the phenomenon where a concrete cover peels off from e column and exposes reinforcing bars to external environment. When high strength concrete is used, spalling tends to occur suddenly and abruptly with debris flying out, leading to a sudden drop of load-carrying capacity and axial stiffness.

As illustrated in Fig. 5.2(a), the loss of concrete cover occurring before the attainment of the concrete compressive strength (f_{co}) is regarded as premature cover spalling. Due to the inherent brittleness nature of high strength concrete, the cover component instantaneously spalls off and completely detaches from the column. Thereafter, the concrete cover is no longer participated in the load-carrying capacity. The compressive stress could thus be assumed to suddenly drop to zero.

As a typical failure mode of concrete structures, cover spalling is also observed in normal strength concrete, however it does not occur in such a brittle manner as that observed in high strength concrete. As presented in Fig. 5.2(b), spalling of normal strength concrete cover is often occurred at the post-peak stage (**Mander et al., 1988**), and does not exert significant influence on the column behavior. In addition, the confinement effect of normal strength concrete is much more significant than that of high strength concrete, which to some extent offsets the adverse effect induced by cover spalling.

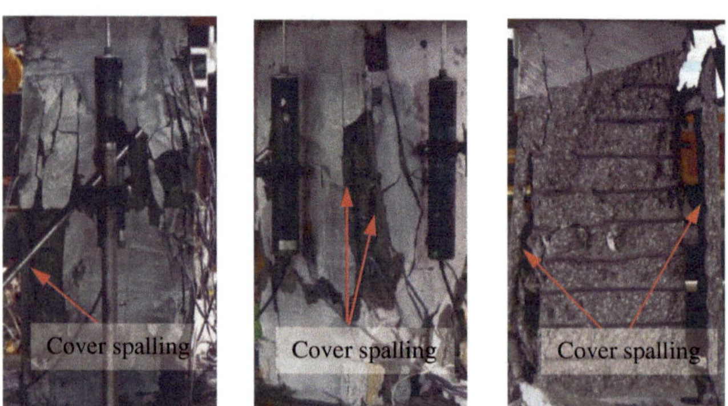

Fig. 5.1 Concrete cover spalling in high strength CES columns.

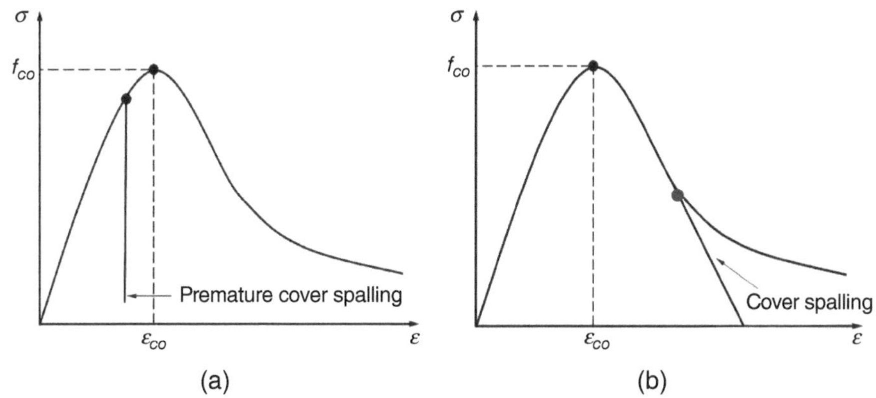

Fig. 5.2 Different cover spalling between high strength and normal strength concrete (a) High strength concrete (b) Normal strength concrete.

5.2.1.2 Cover spalling mechanism

The occurrence of cover spalling is triggered by the propagation of splitting cracks generated along the interface formed by reinforcement cage. As illustrated in Fig. 5.3, when compressively loaded, concrete also expands laterally because of Poisson effect. The core concrete expansion is restrained by the stirrups whereas the concrete cover deforms freely without restraint. This difference in lateral deformation results in tensile stress at the concrete core-cover interface and produces tensile cracks. Splitting crack widens and propagates with compression force increasing, and finally leads to complete cover spalling. In the meantime, the longitudinal reinforcing bar may undergo small lateral deformation which exerts certain pushing force to the concrete cover. Other sources of cover spalling mainly include the differential shrinkage between concrete cover and core concrete **(Collins et al., 1993)**, however it seems no sufficient experimental proof is available to support this conjecture.

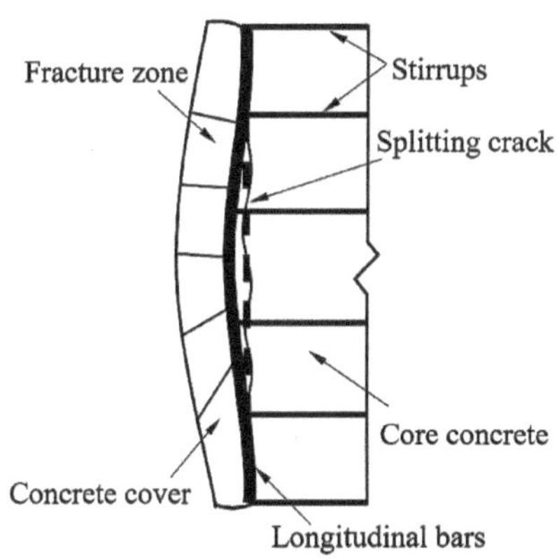

Fig. 5.3 Illustration of concrete cover spalling (Maekawa et al., 2003).

Due to the complex of interaction mechanism between concrete and reinforcing bars, as well as the different properties brought by different raw materials (especially coarse aggregate), and so far, no mature model is well-established to quantifiably predict the spalling behavior. Nevertheless, for design purpose, the cover thickness is recommended to be kept as small as possible but enough to satisfy the requirement of corrosion and fire protection. This is to minimize the loss of axial capacity due to the premature spalling of thick concrete cover. From material perspective, adding steel fiber or other materials with fiber-bridging effect is effective to mitigate or prevent premature cover spalling. If plain concrete is adopted, reduction factor of concrete compressive strength should be incorporated in design as introduced in **Chapter 4**.

5.2.2 Confinement effect of CES section

5.2.2.1 Subdivision of confinement zone

The confinement models of RC columns are well-established involving both normal strength and high strength concrete, however extending these models to CES columns may give rise to underestimation, which is mainly attributed to the ignorance of confinement effect offered by encased steel section. As such, the concrete zone of a CES section can be subdivided according to different confinement degrees **(El-Tawil et al., 1999; Chen & Wu, 2017; Zhao et al., 2019)**.

As shown in Fig. 5.4(a), the concrete zone can be partitioned as Unconfined Concrete (UCC), Partially Confined Concrete (PCC), and Highly Confined Concrete (HCC). UCC refers to concrete cover outside stirrups. The concrete zone enclosed by stirrups and outside steel section is PCC, which is under the same confining condition as that of core concrete in RC columns. The remaining concrete, located within steel section, is regarded as HCC. Fig. 5.4(b) indicates that HCC is confined by the pressures from both stirrups and steel section. Once these two types of pressures are

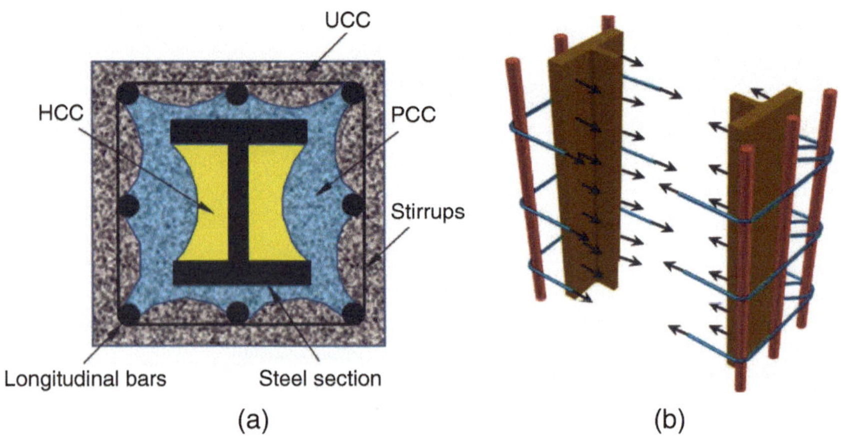

Fig. 5.4 Concrete confinement effect of CES section. (a) Confinement zone (b) Confining pressure acting on HCC.

predictable, the confinement effect of HCC can be estimated based on existing confinement model derived for RC columns (**Zhao et al., 2019**).

5.2.2.2 Steel section-induced confinement

Several attempts have been made to study the confinement effect induced by steel section, including experimental investigation (**Zhao et al., 2014**), numerical simulation (**Jamkhaneh et al., 2020**), and analytical derivation (**Chen & Wu, 2017; Zhao et al., 2019**). Since these studies only focus on pure compression scenario, the confinement effect in the presence of strain gradient is beyond the scope of this chapter.

As illustrated in Fig. 5.5, the steel flange can be regarded as a cantilever beam that deflects outwards freely when pushed by concrete lateral expansion. The confining pressure acting on concrete can be derived according to the force equilibrium between confining pressure and the bending moment generated at flange-web intersection. It is noteworthy that the stress state of the flange-web intersection point is not fully understood so far, which necessitates more experimental and theoretical work for verification and validation.

Another strategy to evaluate steel section-induced confinement effect is based on the biaxial stress state of steel web. As a result of concrete Poisson effect, steel web is subjected to axial compressive stress and lateral tensile stress. The stress along the web thickness direction can be neglected so that the steel web is at a plane stress state. The Von-Mises yield criterion should be employed for examining the actual stress in both directions, and the confining pressure can be derived accordingly once the lateral tensile stress is calculated. Depending on specific steel grade, the yielding condition of steel web differs, requiring more experimental and numerical work to generate a unified knowledge of stress state.

5.2.2.3 Double confinement within steel section

As mentioned above, the Highly Confined Concrete (HCC) is confined by both the transverse reinforcing bars (stirrups) and steel section. Depending on the geometry of the

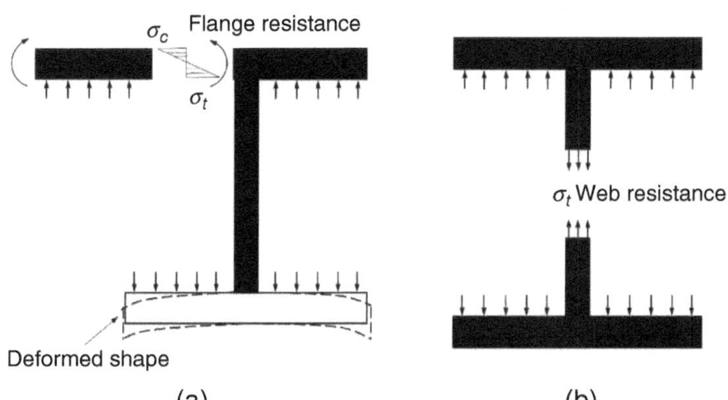

Fig. 5.5 Confinement mechanism of steel section-induced confinement (Lai et al., 2020a). (a) Derive from steel flange (b) Derive from steel web.

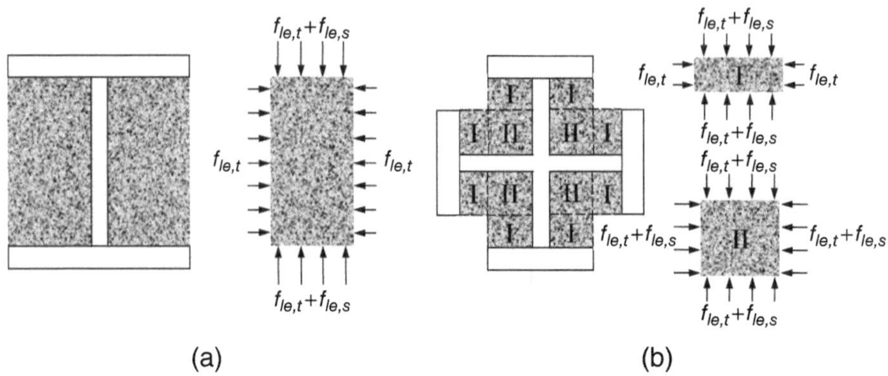

Fig. 5.6 Double confinement effect within steel section (Lai et al., 2021). (a) H-Section (b) Cruciform section.

encased steel section, the confining pressures would be superimposed in different ways. The typical H-section and cruciform section are presented in Fig. 5.6 for elaboration. As illustrated in Fig. 5.6(a), for the confined concrete within H-section, the effective confining pressures are not equal along the two perpendicular directions. Paralleling to the steel web, both the pressure from transverse bars ($f_{le,t}$) and steel section ($f_{le,s}$) act on concrete, whereas only $f_{le,t}$ exists in the perpendicular direction. Similarly, the confining pressure within a cruciform steel section can be discretized in the pattern presented in Fig. 5.6(b).

In view of the inequality of effective confining pressures in the two orthogonal directions of an encased open steel section, the weighted average confining pressure introduced by **Razvi & Saatcioglu (1999)** may be employed to account for this uneven stress distribution. More details of steel-induced confinement derivation can be found in **Lai et al. (2021).**

5.2.3 Behavioral analysis of CES members

5.2.3.1 Fiber section analysis

As an advanced numerical technique, commercial software packages like ABAQUS and ANYSIS implementing Finite Element Analysis (FEA) are commonly embraced to perform simulations of composite columns. However, FEA always consumes tremendous computational effort despite its powerful post-analysis function. For composite beam-columns under static loading, if the focus is placed on general global response or critical cross-section capacity, a more computationally efficient fiber section analysis method is preferable to use. Due to the convenience in programming, fiber section analysis has been widely used to perform numerical study for CFST columns (**Liang, 2011a, b**) and CES columns (**Lai & Liew, 2020 a, c**).

5.2.3.2 Axial force-bending moment diagram

As presented in Fig. 5.7, the composite section is meshed into numerous fiber elements. Assigning these elements with respective material constitutive model, the axial

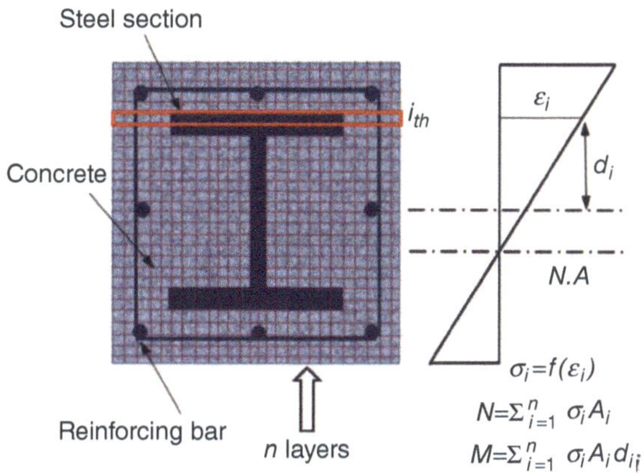

Fig. 5.7 Fiber section mesh of a CES section (Lai & Liew, 2020c).

and flexural resistance generated by the composite section can be computed through the integration of stress resultant, and cross-section capacity can thus be predicted once the failure criterion is properly defined.

Several hypotheses are commonly made: (1) Plane section remains plane after deformation; (2) Steel and concrete is perfectly bonded without slippage so that strain compatibility is satisfied; (3) Concrete tensile resistance and shear deformation are neglected. For the construction of axial force (N)-bending moment (M) interaction diagram, the failure criterion is commonly defined as the crushing of the outermost layer of concrete fiber under compression. It is noteworthy that N-M curve generally corresponds to the ultimate limit state for design purpose, at which state the concrete confinement is not fully activated. Therefore, it may not be prudent to incorporate confinement effect into the N–M curve construction **(Lai et al., 2019d)**.

Due to the adoption of actual stress-strain relation of material components, fiber section analysis gives an accurate prediction of N–M curve, which is applicable for both regular and irregular composite section and eliminates the treatment of steel section partial yielding as that adopted in EN 1994–1–1. The N–M curve of a standard CES section is presented in Fig. 5.8. The upper portion adjacent to pure compression is intercepted by a straight line to account for the premature concrete cover spalling. The axial capacity should be limited to the calculation using Eq. (4.3 (b)). It is noteworthy that concrete confinement effect is also neglected in calculating the axial capacity as its beneficial effect is more prominently in the post-peak stage.

5.2.3.3 Load-deflection response

Similar to the cross-section analysis, the load-deflection curve of CES beam-columns can be predicted based on the same techniques and hypotheses. Since the full range load-deflection response involves all the material nonlinearities, the key attributes including concrete cover spalling, concrete confinement effect, steel section strain-hardening, and rebar buckling need to be properly input into the fiber section model.

The member-level response can be linked to the section-level response by equating the second derivative of lateral deflection to the corresponding cross-sectional

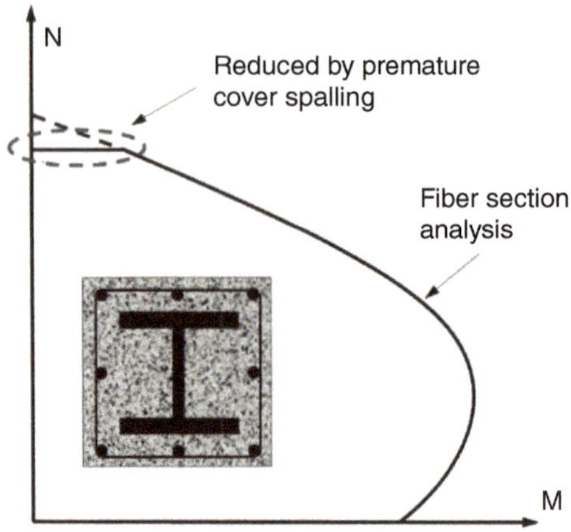

Fig. 5.8 N–M curve of high strength CES composite section.

curvature. The deflection shape of a beam-column is generally approximated as a half sine wave or a parabola. The full range load-deflection curve can be generated through an iterative algorithm implemented in a self-coded program using MATLAB or other platform. The beam-column fiber-section analysis is schematically illustrated in Fig. 5.9. More details of the numerical procedure are described by **Lai & Liew (2020c)**.

The applicability of fiber section analysis has been validated through the comparison between numerical results and experimental results. As evident in Fig. 5.10, good

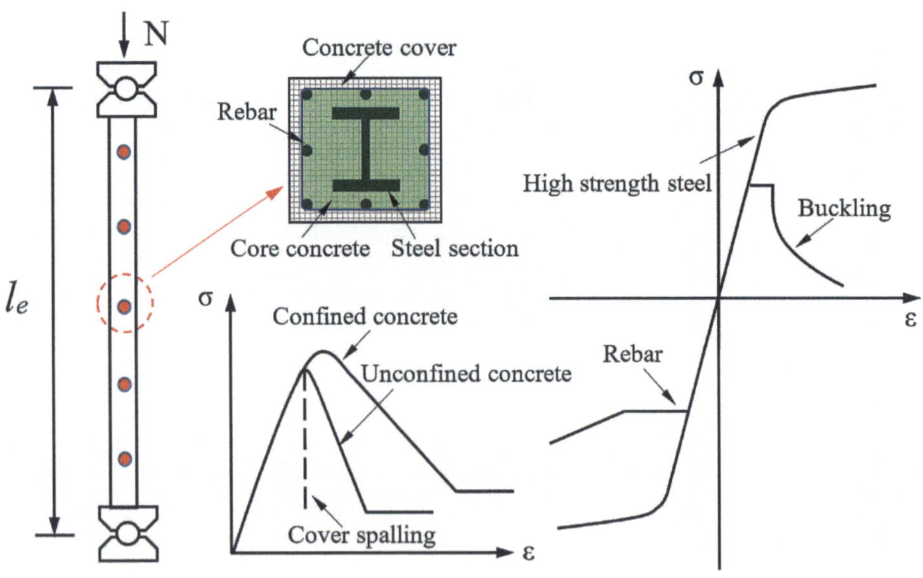

Fig. 5.9 Fiber section analysis in CES beam-columns (Lai & Liew, 2020a).

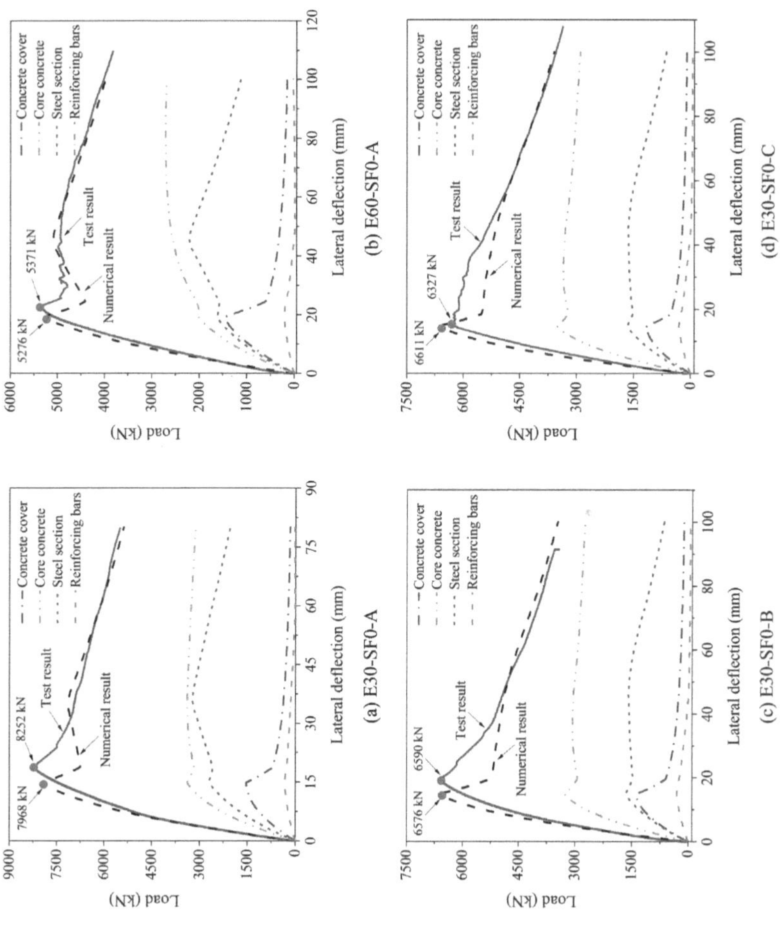

Fig. 5.10 Validation of fiber section analysis (Lai & Liew, 2020a).

agreement is achieved in terms of initial stiffness, load-carrying capacity, and post-peak resistance. In addition, the axial forces sustained by different material components can be extracted from numerical procedure to provide a deeper understanding of composite column behavior. Numerical data generated by this method has been used to supplement the test data to establish the design guide in Chapter 4.

5.3 Concrete filled steel tubular members

The study conducted in this section is limited to concrete filled steel tubular sections made of compact circular or rectangular hollow sections. Welding of the corner joints of a box column made of thick steel plate is normally done by using Submerged Arc Welding (SAW) but special care must be taken to control distortion of the box column caused large heat input. Chapter 7 provides additional information concerning the use of high strength steel plates and high strength concrete. This section focuses on the behavioral aspect of concrete filled steel tubes with special attention given to high strength concrete confined by steel tube to establish a proper constitutive model for implementation in a finite element analysis.

5.3.1 Effect of concrete confinement in CFST columns

With an increase in concrete strength, the compressive behavior of the CFST columns tends to be less ductile. However, it may be improved by an increase in the steel strength of the steel tube due to the confinement effect. The confinement from the steel tube not only improves the ductility of concrete core, but also increases the compressive resistance of the composite section. The higher the confinement pressure acting on the concrete, the higher the concrete compression strength. The level of the confinement can be represented by the ratio of the confining stress over the unconfined compressive strength of concrete (i.e., p/f_{ck}).

Fig. 5.11 shows the force equilibrium and stress status of the steel tube and concrete core at the ultimate limit state of a CFST column under axial compression. Assuming the steel tube and the concrete core have reached their yield strength and the compressive strength at the ultimate limit state, respectively, the confining stress can be calculated based on the following equations.

Considering force equilibrium in the axial direction (see Fig. 5.11(a)):

$$N = A_c f'_{cc} + A_s \sigma_{s,a} \tag{5.1}$$

Taking force equilibrium in the hoop direction (Fig. 5.11(b)):

$$\sigma_{s,h} t_s = p \cdot d_c / 2 \tag{5.2}$$

Based on von-Mises failure criteria, we have

$$\sigma_{s,a}^2 + \sigma_{s,a}\sigma_{s,h} + \sigma_{s,h}^2 = f_y^2 \tag{5.3}$$

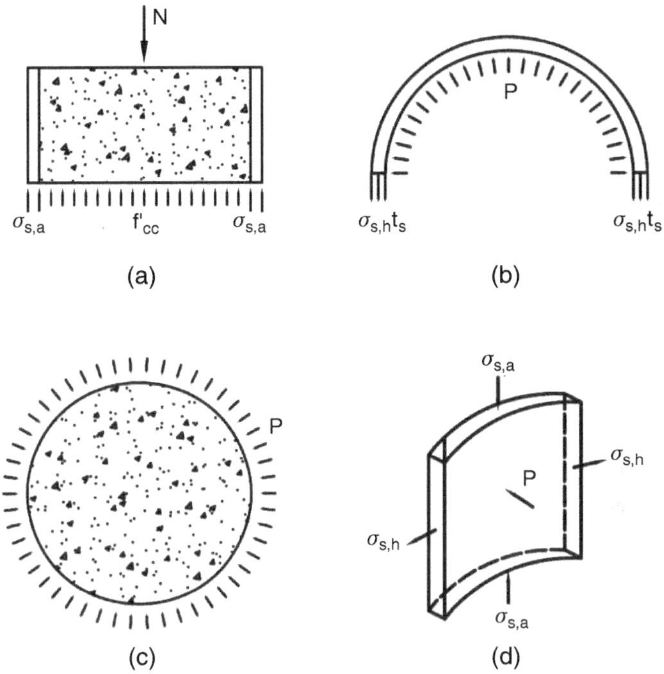

Fig. 5.11 Force equilibrium and stress status of CFST column under axial compression. (a) Force equilibrium in axial direction (b) Force equilibrium in hoop direction (c) Confining stress on concrete core (d) Stress status of steel tube.

The compressive strength of the concrete under confinement (i.e., f'_{cc}) may be calculated from Eq. (5.4) where η_c is a coefficient which can be obtained by calibrating with the test data.

$$\frac{f'_{cc}}{f_{ck}} = 1 + \eta_c \frac{p}{f_{ck}} \tag{5.4}$$

An early study on normal strength concrete (NSC) by Richart et al. (1928) found that the average value of η_c was 4.1, but a different value of 5.6 was proposed by Balmer (1944). Currently, $\eta_c = 4.1$ has been widely used for the NSC. For a type of high strength concrete (HSC) with an unconfined compressive strength of 67 MPa (Lu and Hsu, 2006), the value of η_c was in the range of 5.11 to 3.73 for confinement stress ratio, p/f_{ck}, ranging from 0.05 to 0.84 with an average value of $\eta_c = 4.3$. For a type of ultra-high strength concrete (UHSC) with an unconfined compressive strength of 131 MPa (Williams et al., 2020), the value of η_c was in the range of 6.03 ~ 1.52 for p/f_{ck} ratio ranges from 0.08 to 0.76 with an average value of $\eta_c = 3.4$. The study by Wang et al., (2016) on very higher strength UHSC ($f_{ck} = 212$ MPa) revealed that the value of η_c ranged from 5.25 to 2.58 for the ratio p/f_{ck} in the range of 0.12 ~ 1.89, and the average value of η_c was 3.7. The studies indicate that the η_c value decreases with the increase of confining stress, p, but the confined concrete strength ratio, f'_{cc}/f_{ck}, increases with the value of η_c.

The concrete core in the CFST columns is subject to a passive confinement pressure from the steel tube. For this reason, the EC 4 (EN1994-1–1) approach provides a varying value for the coefficient η_c related to the yield strength of the steel tube as shown in Eq. (5.5), provided the CFST column is circular in shape with a non-dimensional slenderness ratio $\bar{\lambda}$ not greater than 0.5 and a ratio of load eccentricity to column diameter e/d smaller than 0.1 (EN 1994–1–1, 2004). For square or rectangular columns, η_c is taken as 1.0. It is found that the value of η_c is inversely proportioning to the confining stress but is directly related to the yield strength of the steel. However, its relationship with the concrete strength is not clear since the effect of concrete strength is implicitly reflected in $\bar{\lambda}$.

$$\eta_c = \frac{t}{d}\frac{f_y}{p}\left(1-10\frac{e}{d}\right)\left[\min\left(1.0, 4.9-18.5\bar{\lambda}+17\bar{\lambda}^2\right)\right] \tag{5.5}$$

The following equation can be obtained by rearranging Eqs.(5.1) to (5.3) as

$$N = A_c f_{ck}\left[1+(\eta_c-1)\frac{p}{f_{ck}}+\sqrt{\left(\frac{A_s f_y}{A_c f_{ck}}\right)^2-3\left(\frac{p}{f_{ck}}\right)^2}\right] \tag{5.6}$$

The confining stress at the ultimate limit state can be determined by taking dN/dp equal to zero as

$$\frac{p}{f_{ck}} = \frac{A_s f_y}{A_c f_{ck}}\frac{\eta_c-1}{\sqrt{9+3(\eta_c-1)^2}} \tag{5.7}$$

Eq. (5.7) indicates that the level of confinement stress increases with the yield strength of the steel tube, f_y, but decreases with the unconfined compressive strength of the concrete, f_{ck}. It is found that the level of confinement is insignificant for UHSC with $f_{ck} \approx 200$ MPa (Xiong et al., 2017a), and therefore, the confinement effect could be conservatively ignored for the design of CFST columns using UHSC.

Eq. (5.7) also implies that the larger the difference between the steel and concrete strengths, the higher is the level of confinement. From this point of view, the combined use of high strength steel and high strength concrete in a CFST column is advantageous to sufficiently utilize the strengths of the constituent materials. This is valid if local buckling of steel tubes does not happen and thus the steel section must be at least a Class 3 section. However, awareness should be paid to the incompatibility between the HSS and NSC. For a CFST column having the HSS tube and NSC or HSC core, the concrete core is likely to crush earlier than the yielding of the HSS tube. This is illustrated in Fig. 5.12 where basically the C40 concrete or C90 high strength concrete would crush before the yielding of S690 high strength steel. In such cases, the plastic resistance of the composite section, which is the sum of the steel and concrete section resistances, cannot be fully achieved at the ultimate limit state.

Note: The yield strains of steel are calculated based on an Elastic modulus of 210 GPa (EN 1993–1–1, 2005), the peak strains of C40 and C90 concrete are taken

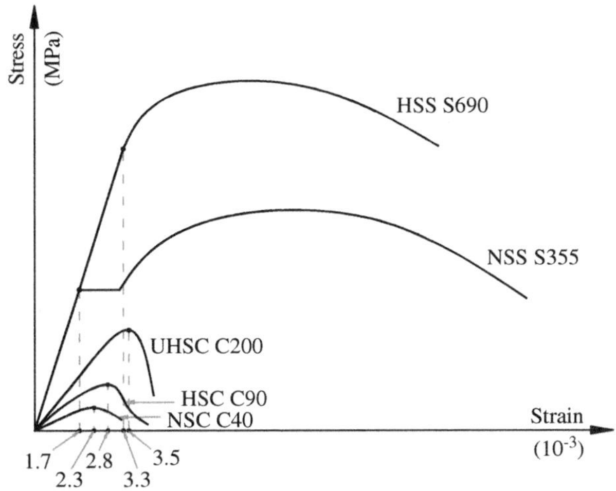

Fig. 5.12 Comparison of stress-strain curves of steel and concrete materials.

from EC 2 (1992–1–1, 2004) and that of C200 are based on the authors' previous work (Xiong et al., 2020).

When incompatible materials are used (refer to Table 3.10 in Chapter 3), the axial and hoop stresses in the steel tube (i.e., $\sigma_{s,a}$ and $\sigma_{s,h}$) at the ultimate limit state is no longer conforming to the von-Mises failure criteria as given in Eq. (5.3). Instead, their relationship should follow the Hooke's law as the steel tube is still in the elastic stage. To avoid such incompatibility, Table 3.10 in Chapter 3 should be followed to ensure the steel grade and concrete classes are matching.

It should be mentioned that the confinement effect on the one hand improves the strength and ductility of the concrete core, but it reduces the limit axial stress in the steel tube (i.e., $\sigma_{s,a}$) at the ultimate limit state according to Eq. (5.3). This is because the hoop stress in the steel tube increases with the dilation of the concrete core under compression. To consider this effect, a reduction factor can be applied to the yield strength of the steel as given in Eq. (5.8) according to EC 4, provided the CFST column is circular in shape with the non-dimensional slenderness ratio, $\bar{\lambda}$, not greater than 0.5 and the eccentricity ratio, *e*/*d*, smaller than 0.1 (EN 1994–1–1, 2004).

$$\eta_a = \min\left(1.0, 0.75 + 0.5\bar{\lambda}\right) + 10\frac{e}{d}\left[1 - \min\left(1.0, 0.75 + 0.5\bar{\lambda}\right)\right] \quad (5.8)$$

Eq. (5.8) indicates that the effects of the steel and concrete strengths on the reduction factor η_a are not explicit (it is implicitly reflected in $\bar{\lambda}$), and this is the reason why it is used in Eq. (4.4) to determine the axial load capacity of the CFST columns, irrespective of the steel and concrete strengths. In other words, the value of η_a given in Eq. (5.8) is applicable to all grades of steel. The validity of it has been established by comparing the test and predicted load capacities in Chapter 3.

5.3.2 Flexural behavior

5.3.2.1 Equivalent stress block of concrete in compression zone

Current research work on flexural behavior of CFST members with high strength materials is rather limited. Among which, Gho and Liu (2004) conducted flexural tests on twelve rectangular CFST beams with high strength concrete of cylinder strength ranging from 56.3 MPa to 87.5 MPa and steel section with yield strength ranging from 409 MPa to 438 MPa. Their investigations showed that the EC 4 approach underestimated the flexural load-carrying capacity by 11% approximately. Guler et al. (2012) experimentally investigated the flexural behavior of nine square CFST beams with high strength concrete up to 120 MPa and normal strength steel tubes, showing both EC 4 (EN 1994–1–1, 2004) and ANSI/AISC (2005) design methods are conservative in predicting the bending moment capacity. The study done by Chung et al. (2013) on six square CFST beams using steel tubes of 325 MPa, 555 MPa, and 900 MPa, and concrete of 82.5 MPa and 119.7 MPa revealed that the AISC-LFRD (1999) design method could underestimate the bending moment capacity. Again, the study by Li et al. (2017) on six square CFST beams with high strength concrete of 98 MPa (cubic strength) and normal strength steel tubes demonstrated that the modern design codes AISC-LFRD (1999), AIJ (1997) and EC 4 (2004) gave rather conservative predictions on the flexural load capacity. Besides, Xiong et al. (2017b) conducted a study on the flexural behavior of eight CFST beams employing ultra-high strength concrete of 200 MPa and S690 steel tubes. Circular and square, single-tube, double-tube, and double-skin (with inner tube not infilled with concrete) CFST beams were tested under the four-point concentrated loads. The comparison with the code predictions indicated that the EC 4 approach is quite conservative.

It is found that the current modern design codes are generally conservative to estimate the bending moment capacity of the CFST beams made of high strength materials. Several researchers suggested that these codes can be safely extended to estimate the flexural resistance of high strength CFST members. However, awareness should be raised on the conservativeness of such method. It can be seen from Table 3.3 in Chapter 3 that the conservativeness, i.e., the safety margin, reduces with the increase of concrete strength (see the unbracketed average values of the Test/EC4 ratios of the circular and rectangular beam-columns). This means the design of such high strength CFST members cannot achieve equivalent reliability with that of their counterparts using normal strength materials. One of the reasons is that the equivalent stress block for normal strength concrete was used in the design calculation, which may not be suitable for high strength or ultra-high strength concretes according to the research done on reinforced high strength concrete members (Oztekin et al., 2003; Ozbakkaloglu and Saatcioglu, 2004; Karthik and Mander, 2011).

For the reinforced concrete (RC) members under pure bending, the concrete strain and stress distributions in the compression zone at the ultimate limit state are shown in Fig. 5.13 where the equivalent rectangular stress block is illustrated. Since the cross-section remains plane under bending, the strain is linearly distributed on the section. As the stresses can be redistributed near the extreme compression fiber after

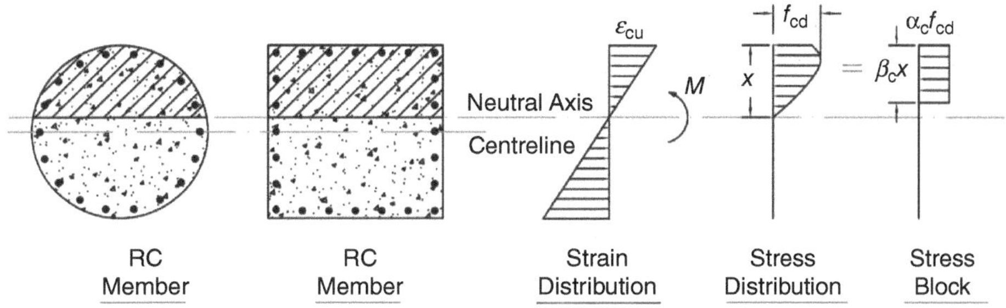

Fig. 5.13 Strain and stress distributions and equivalent stress block of concrete in reinforced concrete members at ultimate limit state.

the peak strength of the concrete has been reached, the ultimate compressive strain (i.e., ε_{cu}) at said fiber can be higher than the peak strain (i.e., ε_{c1}) obtained for a standard cylinder under axial compression. The value of ε_{cu} is taken as 0.0035 for normal strength concrete but decreases with the increase of concrete strength according to EC 2 (EN 1992–1–1, 2004). The stress of the concrete in compression zone is parabolically distributed following the stress-strain relationship of the concrete. The rectangular stress block is obtained when the resultant compression forces in the parabolic and rectangular areas are the same and the resultant moment resistance to the neutral axis are equivalent. The parameters α_c and β_c control the area of the stress block. As the ultimate compressive strains are different between normal strength and high strength concretes, the parameters α_c and β_c of them are different as shown in Eq. (5.9) and Eq. (5.10) according to EC 2.

$$\alpha_c = \begin{cases} 1.0 & \text{for} \quad f_{ck} \le 50\,\text{MPa} \\ 1.0-(f_{ck}-50)/200 & \text{for} \quad 50 < f_{ck} \le 90\,\text{MPa} \end{cases} \tag{5.9}$$

$$\beta_c = \begin{cases} 0.8 & \text{for} \quad f_{ck} \le 50\,\text{MPa} \\ 0.8-(f_{ck}-50)/400 & \text{for} \quad 50 < f_{ck} \le 90\,\text{MPa} \end{cases} \tag{5.10}$$

Different from the RC beams, the CFST members under pure bending should have the better rotational capacity due to the confinement from the steel tubes. Therefore, the cross-section could further rotate beyond the limit state of the RC beams to form a plastic hinge, this means the full plasticity could be reached on the section. Therefore, the plastic design method can be employed for the CFST members employing normal strength materials according to EC 4 (EN 1994–1–1, 2004). The strain and stress distributions in the compression zone of the CFST members at the ultimate limit state are shown in Fig. 5.14. Considering the full plasticity has been achieved on the section and considering the decay part of the stress-strain curve of the NSC is quite smooth (in fact, EC 2 assumes no decay part for the stress-strain curve of NSC in design), the EC 4 approach adopts the value of unity for both the parameters α_c and β_c (Note that EC 4 is only applicable to CFST members with NSC, not applicable to those with HSC and UHSC).

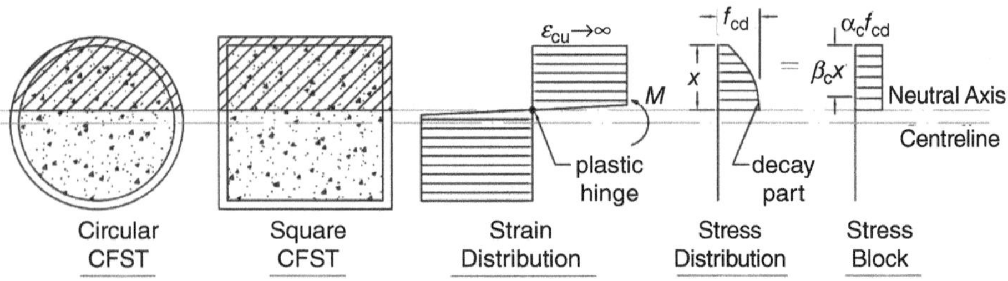

Fig. 5.14 Strain and stress distributions and equivalent stress block of concrete in CFST members at ultimate limit state.

The full plasticity on cross-section can still be achieved in CFST members employing the high strength materials when the strain compatibility between steel and concrete is satisfied (Xiong et al., 2017b&c). Therefore, the parameter β_c can still be taken as unity. However, the decay parts of the stress-strain curves of the high strength and ultra-high strength concretes under compression are much steeper than that of the normal strength concrete due to their brittle behavior, the value of unity is no longer suitable for the parameter α_c. Considering the confinement from the steel tube to the concrete in the compression zone is low for a CFST beam under bending, the value of α_c for the unconfined concrete in an RC beam can be extended for the CFST beam. This explains the use of the reduction factor, η, in Eq. (3.1).

In summary, the parameters α_c and β_c for the CFST members using high strength materials are given in Eq. (5.11) and Eq. (5.12) as

$$\alpha_c = \begin{cases} 1.0 & \text{for } f_{ck} \le 50\,\text{MPa} \\ 1.0-(f_{ck}-50)/200 & \text{for } 50 < f_{ck} \le 90\,\text{MPa} \\ 0.8 & \text{for } f_{ck} > 90\,\text{MPa} \end{cases} \tag{5.11}$$

$$\beta_c = 1.0 \tag{5.12}$$

The α_c value for UHSC (f_{ck} > 90 MPa) is taken as 0.8 which has been justified by the comparisons between the test and predicted results presented in Chapter 3. The parameters α_c (i.e., the reduction factor η) and β_c are also used for the concrete encased steel composite members with the high strength materials due to the good rotational capacity provided by the double confinement from the outer stirrups and the inner encased steel section. Also, the concrete encased steel section possesses very good ductility under the action of pure bending and can achieve full plasticity, and the phenomenon of concrete cover spalling was not observed under flexural load.

By observing the average values of the Test/EC4 ratios of the circular and rectangular beam-columns with flexural loading (i.e., the values in brackets "( )", "{ }" and "| |"), the conservativeness of design is largely improved and the reliability is quite comparable for these members with concrete in different strengths when the parameters α_c (i.e., the reduction factor η) and β_c (implicitly considered in design calculation since it is equal to 1.0) given in Eq. (5.11) and Eq. (5.12) are used.

It should be mentioned that the reduction factor, η, is also used for CFST members without flexural loading, for example the axially loaded columns. This is to consider the brittle nature of the high strength and ultra-high strength concrete core and cater for the fact that the whole cross-section under pure compression would crush immediately upon the reach of plasticity with the limited development of plastic deformation. It is also reasonable to think that the neutral axis of the cross-section under pure compression is in fact infinitely away from the centerline, thus it could be similarly considered as the members with flexural loading. Nevertheless, the use of the reduction factor, η, for CES and CFST members has been justified by the comparisons between the test and predicted results in Chapter 3.

It should be mentioned that, for the CFST beams under bending, the level of confinement to the concrete in compression zone is relatively low. In this regard, the confinement effect is ignored in determining the equivalent stress block. For a more accurate analysis, the stress-block model proposed by Karthik and Mander (2011) for the confined concrete of RC beam may be used.

5.3.2.2 *Effects of steel strength and neutral axis*

It is found from Table 4.3 and Eq. (4.20) that there is a reduction factor (i.e., α_M) applied to the plastic bending moment capacity, $M_{pl,Rd}$, of the composite member. It can avoid unconservative design since it allows for the increase of compressive strain of concrete (which is adverse for the concrete) at the yielding of the steel section when the yield strength is high (Johnson and Anderson, 2004). This reduction factor depends on the grade of structural steel, taken as 0.9 for S235 and S355 steel inclusive and 0.8 for higher steel grades including high strength steel. The value of 0.8 for higher grade of steel section (i.e., 0.8) has been justified by the test results in Chapter 3.

The location of neutral axis is one of the factors affecting the bending moment capacity. The variation of depth of neutral axis of some typical high strength CFST beams is shown in Fig. 5.15 (Xiong et al., 2017b). With the increase of deflection, cracks are formed in the concrete under tension, consequently, the depth of the neutral axis to the centerline is increased. In the nonlinear stage of curve, the depth of the neutral axis is nearly kept constant in the single-tube beams while it is reduced in the double-tube ones and reduced more in the specimen with the inner tube not filled with the concrete. Besides, the depth of the natural axis decreases with the increase of steel strength.

5.3.2.3 *Effective flexural stiffness*

The effective flexural stiffness, i.e., $(EI)_{eff,II}$, is usually used to determine the internal forces and the lateral deformation of the CFST members through second-order analysis. Like the depth of the neutral axis, the effective flexural stiffness of a CFST member also varies with the lateral deflection due to the developments of cracks of the concrete in tension zone and the plastic deformation of the steel sections. In modern design codes (ACI 318, 2019; ANSI/AISC 360, 2018; EN 1994–1–1, 2004), the effective flexural stiffness of the CFST member is usually considered by superimposing

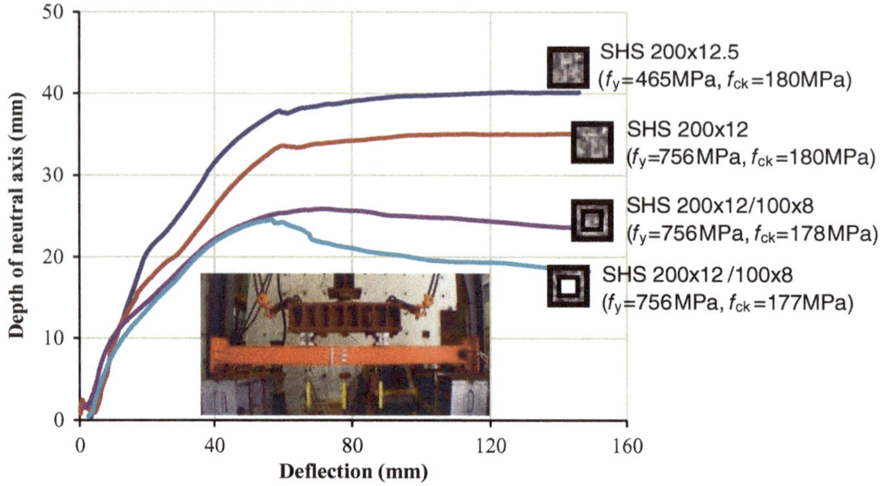

Fig. 5.15 Variation of neutral axis location.

the flexural stiffness of the constituent materials considering the effects of concrete cracking and steel yielding, for example in EC 4, it is expressed as

$$(EI)_{eff,II} = 0.9E_aI_a + 0.45E_{c,eff}I_c \tag{5.13}$$

for the CFST members without reinforcing steel rebars and encased steel section.

Eq. (5.13) indicates that the effective flexural stiffness increases with the increase of concrete strength since the modulus of elasticity of the concrete increases with the compressive strength, but independent on the steel strength (the elastic modulus of steel is irrespective of the steel grade). A study carried out by Xiong et al.(2007b) on the flexural behavior of eight CFST beams with UHSC of 200 MPa and S690 HSS tubes justified that the effective flexural stiffness given in Eq. (5.13) approximated the secant flexural stiffness of the moment-curvature curves of the CFST beams up to 0.7 times the ultimate bending resistance, which was defined herein as the serviceability limit state. It should be mentioned that the effective flexural stiffness provided in Eq. (4.17), i.e., $(EI)_{eff}$, is generally used for determining the buckling resistance of the composite member (EN 1994–1–1, 2004), which is recalled herein as Eq. (5.14).

$$(EI)_{eff} = E_aI_a + 0.6E_{c,eff}I_c \tag{5.14}$$

Li et al. (2020) conducted cyclic pure bending tests on six square CFST beams with high strength concrete of 95 MPa (cubic strength) and normal strength steel tubes, and the effective flexural stiffness calculated by Eq. (5.14) and that specified in BS 5400 (2005) as Eq. (5.15) were used to compare with the test flexural stiffness (taken as the secant stiffness up to 0.6 times the ultimate bending moment).

$$(EI)_{eff,II} = 0.95E_aI_a + 0.45E_{c,eff}I_c \tag{5.15}$$

It is found that the use of Eq. (5.14) overestimated the test flexural stiffness, but Eq. (5.15) yielded reasonable estimations on the said secant stiffness. Eq. (5.13) is quite similar to Eq. (5.15), a re-calculation based on Eq. (5.13) by the authors indicated that Eq. (5.13) is suitable to estimate the secant stiffness up to 0.6 times the ultimate bending moment of the high strength CFST members.

The effective flexural stiffness may not be degraded under cyclic loads at a low axial load level (Varma et al., 2002), but could be degraded under the high axial load levels, for example under an axial load level of 0.7 as observed in Xiong et al. s study (2017b). Also, such degradation increases with the increase of concrete strength (Wang et al., 2019; Zhang et al., 2020). In the proposed design method in Chapter 4, the effect of the degradation is implicitly considered for high strength composite members by the reduced elastic modulus of concrete as calculated by Eq. (3.2), which reduces the values of $(EI)_{eff}$ and $(EI)_{eff,II}$ in a conservative way.

5.4 Numerical models for high strength CFST members

5.4.1 Constitutive model of steel tube confined concrete

To analyze the structural behavior of the CFST members employing advanced finite element methods, the constitutive laws of the steel tube and the confined concrete should be known. For slender beams and columns failed in global bending and buckling, respectively, or the short square/rectangular columns failed by reaching the N–M interaction strength, the confinement effect may be ignored, and the uniaxial stress-strain relationship of the steel and unconfined concrete can be used in the finite element analysis (FEA) (Xiong et al., 2017c). For the short circular columns subject to the cross-section failure, the confinement effect should be considered. Hereinafter, a general procedure to determine the constitutive law of the steel tube confined concrete is introduced.

Concrete under laterally confining pressure from the steel tube can undergo pronounced inelastic axial deformation prior to reaching the failure load. The constitutive model based on the plasticity theory is appropriate to describe the material response of such concrete. Due to the difficulties in measuring the lateral dilation of concrete and the confining pressure on steel-concrete interface, the constitutive models of the confined concrete could be developed by matching the numerical analysis and test results of the CSFT columns (Hu et al., 2003; Ellobody et al., 2006; Dai and Lam, 2010; Tao et al., 2013). However, these models are case-oriented, and should be re-examined with the test results in further use.

Fig. 5.16 shows a CFST column under compression. At the initial stage of elastic deformation, the inner surface of the steel tube separates from the concrete core because the Poisson's ratio of the steel is higher than that of the concrete. Therefore, there is no confinement between them. With the increase of the compression load, micro-cracks are initiated and propagated in concrete, resulting in fast dilation of concrete and then it contacts with the steel tube. Such contact induces hoop stresses in the steel tube and consequently enhances the hardening/softening behavior of concrete. In view of this, the flow rule and hardening/

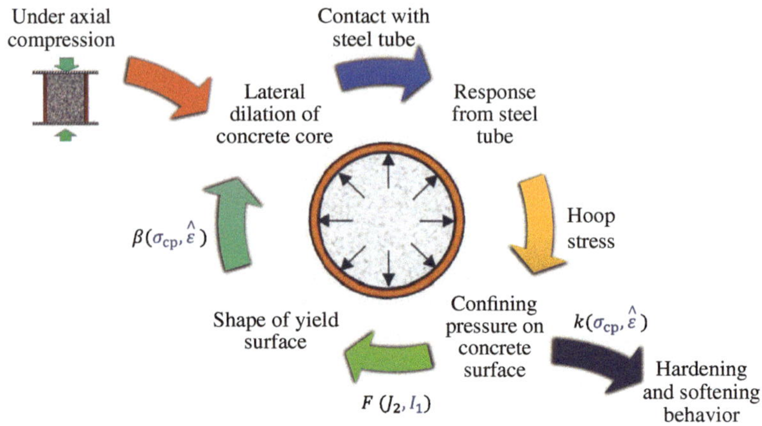

Fig. 5.16 Interaction between steel tube and concrete core.

softening rule should be properly developed to describe such enhancement. Since the flow rule and hardening/softening rule are a function of the plastic deformation, a non-associated flow rule with a variable dilation angle is required to reflect the influences of the confining pressure and the plastic deformation. Various strain hardening/softening rules are needed to cater for the varying confining pressure during loading. Besides, the yield surface is sensitive to the hydrostatic pressure (confining and axial pressures) and the lode angle (the third invariant of the deviatoric stress tensor). Overall, to obtain a generalized constitutive model for the steel tube confined concrete that is not cast-oriented, the following characteristics of concrete should be addressed:

(a) A yield criterion related to hydrostatic pressure ($I_1/3$) and lode angle, $F(I_1, J_2, J_3)$;
(b) A non-associated flow rule with a dilation angle β that is a function of the confining pressure and the equivalent plastic strain, $G \neq F$ and $\beta \neq$ constant;
(c) A hardening/softening rule related to the confining pressure and the equivalent plastic strain, $k(\sigma_{cp}, \hat{\varepsilon})$;

where F is the yield function, I_1 is the first invariant of the stress tensor, J_2 and J_3 is the second invariant and the third invariant of the deviatoric stress tensor, G is the flow potential function, β is the dilation angle, and k is the hardening/softening parameter.

(a) Yield criterion

The yield criterion describes the shape of yield surface and subsequent loading surfaces in the stress space. Various yield functions have been established to describe the pressure-sensitive characteristics of the concrete materials, such as the well-known Drucker-Prager criterion (2 parameters), Ottosen criterion (4 parameters), and the William-Warnke criterion (5 parameters). The Drucker-Prager criterion is commonly employed in the numerical modeling of the confined concrete due to its simplicity compared with the other yield functions. In the material library of the commercial FEA software ABAQUS, the Drucker-Prager yield criterion has been extended by introducing a new parameter related with the J_3 value in the yield function to offer different descriptions of the tensile and compressive meridians but with a smoothed noncircular yield surface in the deviatoric plan (Wang and Liew, 2016). The linear extended Drucker-Prager yield criterion has the form

$$F = t(K)\sqrt{J_2} - \tan\varphi \frac{I_1}{3} - k = 0 \tag{5.16}$$

where

$$t(K) = \frac{\sqrt{3}}{2}\left[1 + \frac{1}{K} - \left(1 - \frac{1}{K}\right)\frac{3\sqrt{3}}{2}\left(\frac{J_3}{J_2^{3/2}}\right)\right] \tag{5.17}$$

$$I_1 = \sigma_1 + \sigma_2 + \sigma_3 \tag{5.18}$$

$$J_2 = \frac{1}{2}s_{ij}s_{ji} = \frac{1}{6}\left[(\sigma_1 - \sigma_2)^2 + (\sigma_2 - \sigma_3)^2 + (\sigma_3 - \sigma_1)^2\right] \tag{5.19}$$

$$J_3 = \frac{1}{3}s_{ij}s_{jk}s_{ki} \tag{5.20}$$

where σ_i is the principal stresses, $I = 1$ represents the axial direction, $I = 2$ or 3 represents the radial direction, and s_{ij} is the deviatoric stress tensor. The yield surface is defined by the three material parameters where φ is the friction angle representing the slope of the compressive meridian, k is the softening/hardening parameter being the interception of the compressive meridian to the hydrostatic pressure axis, and K is the flow stress ratio being an additional parameter in the extended Drucker-Prager criterion to control the shape of the yield surface in the deviatoric plane with $0.778 \leq K \leq 1$. When $K = 1$, Eq. (5.17) reduces to the classic Drucker-Prager yield criterion with a circular cross-section in the deviatoric plane.

(b) Flow rule

In the stage with inelastic deformation, the total strain is the summation of the elastic strain and plastic strain

$$\varepsilon_{ij} = \varepsilon_{ij}^{e} + \varepsilon_{ij}^{p} \tag{5.21}$$

where

$$\varepsilon_{ij}^{p} = \lambda \frac{\partial G}{\partial \sigma_{ij}} \tag{5.22}$$

ε_{ij} is the strain tensor, σ_{ij} is the stress tensor, λ is a non-negative scalar parameter, and G is the plastic potential function. As said, a non-associated flow rule ($G \neq F$) can be adopted to determine the direction of plastic strain flow in the form of the Drucker-Prager yield function

$$G = \sqrt{3J_2} - \tan\beta \frac{I_1}{3} - c \tag{5.23}$$

where β is the dilation angle and c is a constant. As the flow potential function presents in the form of a partial derivative of the stress tensor as shown in Eq. (5.22), there is only one valid parameter, dilation angle β, controlling the direction of the plastic strain flow. It should be mentioned that a constant dilation angle is incapable to describe the plastic dilation of the concrete under various confining pressures (Imran and Pantazopoulou 1996; Ansari and Li 1998; Lim and Ozbakkaloglu 2014a). The dilation angle, β, is a function of the confining pressure and plastic deformation,

$$\beta = \beta(\sigma_{cp}, \hat{\varepsilon}^{p}) \tag{5.24}$$

where σ_{cp} is the confining pressure and $\hat{\varepsilon}^{p}$ is the equivalent plastic strain. The increment of the equivalent plastic strain is given by,

$$d\hat{\varepsilon}^{p} = \sqrt{d\varepsilon_i^{p} d\varepsilon_i^{p}} = \sqrt{d\varepsilon_1^{p} d\varepsilon_1^{p} + d\varepsilon_2^{p} d\varepsilon_2^{p} + d\varepsilon_3^{p} d\varepsilon_3^{p}} \tag{5.25}$$

For uniformly confined concrete, the plastic strains, ε_i^{p} can be derived from Eq. (5.21) and the Hooke's law. As $\sigma_2 = \sigma_3 = -\sigma_{cp}$, we have

$$\varepsilon_1^{p} = \varepsilon_1 - \frac{1}{E_c}(\sigma_1 - 2\nu\sigma_3) \tag{5.26}$$

$$\varepsilon_2^{p} = \varepsilon_3^{p} = \varepsilon_3 - \frac{1}{E_c}\left[(1-\nu)\sigma_3 - \nu\sigma_1\right] \tag{5.27}$$

(c) **Hardening/softening rule**

The hardening/softening rule defines the evolution of the loading surface due to the development of the plastic strain, including the hardening part from the initial yield surface to the ultimate yield surface (failure surface), the following softening part, and the residual surface. To simulate a CFST short column subject to monotonic compression, the isotopic hardening can be adopted. Existing experiments on NSC, HSC and UHSC has indicated that the development of the subsequent yielding surface is sensitive to the plastic deformation and the confining pressure (Imran and Pantazopoulou 1996; Ansari and Li 1998; Lim and Ozbakkaloglu 2014b; Wang et al. 2016). Therefore, the hardening/softening parameter should be calibrated under different confining pressures. The hardening/softening parameter is defined as

$$k = k(\sigma_{cp}, \hat{\varepsilon}^{p}) \tag{5.28}$$

Solving Eq. (5.16) for k yields

$$k = t(K)\sqrt{J_2} - \tan\varphi \frac{I_1}{3} \tag{5.29}$$

Noted that $\sigma_2 = \sigma_3 = -\sigma_{cp}$, Eq. (5.29) can be rewritten as

$$k = (\sigma_3 - \sigma_1) - \frac{\tan\varphi}{3}(\sigma_1 + 2\sigma_3) \tag{5.30}$$

Table 5.1 Peak stress and yield stress under triaxial compression.

σ_{cp} MPa	$f'_{cc}(f'_{c0})$ MPa	Ultimate yield surface		Initial yield surface	
		$I_1/3$	$\sqrt{J_2}$	$I_1/3$	$\sqrt{J_2}$
0	212.0	70.7	212.0	21.2	63.6
25	345.0	131.7	320.0	59.1	102.4
50	418.4	172.8	368.4	97.1	141.2
100	546.9	248.1	444.3	172.9	218.7
200	860.5	420.2	660.5	324.6	373.9
400	1244.5	681.5	844.5	628.1	684.2

5.4.2 Parameter calibration

To illustrate how to calibrate the parameters involved in the determination of yield criterion, flow rule, and hardening/softening rule, the test results including the axial stress-strain and lateral strain-to-axial strain curves under triaxial compression with confining pressure ranging from 25 to 400 MPa are used (Wang et al., 2016). This test was conducted on a type of UHSC with the uniaxial compressive strength of 212 MPa (cylinder specimens with a diameter of 50 mm and a height of 100 mm).

To calibrate parameters for the yield surface of the UHSC, tests under at least three different multiaxial stress states are generally required, having two stress states located on the compressive meridian and one on the tensile meridian. The uniaxial and triaxial compression test results under five different confining pressures for the compressive meridian are given in Table 5.1 where f'_{c0} and f'_{cc} are the unconfined and confined compressive strength, respectively. The compressive meridian for the ultimate yield surface (failure surface) is determined through the least square fit to the test data as shown in Fig. 5.17. The friction angle φ is 45.7° with $R^2 = 0.98$. For the initial yield surface, the same friction angle of 45.7° with a reduced size corresponding $0.3f'_{c0}$ is assumed. Therefore, the interception of the initial yield surface to the hydrostatic pressure axis is $k_0 = 41.9$. On the other hand, the biaxial compressive test data reported by Curbach and Speck (2008) are used to determine the dependence of the yield surface on the intermediate principal stress. It is found that the parameter K for the UHSC is 0.935. The shape of the yield surface on the deviatoric plane is nearly a circular, as shown in Fig. 5.18, with the comparison of the surface with $K = 1$ and $K = 0.8$.

To calibrate the dilation angle $\beta(\sigma_{cp}, \hat{\varepsilon})$ for the flow rule, the radial-to-axial plastic strain curves are converted into the dilation angle versus equivalent plastic strain curves under different confining pressures. Substituting Eq. (5.23) into Eq. (5.22) yields the plastic strain increment,

$$d\varepsilon_1^{p} = d\lambda\left(\frac{\sqrt{3}}{2\sqrt{J_2}}\frac{2(\sigma_1 - \sigma_3)}{3} + \frac{\tan\beta}{3}\right) \tag{5.31}$$

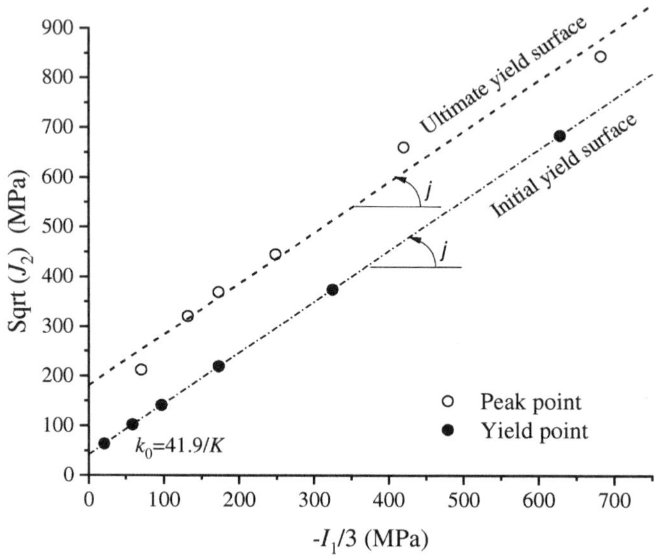

Fig. 5.17 Compressive meridian of UHSC.

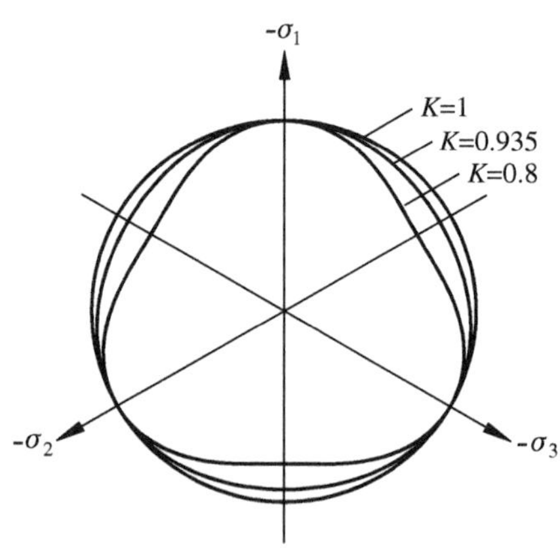

Fig. 5.18 Yield surface on the deviatoric plane.

$$d\varepsilon_2^{\mathrm{p}} = \mathrm{d}\varepsilon_3^{\mathrm{p}} = \mathrm{d}\lambda\left(\frac{\sqrt{3}}{2\sqrt{J_2}}\frac{2(\sigma_3 - \sigma_1)}{3} + \frac{\tan\beta}{3}\right) \tag{5.32}$$

By rearranging Eqs.(5.31) and (5.32), the dilation angle is calculated as

$$\tan\beta = \frac{1}{2}\frac{d\varepsilon_1^{\mathrm{p}} + 2d\varepsilon_3^{\mathrm{p}}}{d\varepsilon_1^{\mathrm{p}} - d\varepsilon_3^{\mathrm{p}}} = \frac{\sqrt{3}}{2}\frac{dI_1^p}{\sqrt{J_2^p}} \tag{5.33}$$

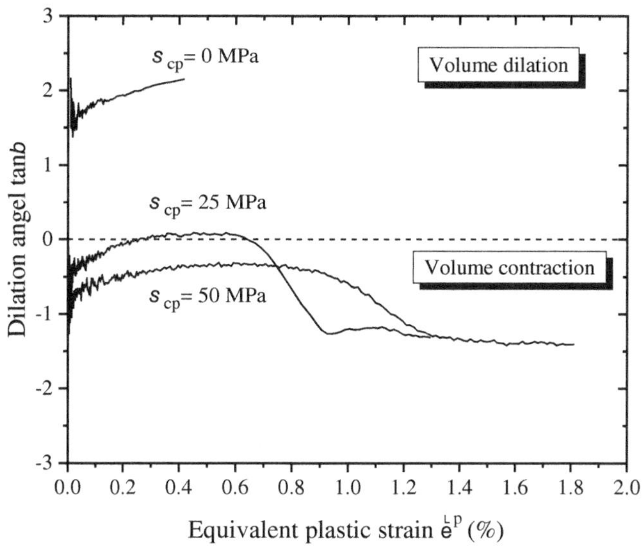

Fig. 5.19 Dilation angle versus equivalent plastic strain curves of unconfined and confined UHSC.

where

$$dI_1^p = d\varepsilon_1^p + d\varepsilon_2^p + d\varepsilon_3^p \tag{5.34}$$

$$dJ_2^p = \frac{1}{6}\left[(d\varepsilon_1^p - d\varepsilon_2^p)^2 + (d\varepsilon_2^p - d\varepsilon_3^p)^2 + (d\varepsilon_3^p - d\varepsilon_1^p)^2\right] \tag{5.35}$$

Fig. 5.19 shows the dilation angle versus equivalent plastic strain curves under different confining pressures. The positive value of tanβ represents the volumetric dilation and the negative one represents the volumetric contraction. It is found that the dilation angle is a function of not only the equivalent plastic strain but also the confining pressure. For the unconfined UHSC, the initial value of the dilation angle is 58.3°. With the increase of equivalent plastic strain, the dilation angle increases to 66.1° at the peak load. The UHSC cylinder under a confining pressure of 25 MPa starts from a volumetric contraction with the dilation angle of −22.0°. The dilation angle increases with the equivalent plastic strain and reaches the maximum value of 13.1 (volumetric dilation) near the peak load. Then, it turns into the volumetric dilation during the softening branch and levels off with the dilation angle of −51.3°. The UHSC cylinder under confining pressure of 50 MPa has the similar $\beta - \hat{\varepsilon}^p$ curve as that of σ_{cp} = 25MPa, however, with the increased ductility in the softening branch.

To calibrate the hardening/softening parameter k as Eq. (5.30), the friction angle tanφ should be determined first during the yield surface calibration, and the stresses σ_1 and σ_3 are obtained from the triaxial compression tests as shown in Fig. 5.20. In

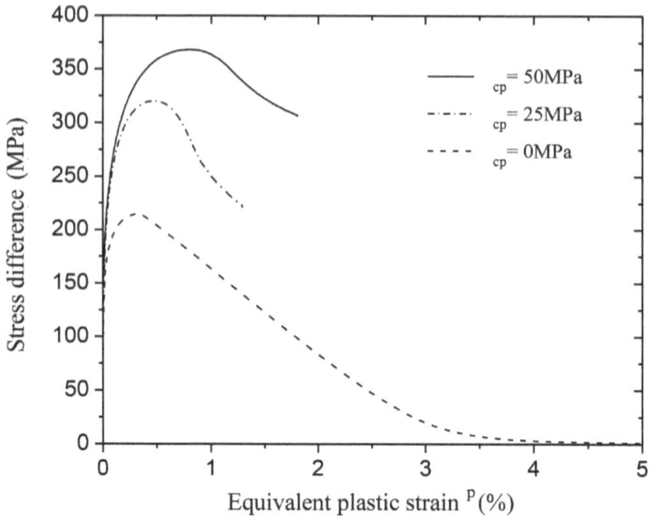

Fig. 5.20 Measured stress difference-equivalent plastic strain curves of unconfine and confined UHSC.

numerical analysis (e.g., in analysis via ABAQUS), the hardening/softening parameter is usually inputted as an equivalent uniaxial stress-plastic strain curve. In this regard, it is better to convert the hardening/softening parameter k under different confining pressures into the equivalent uniaxial stress $\hat{\sigma}_1$ like the unconfined concrete. Noting k is a known variable and $\sigma_2 = \sigma_3 = 0$, substituting k, σ_2 and σ_3 into Eq. (5.30) yields the equivalent uniaxial stress as

$$\hat{\sigma}_1 = \frac{-k}{1 - \frac{\tan\varphi}{3}} s \tag{5.36}$$

5.4.3 Implementation in finite element analysis

The above determined constitutive law of the steel tube confined UHSC is implemented by analyzing the compressive behavior of three short CFST columns tested by Xiong et al. (2017a). The analysis is done by using the commercial FE software ABAQUS through the available user subroutine interface VUSDFLD. Two field variables, i.e., the confining pressure and the equivalent plastic strain, are defined and directly added at the integration points of the finite elements. FORTRAN source file is programmed to obtain the stress and strain status from the material points and the path values to the field variables of the user subroutine.

The test and predicted load-axial strain and load-lateral strain curves are compared as shown in Fig. 5.21. The predicted initial stiffness, peak load and hardening regions match well with the experimental curves except for the residual resistance after the

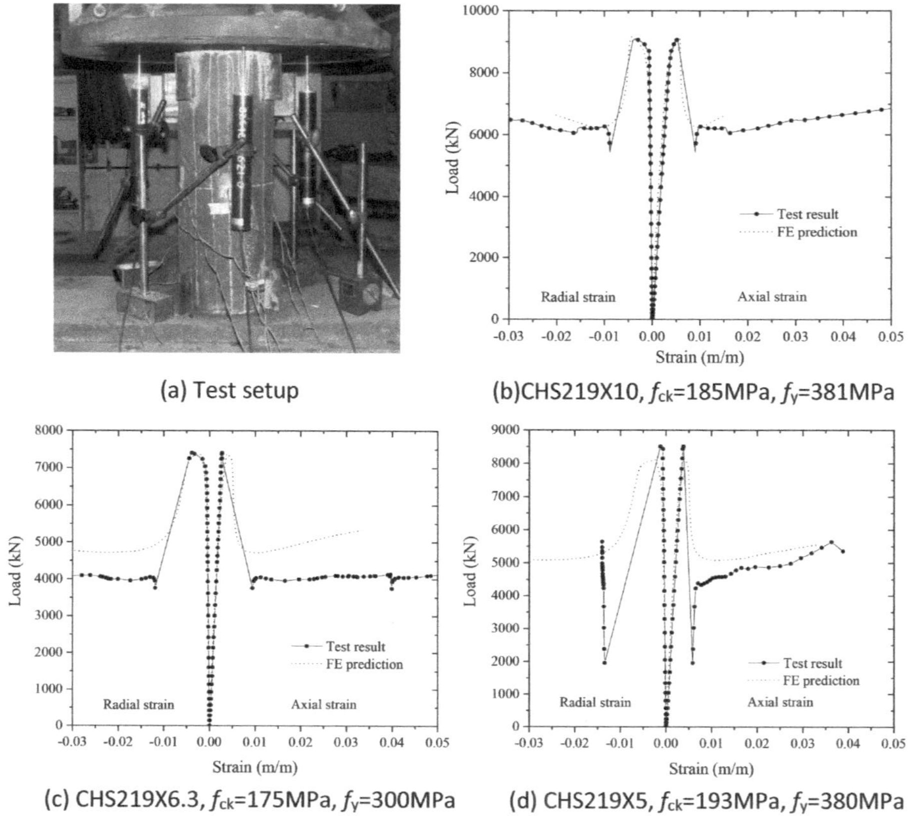

Fig. 5.21 Comparison of predicted and experimental load-axial strain and load-lateral strain curves. (a) Test setup (b) CHS219 × 10, f_{ck} = 185 MPa, f_y = 381 MPa (c) CHS219 × 6.3, f_{ck} = 175 MPa, f_y = 300 MPa (d) CHS219 × 5, f_{ck} = 193 MPa, f_y = 380 MPa.

load drop from the peak, indicating a further investigation on the behavior of UHSC under triaxial compression with the low confining pressures is necessary. Overall, a good agreement has been reached between the test and predicted curves, showing the generalized constitutive law of the steel tube confined concrete is suitable for behavioral analysis of the high strength composite members.

Fire resistant design

6

Outline

6.1 General

Fire resistance reflects the duration for which a structural member can withstand a standard fire exposure. In structural design for fire, the fire resistance is often evaluated to achieve a certain fire rating for various occupancies (SCDF fire code, 2018). There are three aspects of determining the fire resistance of a structural member, which are the selection of design fire scenarios, the heat transfer analysis, and the thermal-stress analysis.

The selection of design fire scenarios concerns the fire development and spread in a structure, especially in a compartment of the structure. Simplified and advanced calculation models have been provided in modern design codes to determine the temperature of fire. This information can be found in Section 6.2.

The heat transfer analysis determines the temperature distribution in a structural member being affected by the fire. It needs the thermal properties of the constituent materials of the structural member such as thermal conductivity, specific heat, and mass loss to be known. These aspects are discussed in Section 6.3.1. Normally, the stress condition related with the thermal expansion, water migration, and external

Design of Steel-Concrete Composite Structures Using High Strength Materials. DOI: 10.1016/C2019-0-05474-X

loading does not affect the temperature distribution, but it does affect the spalling behavior of high strength concrete covered in Section 6.3.2. The detailed calculation procedure for the temperature distribution is provided in Section 6.4.

The thermal-stress analysis evaluates the load-carrying capacity of a structural member under a fire. The empirical and prescriptive method with the tabulated data as shown in Section 6.5 provides a simple way to determine passive protection to fulfil the required fire rating. Alternatively, the fire engineering approaches and the advanced calculation models introduced in Sections 6.6 and 6.7 may be used to determine the fire resistance of a structural member. The fire engineering approaches use the design-oriented stress-strain relationship of the constituent materials where only some characteristic points on the stress-strain curves need to be known, for example, the compressive stresses, peak strains, and elastic modulus at elevated temperatures. The fire engineering approaches can be implemented in a step-by-step manner on a numerical analysis platform such as MATLAB. The calculation procedure is quite straightforward and easy for error scrutiny. On the other hand, the advanced calculation models use the analysis-oriented stress-strain relationship of the constituent materials where their full-spectrum development at the elevated temperatures should be known. The implementation of the advanced calculation models practically resorts to the finite element analysis platforms such as ABAQUS and ANSYS. The design- and analysis-oriented stress-strain relationships of the constituent materials of composite columns can be found in Section 6.3.3.

6.2 Design fire scenarios

Fire is the rapid oxidation of a combustible material releasing heat, light, and various reaction products such as carbon dioxide and water. Fire starts when a combustible material with an adequate supply of oxygen is subjected to enough heat and is able to sustain a chain reaction. Combustible material, oxygen and heat are commonly called as the fire tetrahedron. Fire can be extinguished by removing any one of the elements of the fire tetrahedron. To protect structures from fire hazards, both active and passive fire protections could be utilized. The active fire protection involves manual and automatic detection and suppression of fires, such as fire sprinkler system, fire extinguisher, fire alarm and smoking detector; whereas the passive fire protection involves fire-resistance rated walls and floor assemblies to form the fire compartments that limit the spread of fire. Fire insulation materials are also needed to protect the structural members. In addition to the active and passive fire protections, the structural members also possess inherent fire resistance contributed by their constituent materials.

To consider the effect of fire in structural design, the temperature of fire as a function of time should be known. Two types of temperature-time relationship are introduced in Eurocode 1, i.e., nominal temperature-time curves and natural fire models (EN 1991–1–2, 2002). Among the nominal temperature-time curves, the standard

ISO-834 fire curve, given by Eq. (6.1), is generally used for fire resistant design of structural members in building structures, while the external fire curve and hydrocarbon fire curve are available for other specific uses.

$$\Theta_g = 20 + 345\log_{10}^{(8t+1)} \left({}^{\circ}C\right) \tag{6.1}$$

where

t is the time in minute

Θ_g is the gas temperature in the fire compartment

In terms of performance-based fire resistant design, natural fire models can be used which are categorized into two groups, i.e., simplified fire models and advanced fire models. The simplified fire models comprise of compartment fires and localized fires. The former adopts a uniform temperature distribution as a function of time inside the compartments, whereas the latter considers a non-uniform temperature distribution. Since these models are determined based on specific physical parameters, they have limited field of application. The compartment fires are determined according to the compartment dimensions, opening area and fire load density. The fire load density is related with the classification of occupancy, calorific values of combustible materials inside the compartment, and the fire detection and suppression systems. The standard ISO-834 and typical compartment fire temperature-time curves are compared in Fig. 6.1. The standard ISO-834 fire curve features a rapid rise of temperature at the early stage of heating, especially within the first minute, the heating rate could reach up to 329 °C/min, then, the heating rate gradually drops down to 5 °C/min after half-hour heating and 2 °C/min after one-hour heating. The compartment fire curve features a relatively slow rise of temperature and a decay stage that is resulted from the gradual consumption of combustible materials or oxygen with time. Mostly, the compartment fire temperature is lower than the ISO-834 fire temperature. Hence, the fire resistant design based on the compartment fire temperature would in general lead to some saving in fire protection, even in some cases, it may be not necessary to apply fire protection if the fire resistance is adequate.

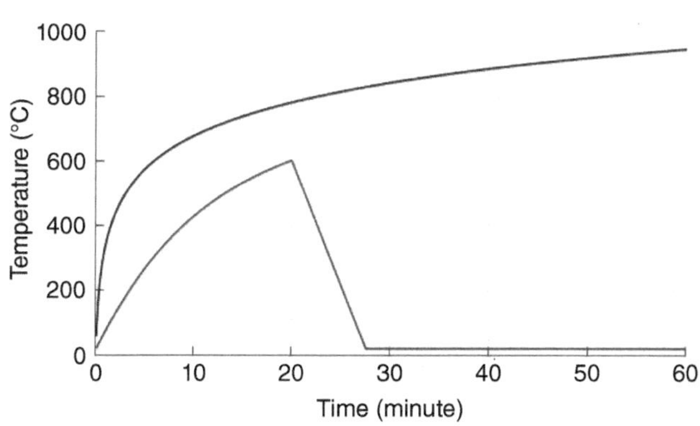

Fig. 6.1 Comparison between standard ISO and typical parametric fire temperatures.

The advanced fire models consider not only the gas properties and dynamic exchanges of mass and energy, but also the possible locations of fire which are determined based on risk assessment. The advanced fire models could be the computational fluid dynamic (CFD) models that give a completely time-dependent and space-dependent temperature evolution in the compartment. Fig. 6.2 shows a CFD model to determine the temperature distribution of fire in the roof structure at airport terminal.

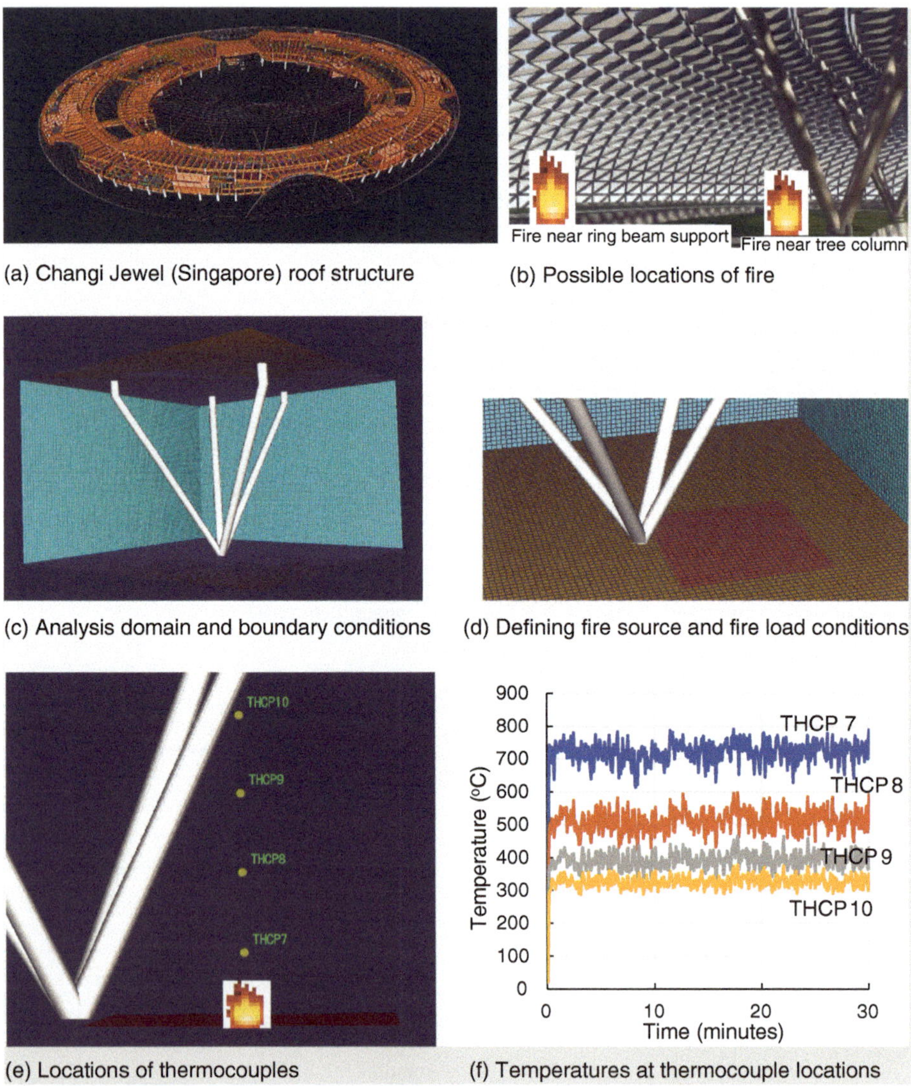

Fig. 6.2 CFD model to determine fire temperatures in Jewel roof structure.

6.3 Fire performance of materials

6.3.1 Thermal properties

Thermal properties involved in heat transfer analysis of a steel-concrete composite column are the temperature dependent thermal conductivity, specific heat, and density (mass loss) of the steel and concrete materials, respectively. The specific heat and density are usually combined together in the form of thermal capacity or volumetric specific heat. The thermal properties of high strength steel and concrete could be much different from those of their normal strength counterparts due to the different micro-structures. Such differences affect temperature distribution and then the fire resistance of the columns. Generally, the thermal properties of steel and concrete can be obtained based on standard test procedures (ASTM E1461, 2013; ASTM E1269, 2018; ISO 22,007–2, 2015). Alternatively, they may be determined based on the additivity theorem which approximately calculates the thermal properties by summing up those of the components of the material in terms of volume or weight fractions (Harmathy, 1970).

6.3.1.1 High- and ultra-high strength concretes

There are many data on the thermal properties of NSC (Naus, 2005), but less data on those of HSC and UHSC in the available literature (Toman and Černý, 2001; Shin et al., 2002; Kodur and Sultan, 2003; Bamonte and Gambarova, 2010; Kodur and Khaliq, 2011; Khaliq and Kodur, 2011; Li et al., 2020; Yang and Park, 2020; Kodur et al., 2020). Nevertheless, useful calculation models have been proposed to determine the thermal properties of NSC (EN 1992–1–2, 2004), HSC (Kodur and Khaliq, 2011) and UHSC (Kodur et al., 2020) for fire resistant design, respectively.

1. Thermal conductivity
   Thermal conductivity is a measure of ability to conduct heat. Results presented in literature indicate that the thermal conductivity of concrete is mainly influenced by the curing age (hardened cement paste), moisture content, type of coarse aggregate, microcracks, pore size and distribution, and the temperature. It slightly increases with an increase in curing age (Yang and Park, 2020), but significantly and almost linearly increases with the increase of moisture content (Kim et al., 2002). Concrete with siliceous aggregates has a higher thermal conductivity than that with carbonate aggregates, indicating the presence of carbonate aggregates in concrete can improve the fire resistance of composite columns, but such effect of type of coarse aggregate is not considered in the Eurocode 2 calculation model for the thermal conductivity of concrete (EN 1992–1–2, 2004).
   Fig. 6.3 shows a typical comparison between the thermal conductivities of NSC, HSC and UHSC. The lower and upper limit values are calculated according to the Eurocode 2 model (EN 1992–1–2, 2004). The lower limit is usually used for NSC and the upper limit can be used for HSC. The other values are taken from the test results of Kodur and Khaliq (2011) on HSC with a compressive strength of 90 MPa, and Kodur et al. (2020) on UHSC having a compressive strength of 164 MPa. Fig. 6.3 shows that the thermal conductivities of HSC and UHSC are higher than those proposed by Eurocode 2. This is because HSC and UHSC have denser and less porous microstructures than NSC, consequently, the air in the voids is quite limited. Considering the density and thermal conductivity of air are much smaller than those of hydrated cement paste, HSC and UHSC exhibit higher thermal

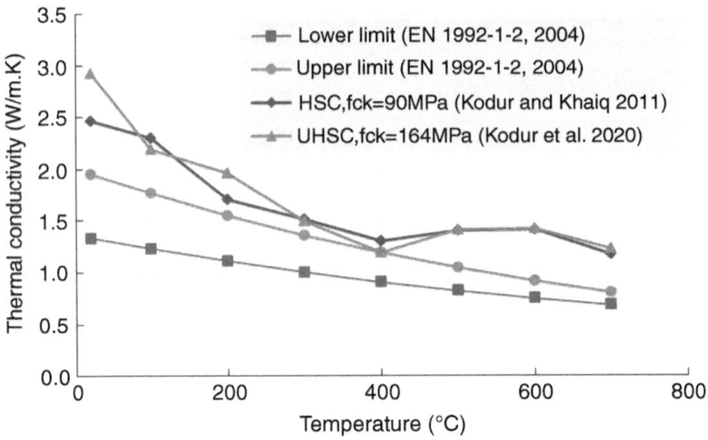

Fig. 6.3 Variations of thermal conductivity of concrete with compressive strength and temperature.

conductivity than NSC based on the additivity theorem. However, the differences in the thermal conductivities between HSC and UHSC are minimal. This may be due to the fact that the fine aggregates were partially substituted by coarse aggregates in UHSC, leading to the similar porous microstructures between HSC in Kodur and Khaliq's test (2011) and UHSC in Kodur et al.s test (2020) ($\rho_{HSC} = 2462$ kg/m^3, $\rho_{UHSC} = 2549$ kg/m^3).

Fig. 6.3 also shows that the thermal conductivities of HSC and UHSC generally decrease with increasing temperature due to the evaporation of free water, decomposition of hydrated cement paste, and the thermal cracks. A small increase can be seen in the range of 400 °C ~ 500 °C due to the release of bound water (Kodur and Khaliq, 2011; Laneyrie et al., 2016), but this small increase of thermal conductivity is not reflected in the Eurocode 2 calculation model. At the low temperatures, concrete with moisture content could have higher thermal conductivity than that at normal temperature (Schneider, 1982).

2. Thermal capacity

Thermal capacity represents the ability of a material to absorb or store heat. It is defined as the product of specific heat and density and stands for the amount of heat required to change the temperature of unit volume of material (1 m^3) by unit temperature (1 °C). The thermal capacity of concrete is influenced by moisture content, type of coarse aggregate, and density of concrete (Kodur and Khaliq, 2011), among which the moisture content plays the leading role. This can be seen from a typical comparison between the thermal capacities of NSC, HSC and UHSC in Fig. 6.4. There is an abrupt increase of thermal capacity for NSC with moisture content in weight varying from 0% (dry concrete) to 3%. This is because water has higher thermal capacity ($\approx$ 4200 kJ/m^3. °K) than the concrete. The thermal capacities of dry NSC, HSC and UHSC are similar due to their low moisture contents.

Fibers such as the steel and polypropylene fibers are usually added in the mixes of HSC and UHSC to improve strength and ductility as well as prevent spalling under fire. Steel fibers have higher thermal conductivity and thermal capacity than those of concrete (EN 1992–1–2, 2004; EN 1993–1–2; 2005); whereas polypropylene fibers have a very low thermal conductivity (0.1 ~ 0.2 W/m °C) and melt at around 170 °C (Xiong and Liew, 2015). The melting of polypropylene fibers increases the pore sizes in the concrete matrix and also leaves empty channels for moisture to escape, which reduces the thermal conductivity and thermal capacity of concrete. However, due to their low dosages in the mix (usually lower

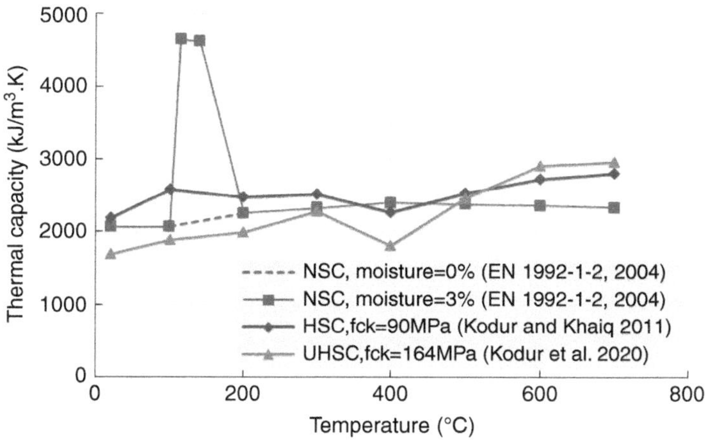

Fig. 6.4 Variations of thermal capacity of concrete with compressive strength and temperature.

than 0.5% in volume to ensure workability and pumpability of concrete), the addition of steel and/or polypropylene fibers has only a marginal influence on the thermal properties of HSC and UHSC (Kodur and Khaliq, 2011; Kodur et al., 2020). In this regard, their influence could be ignored in heat transfer analysis.

6.3.1.2 High strength steels

The strength of steel can be improved by cold-forming process or heat-treatment process. The heat-treatment process includes such as quenching and tempering (QT) process and thermo-mechanically controlled process (TMCP). There are very limited studies on the thermal properties of cold formed or heat-treated steel at elevated temperatures in the literature (Choi et al., 2014; Craveiro et al., 2016; Steau et al., 2020). Fig. 6.5 and Fig. 6.6 show the comparisons of specific heat and thermal conductivity among the cold-formed S280 Gd+Z steel (Craveiro et al., 2016), the heat-treated

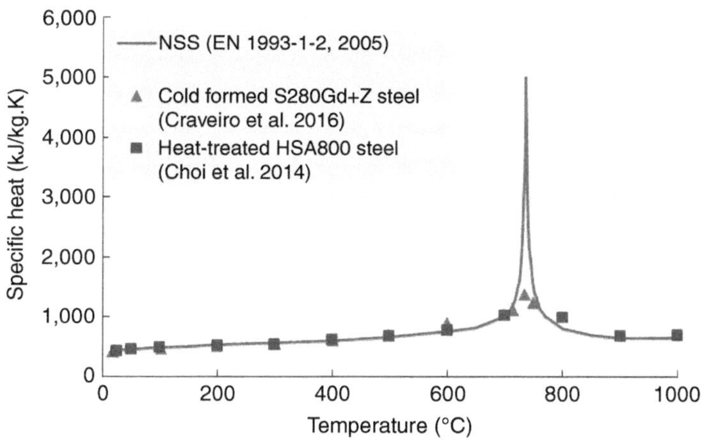

Fig. 6.5 Comparison between specific heat of NSS, cold-formed and heat-treated steels.

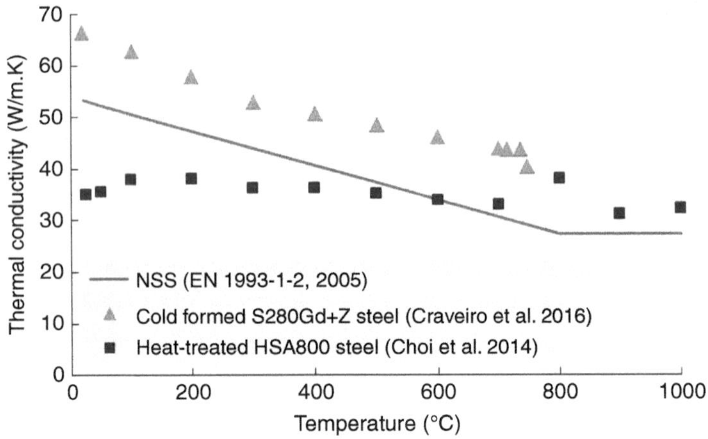

Fig. 6.6 Comparison between thermal conductivity of NSS, cold-formed and heat-treated steels.

high strength HSA800 TMCP steel (Choi et al., 2014), and the hot-rolled mild steel (NSS). The thermal properties of NSS are calculated based on the Eurocode 3 model (EN 1993–1–2, 2005). The comparisons indicate that both the cold-forming and heat-treatment processes influence little on the specific heat before phase change where the ferrite and cementite structures are transformed into the austenite structure at around 735 °C. At the phase change temperature, the specific heat of NSS increases abruptly up to 5000 kJ/kg°K, but those of the cold-formed and heat-treated steels increase slightly.

The cold-forming process could improve the thermal conductivity by 20% according to Fig. 6.6, but the heat-treatment process reduces it when temperature is below 400 °C. This is because the microstructures of the heat-treated HSS are mainly characterized by the martensite or bainite structures which have quite similar thermal conductivity at elevated temperatures (Lyassami et al., 2018). The thermal conductivity of them is higher than that of the hot-rolled NSS featuring the ferrite and cementite structures (Wilzer et al., 2013). When the temperature is higher than 400 °C, the thermal conductivity of the heat-treated HSS is quite similar to that of NSS. This is due to the fact that the microstructure of the heat-treated HSS gradually transforms back to that of the hot-rolled NSS under increasing temperatures (Xiong and Liew, 2016a & 2020).

6.3.2 *Spalling of concrete at elevated temperatures*

HSC is prone to spall under fire. Technically, spalling occurs when the tensile stress in the concrete exceeds its tensile strength. The increase in the tensile stress can be caused by the thermal stresses due to temperature gradient during heating, and the pore pressure due to evaporation of water. There are many factors governing the spalling of HSC, including permeability, moisture content, aggregate type and size, pore size and volume, specimen shape and size, heating rate, and external loading, etc. (Shah et al., 2019). The addition of steel fibers and/or polypropylene fibers could be effective to prevent such spalling of HSC (Zeiml et al., 2006; Hadi, 2007). This is because the addition of steel fibers can increase the tensile strength of concrete. On

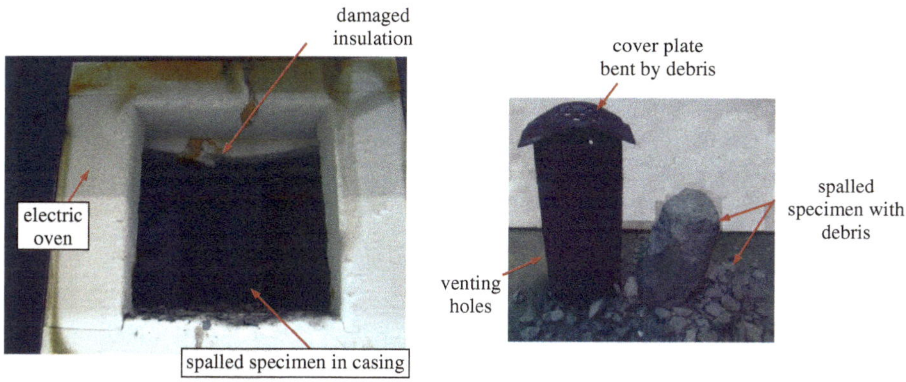

Fig. 6.7 Spalling of UHSC (f_{ck} = 167 MPa) without fibres at high temperature.

the other hand, polypropylene fibers help to reduce the pore pressure after melting at around 170 °C by creating the interconnected channels for water vapor to escape. (Xiong and Liew, 2015).

It is believed UHSC is more likely to spall under fire than HSC owing to the denser structure. Fig. 6.7 shows the spalling of a typical UHSC with a compressive strength of 167 MPa and without any fiber reinforcement (Xiong and Liew, 2015). The spalling occurred when the cylinder specimen ($d \times h$ = 100 mm × 200 mm) was heated up to 490 °C (the heating rate was 5 °C/min), causing the outer layer of the concrete cylinder breaking into pieces and the cover plate of the steel casing bent by the flying debris. The insulation of the oven was also damaged by the spalling.

After adding the steel fibers with a dosage of 1.0% in volume in the mix of UHSC, the spalling still happened as shown in Fig. 6.8(a), but the outer layer of the concrete did not scale down. After adding the polypropylene fibers with a dosage of 1.0% in volume, the spalling was completely prevented as shown in Fig. 6.8(b). Hence, the

(a) UHSC with steel fibres of 1.0% in volume after heating (f_{ck}=170 MPa)

(b) UHSC with polypropylene fibres of 1.0% in volume after heating (f_{ck}=144 MPa)

Fig. 6.8 Spalling behavior of UHSC cylinders after adding steel or polypropylene fibers and heating.

addition of polypropylene fibers is effective to prevent spalling of the UHSC. It is worth noting that the addition of polypropylene fibers could reduce the compressive strength because of the reduced workability of fresh concrete (flowability, pumpability, etc.). To ensure the compressive strength and workability of UHSC were not reduced significantly, further trial tests have been carried out with varied dosage of polypropylene fibers. It is found that the polypropylene fibers with a dosage of 0.1% in volume is sufficient to prevent the spalling of UHSC in the meanwhile maintaining a good workability that allows the UHSC to be pumped into the hollow steel tubes (Xiong and Liew, 2016b&2020b).

EN1992-1–2 (2004) specifies that, for concrete grades C55/67 to C80/95, the spalling is unlikely to occur when the moisture content is less than 3% and the maximum content of silica fume is less than 6% by weight of cement. Above these limit values, a more accurate assessment of moisture content, silica fume content, type of aggregate, permeability of concrete and heating rate should be considered. The assessment could be based on laboratory trial or specialist advices.

For concrete grades higher than C80/95, may occur when the concrete is directly exposed to fire, regardless of the content of moisture and silica fume. Therefore, at least one of the following methods should be provided:

- Include in the concrete mix more than 2.0 kg/m^3 of monofilament propylene fibers.
- A type of concrete for which it has been demonstrated by local experience or by testing that no spalling occurs under fire exposure.

As mentioned above, the addition of propylene fibers affects the workability of concrete. In cases where HSC and UHSC are pumped in CFST columns, the pumpability should be checked specifically. Alternatively, the specialist advice should be consulted, regarding the pumping height, pumping rate, etc.

Furthermore, for the release of vapour from the CFST columns, the hollow steel section shall contain steaming holes with a diameter of not less than 20 mm located at least one at the top and one at the bottom of the column in every storey, as shown in Fig. 6.9. The spacing of these holes should not exceed 5.0 m along the column length. The steaming holes should be sealed during casting of concrete in case of leaking, but removed after hardening.

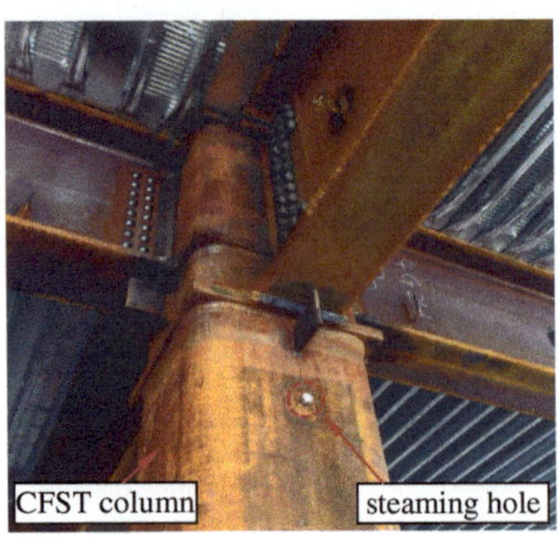

Fig. 6.9 Steaming hole at top of a CFST column.

6.3.3 Mechanical properties

To evaluate the fire resistance of steel-concrete composite columns using fire engineering approaches introduced in Section 6.6 (EN 1994–1–2, 2005; Xiong and Liew, 2018), the mechanical properties of high strength materials, such as the compressive strength and elastic modulus at elevated temperatures, are required. To obtain the full-spectrum load-deformation response of the composite column under fire using the advanced calculation models introduced in Section 6.7 (EN 1994–1–2, 2005), their stress-strain relationships are needed. The advanced calculation models also need the thermal expansions of the constituent materials to be known to determine the thermal stresses in the composite columns. The thermal expansion is often defined as one type of the thermal properties of materials (EN 1992–1–2, 2004; EN 1994–1–3, 2005), but it is covered in this part with the other mechanical properties. Abovementioned mechanical properties of materials could be investigated conforming to a series of test standards in the literature (RILEM 2007; ISO 6892–2, 2018; ASTM E 21, 2020).

6.3.3.1 High- and ultra high- strength concretes

1. Test methods
   Unstressed and stressed test methods can be used to investigate the mechanical properties of concrete at elevated temperatures (RILEM 2007), which form the upper and lower bounds of such properties, respectively. A typical set-up of unstressed or stressed test is shown in Fig. 6.10. For the unstressed method, the specimen is loaded to failure under a constant temperature. This is different from the stressed method whereby the specimen is heated to failure under a constant load level. The unstressed method is mostly used due to the convenience to obtain the stress-strain curves directly. However, this is difficult for the stressed method as the measured strain includes not only the mechanical strain but also the thermal and transient creep strains. Therefore, supplementary tests are usually required to

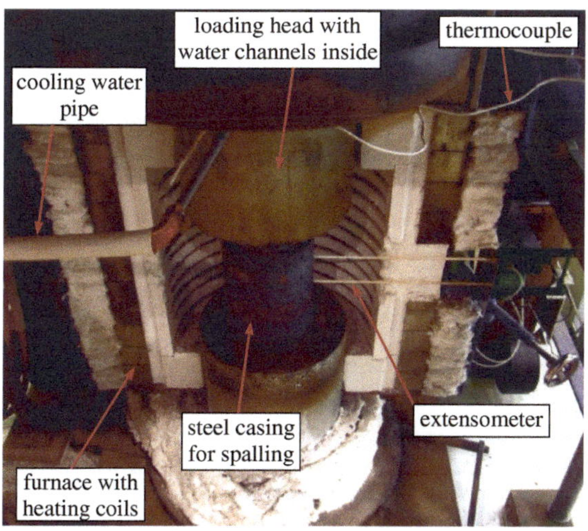

Fig. 6.10 Typical setup of standard compression test on mechanical properties of concrete at elevated temperatures.

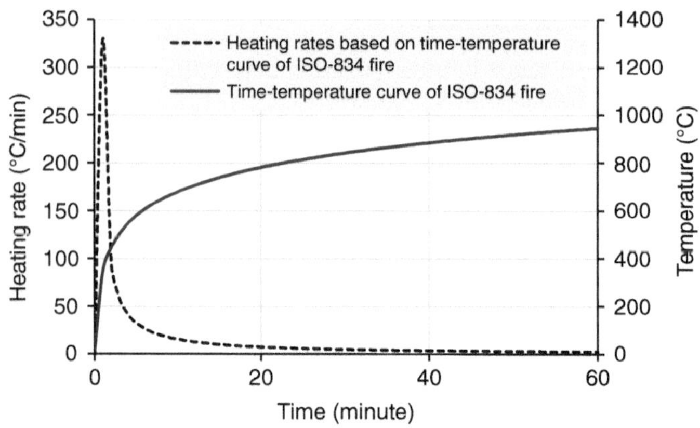

Fig. 6.11 Time-temperature curve of ISO-834 fire and heating rates.

separate them. Their main difference is that the stressed method could capture the transient creep strain. For a composite column under fire conditions, ignoring the transient creep strain could overestimate the buckling resistance of the composite column, however the overestimation may not be large due to the existence of non-uniform temperature distribution through its cross-section (Huang and Burgess 2012).

Heating rate affects the mechanical properties of concrete significantly. In fact, the heating rate varies when a structural member is subjected to a realistic fire. However, it is difficult to consider the varying heating rate in the compression test of concrete at the elevated temperatures. RILEM recommends a heating rate of 1 °C/min for such material test, but such a heating scheme leads to a very long time duration for testing one specimen. By observing the heating rate of the standard ISO-834 fire shown in Fig. 6.11, it is found that, within the first minute, the heating rate can go up to 329 °C/min, but it rapidly drops to 5 °C/min after 25 min and 2 °C/min after 60 min. Especially, when the concrete is infilled in the steel tubes for CFST columns, the heating rate to the concrete core would be further lower due to the heat sink effects of the steel tube and the fire protection if any. Therefore, taking a heating rate of 5 °C/min in the standard compression test of concrete (unstressed or stressed) is reasonable to investigate its mechanical properties. The heating rate of 5 °C/min will result in a conservative design because it is higher than the RILEM specified value of 1 °C/min.

2. Thermal expansion

Cement paste and aggregate are two major components of NSC. The thermal expansion of NSC depends mainly on the type of aggregate because the aggregate constitutes a major proportion of the mix compared with the cement paste. In general, NSC with siliceous aggregates such as the flint aggregates has a higher thermal expansion than that of NSC with calcareous aggregates such as limestones. Since very limited amount of coarse aggregates are used in HSC and UHSC, their thermal expansion properties can be different from that of NSC

Fig. 6.12 shows a comparison between the thermal expansion of NSC, HSC and UHSC. The values of NSC are calculated from the Eurocode 2 model (EN 1992–1–2, 2004), whereas the others are test values. The limestone coarse aggregates were used to partially replace the fine aggregates in the mix of UHSC reported by Kodur et al. (2020), whereas no coarse aggregates were used in the mix of UHSC of Li et al. (2020). Generally, the thermal expansion increases with the temperature up to 600 °C. Above 600 °C, there is a drop in the

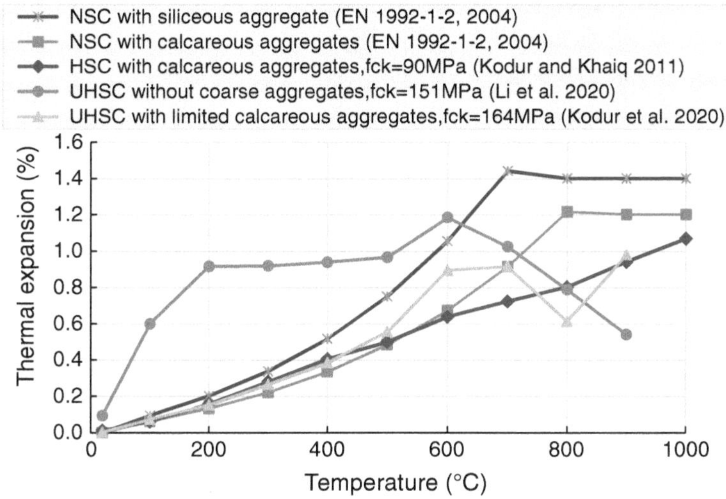

Fig. 6.12 Comparison between thermal expansion of typical NSC, HSC and UHSC

thermal expansion of the UHSC mainly due to the higher shrinkage occurred). With the use of the same type of aggregate, e.g., the calcareous coarse aggregates, the thermal expansions of NSC, HSC and UHSC are similar (EN 1992–1–2, 2004; Koudur and Khaiq, 2011; Kodur et al., 2020). Without the use of coarse aggregates, the thermal expansion of UHSC (Li et al., 2020) is higher than that of the counterpart UHSC using the coarse aggregates (Kodur et al., 2020) before 600 °C. Above 600 °C, their thermal expansions similarly drop with an increase in the temperature.

3. Compressive strength

The compressive strength of concrete at elevated temperatures is significantly affected by the type of aggregate. Fig. 6.13 shows a comparison between the retention factors of

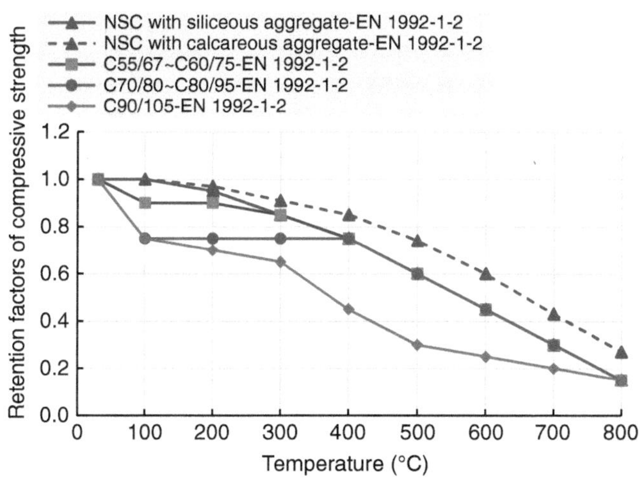

Fig. 6.13 Retention factors of compressive strength of NSC and HSC provided in EN 1992–1–2.

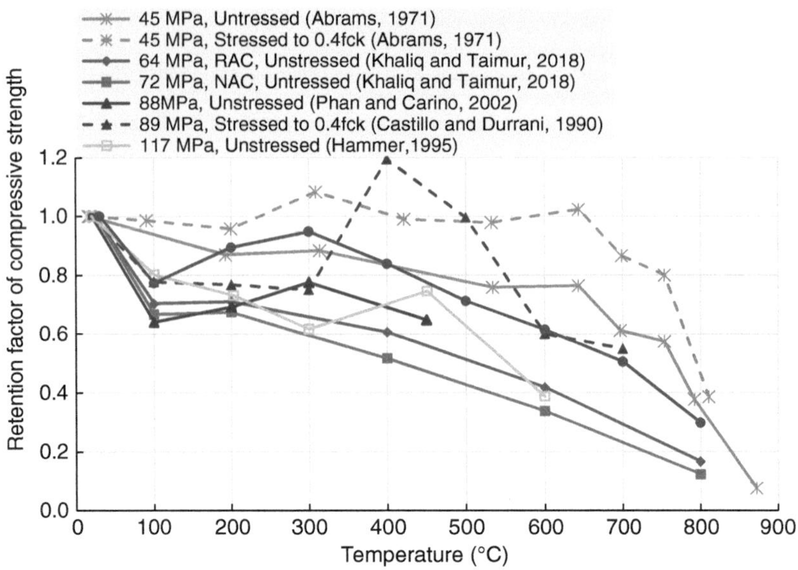

Fig. 6.14 Retention factors of compressive strength of NSC, HSC and UHSC in literature.

the compressive strength of NSC and HSC at elevated temperatures calculated from the Eurocode 2 model (EN 1992–1–2, 2004). The retention factor is defined as the ratio of the compressive strength at elevated temperatures and that at normal temperature. The compressive strength of concrete made with siliceous aggregates reduces faster that made with calcareous aggregates at elevated temperatures. Moreover, the compressive strength of HSC reduces faster than that of NSC at elevated temperatures.

Similar comparison between the retention factors of compressive strength of NSC, HSC and UHSC at elevated temperatures is shown in Fig. 6.14. The values are based on test results in the literature. For concretes having similar strengths (Abrams, 1971; Castillo and Durrani, 1990; Phan and Carino, 2002), the compressive strength measured in the stressed tests is reduced less than that measured in the unstressed tests. This is because of the slower development of cracks under the imposition of a preload (Cree et al. 2017). HSC with the recycled coarse aggregates (RCAs) has the smaller retention factors than that with the natural coarse aggregates (NCAs). This is owing to the more porous microstructure of HSC with RCAs that allows easier moisture migration and produces fewer thermal cracks at elevated temperatures (Khaliq and Taimur, 2018).

It is worth noting in Fig. 6.14 that the compressive strength sharply reduces at 100 °C ~ 200 °C, and slightly recovers at 300 °C. It is believed that the chemical composition of the cement paste is not noticeably changed around 100 °C ~ 200 °C. Hence, the sharp reduction of strength could be either due to the built-up internal pressure by the evaporation of free water or the expansion of water between the C-S-H structures causing a decrease in the surface forces (Xiong and Liew, 2016b). For the recovery of strength at 300 °C, it could be attributed to the general stiffening of the cement gel by shrinkage. In other words, the increase of surface forces (Van der Walls forces) between the gel particles due to the removal of water (Behnood and Ziari, 2008). The temperature at which water is removed and the strength begins to recover depends on the porosity of the concrete (Chen and Liu, 2004). Above 300 °C, the compressive strength decreases due to the decomposition of hydration products such as C-S-H and $Ca(OH)_2$, the deterioration of aggregates, and the cracks due to thermal incompatibility between aggregates and cement pastes that leads to stress concentration.

4. Elastic modulus
The elastic modulus of concrete at elevated temperatures is mainly governed by the type of aggregate and the water/cement ratio (Naus, 2006; Bahr et al., 2013). Similar to the effect of aggregate type on the compressive strength, the concrete with siliceous aggregates such as quartz aggregates has greater reduction in elastic modulus than that with the calcareous aggregates such as limestone aggregates. The reduction of elastic modulus increases with an increase in the water/cement ratio at elevated temperatures. Regarding the test methods, the stressed tests generally produce higher elastic modulus than the unstressed tests (Schneider et al., 1982).
Fig. 6.15 shows a comparison between the retention factors of elastic modulus of NSC, HSC and HSC based on the unstressed test results. The elastic modulus decreases with an increase in the temperature. The sharp reduction and recovery still can be found in UHSC but not noticeable for the NSC and HSC

5. Stress-strain relationship
Experimental works were carried out on the stress-strain relationship of HSC at elevated temperatures (Diederichs et al., 1988; Castillo and Durrani, 1990; Felicetti et al., 1996; Cheng et al., 2004). Stress-strain models have been proposed for numerical analysis. A comparison between the stress-strain curves of a 100 MPa concrete calculated based on the Eurocode 2 model (EN 1992–1–2, 2004) and the Cheng et al.s model (2004) is shown in Fig. 6.16. Up to 400 °C, the compressive strengths in both models are quite similar, but the elastic modulus based on Eurocode 2 model is smaller than that based on the Cheng et al.s model. Overall, the compressive stress after peak according to the Cheng et al.s model drops faster than that according to the Eurocode 2 model, indicating a more brittle behavior.

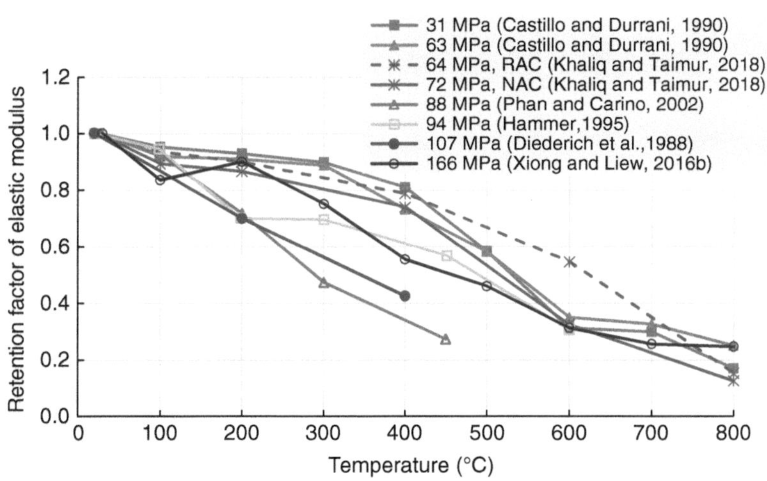

Fig. 6.15 Retention factors of elastic modulus of NSC, HSC and UHSC in literature.

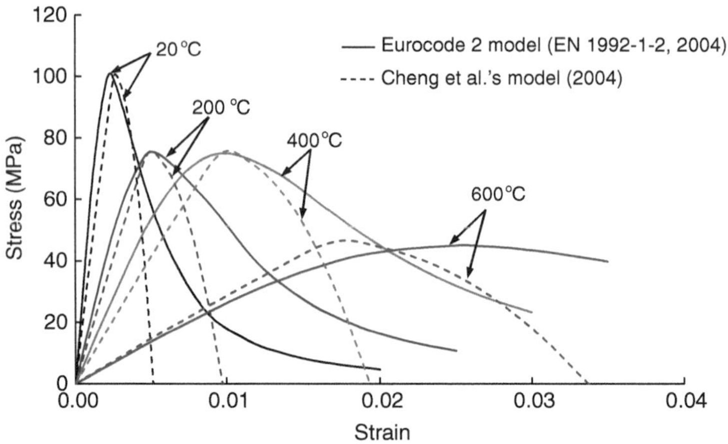

Fig. 6.16 Stress-strain curves of HSC at high temperatures.

6.3.3.2 High strength steels

1. Test methods
   Similar to the test methods of concrete under axial compression, there are also stressed and unstressed methods widely adopted for steel tested under tension at elevated temperatures, which are transient-state method and steady-state method, respectively. In the steady-state tests, the coupon specimen is firstly heated up to a target temperature and then loaded to fail. The stress-strain curves at the target temperatures can be directly obtained during loading. In the transient-state tests, the coupon specimen is heated under a constant pre-load until it fails. The temperature-strain curves are recorded directly during heating instead of the stress-strain curves.
   The temperature-strain curves can be converted into the stress-strain curves as shown in Fig. 6.17. Since only very limited data points on the temperature-strain curves can be used to form the stress-strain curves, the stress-strain curves are not smooth. A large number of pre-load levels are required to make the stress-strain curves smooth as the temperature-strain curves. This is costly and time-consuming. It should be mentioned for the conversion

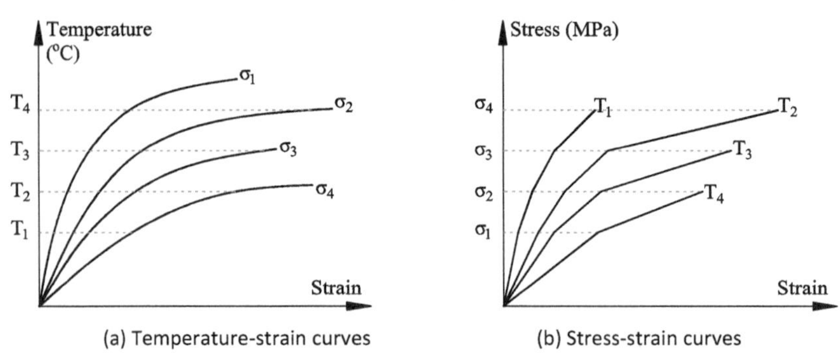

Fig. 6.17 Conversion from temperature-strain curves to stress-strain curves of steel at high temperatures.

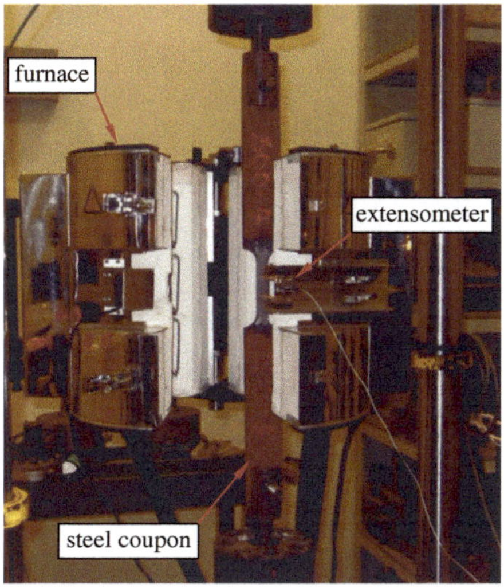

Fig. 6.18 Typical test setup of steel under tension at elevated temperatures.

that the thermal strain should be subtracted from the total strain recorded in the transient-state test. By doing so, a separate heating test for the thermal strain is usually required. Generally, the steady-state tests are easier to be carried out than the transient-state tests. Therefore, they are more widely used (Outinen et al., 2001). However, the transient-state tests give more realistic experimental results since they simulate the true loading condition and capture the thermal creep effect of the load-sustained structural members exposed to fire. Previous studies on the temperature dependent mechanical properties of mild steels indicate that the transient-state tests are preferred, although the steady-state tests may yield similar results at small strain levels of steel (Kirby and Preston, 1988).

The preparation of coupon specimen and test procedure could refer to ISO 6892–2 (2018) and ASTM E 21 (2020). A typical test setup of steel coupon under tension is shown in Fig. 6.18. Due to the higher thermal conductivity and the smaller section size compared with those of concrete specimen, the heating rate for steel coupon could go up to 10 °C/min (Xiong and Liew, 2016a). The loading rate may be taken as a nominal strain rate of 0.006/min (gauge length = 25 mm) in the steady-state tests, and 1.5 kN/min in the transient-state tests.

2. Effects of heat-treatment process

There are two major ways of producing HSS. One is adding alloy elements such as carbon and manganese elements, but it worsens the fabrication & installation properties of HSS, in particular, the weldability. The other way is by means of the heat-treatment processes that generally produce the finer-grained microstructures. As mentioned above, there are typically two heat-treatment processes, i.e. QT and TMCP, to produce the HSS. Compared with the hot-rolling process for manufacturing NSS, QT process consists of additional rounds of heating, quenching, and tempering process, whereas TMCP process involves accelerated cooling and tempering after hot-rolling. As the heat-treatment processes are different, the QT and TMCP HSS may have different physical and mechanical properties. It is reported that the TMCP S960 produced significantly higher residual stresses than the QT S960 under the same welding conditions (Schaupp et al., 2017), and the TMCP steel exhibited better performance in terms of toughness than QT steel when the yield strength exceeded 1000 MPa (Yu et al., 2011).

Table 6.1 Chemical compositions of typical NSS and HSS (wt percent).

	C	Si	Mn	S	P	Cr	Mo	Nb
MS-S355	0.15	0.29	1.02	0.022	0.024	0.080	0.008	0
QT-S690	0.15	0.42	1.40	0.002	0.018	0.020	0.004	0.036
TMCP-S690	0.15	0.34	1.21	0.002	0.014	0.070	0.110	0
	V	**Ni**	**Ti**	**N**	**Cu**	**Al**	**B**	C_{eq}
MS-S355	0	0.060	0	0.008	–	–	–	0.216
QT-S690	0.068	0.016	0.026	0.003	0.0029	0.039	0.002	0.252
TMCP-S690	0	–	0.010	–	0.0100	0.026	0.001	0.238

Note: 1) "-" Not provided in the mill certificates.
2) C_{eq}:= C+Si/30+Mn/20+Cu/20+Ni/60+Cr/20+Mo/15+V/10+5B (WES 3001, 2013).

For the heat-treated HSS, different chemical compositions and microstructures should be investigated to understand their different physical and mechanical properties at elevated temperatures. A typical comparison between the chemical compositions of NSSMS-S355 (Yan et al., 2014), HSSQT-S690 (Corus, 2006) and HSSTMCP-S690 (Endo and Nakata, 2015) are given in Table 6.1. The nominal thickness of plates is 12 mm. HSS and NSS have similar carbon contents although they may be slightly different when the plate thicknesses are different. The main difference between the NSS and HSS is the contents of alloy elements, such as the Niobium, Titanium, and Vanadium, etc. The contents of such alloy elements are low, but they are essential for HSS to suppress the formation of soft ferrite and pearlite structures and facilitate the growth of hard bainite and martensite structures after rapid cooling (Llewellyn and Hudd, 1998). Compared with QT-S690, TMCP-S690 contains fewer alloy elements. Sometimes there are no Niobium and Vanadium. Therefore, the high strength of TMCP-S690 is more likely to be achieved by the accelerated cooling process with the cooling rate close to the theoretical limit as reported by JFE (Shikanai et al., 2008). The accelerated cooling could extremely hinder the solid-state transformation of soft ferrite and pearlite (Bhadeshia and Honeycombe, 2017). Obviously, the carbon equivalent of NSSMS-S355 is the smallest, whereas that of TMCP-S690 is smaller than QT-S690 mainly due to the fewer alloy elements added, leading to the better weldability.

Fig. 6.19 shows the heat-treatment process of QT-S690 plates (Corus, 2006). The hot-rolling process is finished above the A_{c3} temperature, then the plate is slowly annealed to room temperature with the microstructure characterized by the ferrite and cementite structures. After that, it is reheated above the A_{c3} temperature with the microstructure fully austenitized and most alloy elements completely dissolved. It is then quenched at a rapid cooling rate to form the martensite structure. As the martensite is hard and brittle, the plate is finally tempered below the A_{c1} temperature where the austenite structure starts to form. The tempering process is to modify the mechanical behavior of HSS by achieving a good balance between the strength and the toughness.

The different heat-treatment process of TMCP-S690 plates (Endo and Nakata, 2015) is shown in Fig. 6.20. There is no annealing process and the hot-rolling process is terminated in a non-recrystallized temperature range below A_{c3} in order to produce the plastic strains in austenite. Accelerated cooling is then applied to the plates but interrupted in the temperature range between the bainite transformation start temperature (B_s) and the finish temperature (B_f). At this time, the microstructure is characterized by the bainite and retained austenite. Following that, the plates are rapidly heated to a temperature close to

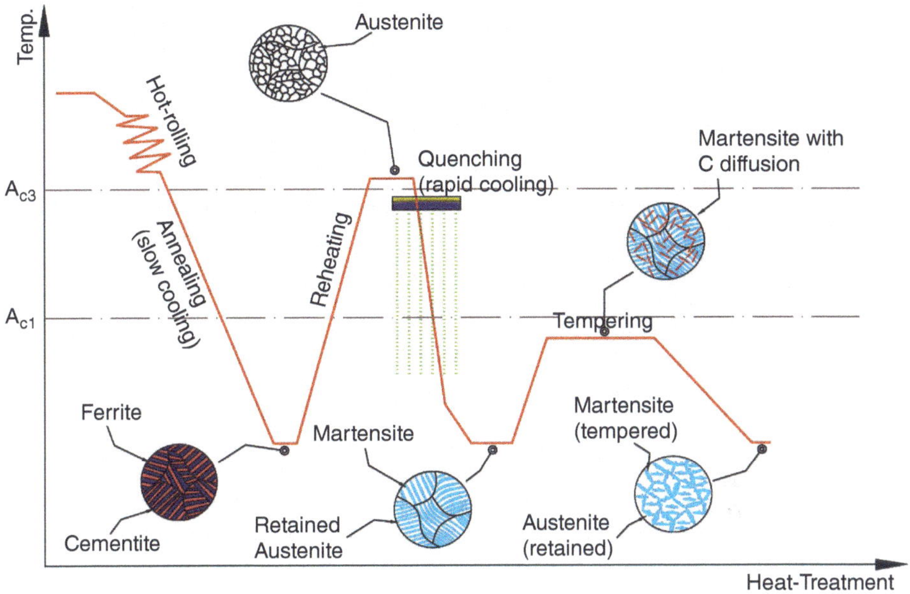

Fig. 6.19 Heat-treatment process of QT-S690 steel.

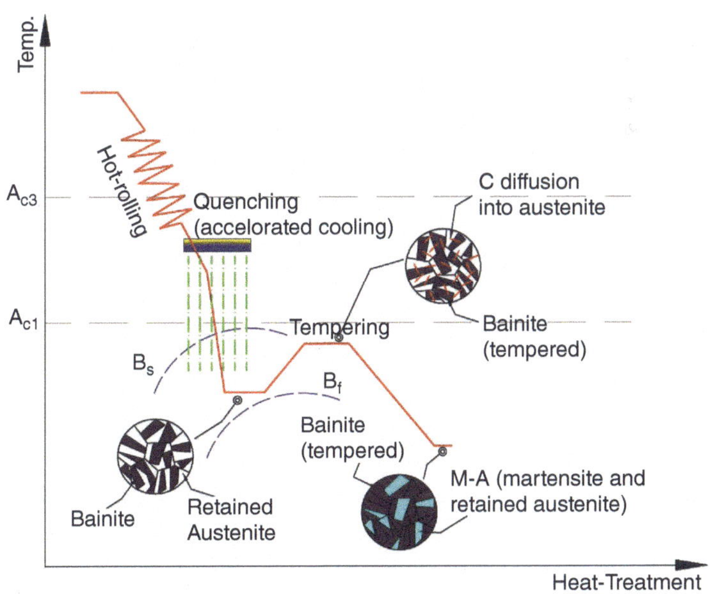

Fig. 6.20 Heat-treatment process of TMCP-S690 steel.

A_{c1} where the supersaturated carbon in bainite diffuses into the retained austenite therefore more bainite recovers. The highly hardenable retained austenite subsequently transforms into an M-A structure due to the high carbon solution when the plates cool to room temperature. Therefore, the microstructure is characterized by a hard phase structure (M-A) and a relatively soft phase structure (tempered bainite).

Through the comparisons between the two typical types of QT and TMCPHSS, it is found their chemical compositions and microstructures are different. This would lead to different physical and mechanical properties at elevated temperatures which are introduced hereinafter.

3. Thermal expansion

Thermal expansion of steel is a tendency to change the separation between the atoms with the changed temperature. It is dominated by the inter-atom forces. The thermal expansions of mild steel NSS, QT-S690 HSS and TMCP-S690 HSS are shown in Fig. 6.21. The values of mild steel NSS are calculated from the Eurocode 3 model (EN 1993–1–2, 2005), whereas those of QT-S690 (Xiong and Liew, 2016a) and TMCP-S690 (Xiong and Liew, 2020a) are taken from the test results. The thermal expansions of TMCP-S690 and QT-S690 are smaller than those of mild steel, and the differences are in the range of 1.9% ~ 36.2% and more noticeable when the temperatures are low. The differences are mainly due to the different microstructures of HSS and mild steel. It is well known that an object expands when it is heated. This is because the heat-induced kinetic energy causes atomic vibration. As a result, each atom will take up more space when it is moving. As mentioned above, HSS features mainly the martensite or bainite structures having high interatomic forces, therefore it is difficult for atoms to move. The mild steel is mainly characterized by the weaker structures such as ferrite and cementite, therefore, the atoms are free to move more. In this regard, the HSS tends to have a lower thermal expansion than the mild steel.

4. Elastic modulus

The elastic modulus of steel is a measure of the steel's resistance to elastic deformation under an applied force. To illustrate the effects of different test methods, a comparison between the retention factors of elastic modulus of TMCP-S690 from the steady- and transient-state tests is shown in Fig. 6.22 (Xiong and Liew, 2020a). For temperatures lower than 300 °C, the retention factors from the transient-state tests are comparable with the steady-state ones, and their differences are insignificant with a value of 6%. When the temperature reaches 300 °C and above, a noticeable difference can be observed. The transie

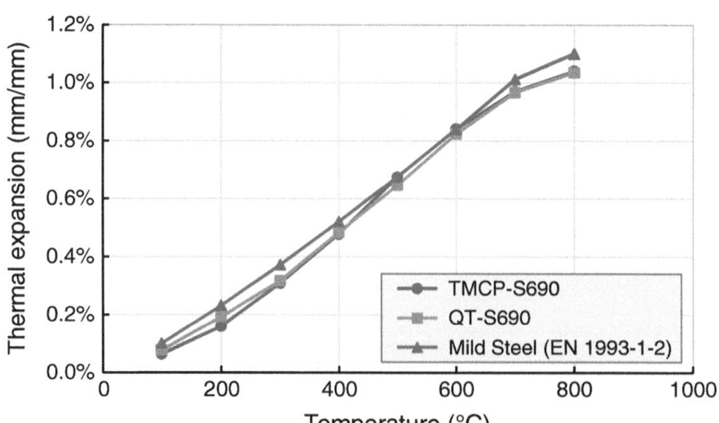

Fig. 6.21 Thermal expansions of NSS and HSS at elevated temperatures.

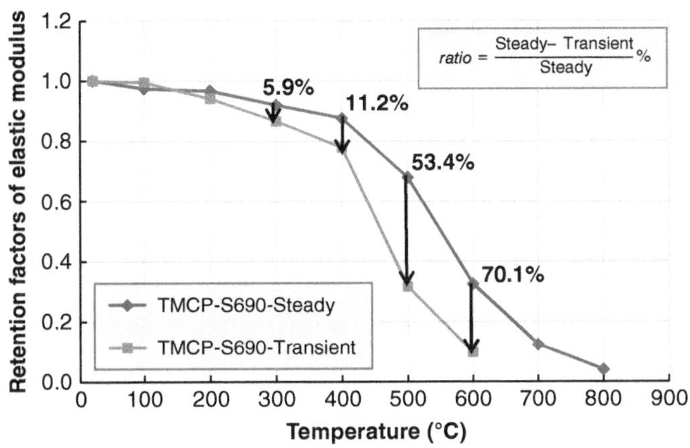

Fig. 6.22 Comparison of retention factors of elastic modulus of TMCP-S690 between steady-state and transient-state tests.

nt-state tests give 5.9% ~ 70.1% lower retention factors as compared to the steady-state tests. The remarkable drop in elastic modulus in the transient-state tests is because of the thermal creep strain involved. The thermal creep can be defined as a time-dependent behavior during which the strain increases continuously while the stress and temperature are kept constant. It has been reported that, at high temperatures, the thermal creep and inelastic deformation can happen even in the elastic region, and the deformation decreases the slope of the elastic region (Sawada et al., 2005). Generally, the creep deformation of steel becomes significant at temperature above 400 °C which is about one third the melting point of steel (Xiong and Liew, 2016a), but it has been experimentally proven that, at high stress levels, the thermal creep can happen at temperature even lower than 400 °C if the applied stress is relatively high (Huang et al., 2006). The comparison shown in Fig. 6.22 also agrees with this statement, as the significant drop in elastic modulus is observed from 300 °C onwards.

To illustrate the effects of steel strength and heat-treatment process, a comparison between the retention factors of elastic modulus of NSS and HSS is shown in Fig. 6.23. For temperatures lower than 400 °C, the retention factors of TMCP-S690 and QT-S690 are slightly higher than the mild steel due to the finer grain size. As the microstructure is coarsening at the higher temperatures, the retention factors of HSS are noticeably smaller than those of the mild steel, indicating a higher thermal creep strain in HSS. This is reasonable since the thermal creep depends on the magnitude of stress sustained. The higher the stress sustained in the HSS, the greater is the thermal creep strain. Nevertheless, the heat-treatment process has little effect on the elastic modulus of HSS.

5. Effective yield strength

Steel does not exhibit distinctive yield plateau at elevated temperatures. Therefore, effective yield strengths which correspond to different strains may be used in different situations for fire resistant design of steel structures. According to BS 5950–8 (2003), the effective yield strengths corresponding to 0.5% ~ 1.5% total strains are used for members in compression depending on the member slenderness ratios. For 3-side heated steel beams, the effective yield strength corresponds to a strain of 2.0% is adopted. The effective yield strengths correspond to 0.5% ~ 2.0% strains are applied for composite beams based on whether it is protected or not. Different from BS 5950–8 (2003), Eurocode 3 (EN 1993–1–2, 2005) recommends a total strain of 2.0% to determine the effective yield strength of steel for the fire resistant design of steel structures in all situations.

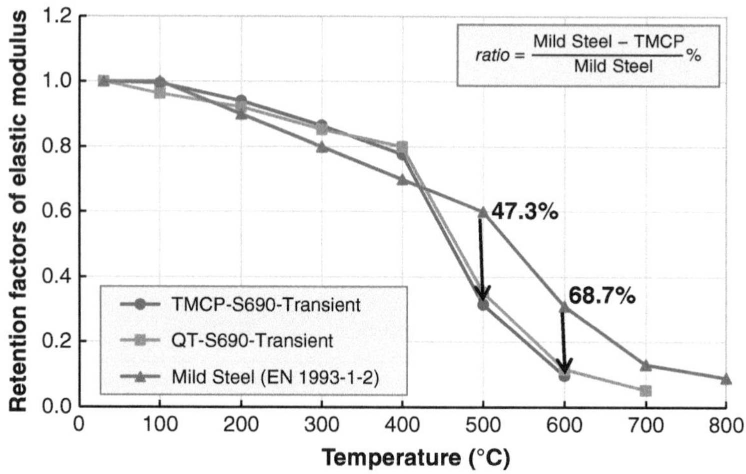

Fig. 6.23 Comparison of retention factors of elastic modulus between mild steel, TMCP-S690 and QT-S690 based on transient-state results.

Stick to the definition given by Eurocode 3, the temperature dependent retention factors of effective yield strength of HSS based on different test methods are compared in Fig. 6.24 (Xiong and Liew, 2020a). When the temperature is lower than 400 °C, their difference is not noticeable since the thermal creep is insignificant. When the temperature is higher than 400 °C, the retention factors obtained from transient-state tests are significantly smaller than those obtained from steady-state tests by up to 47.2%. The higher the temperature is, the larger is the difference. The difference could be attributed to the creep effect involved in the transient-state tests.

The effects of steel strength and heat-treatment process are shown in Fig. 6.25. The strength retention factors of TMCP-S690 and QT-S690 are smaller than those of the mild steel, and the

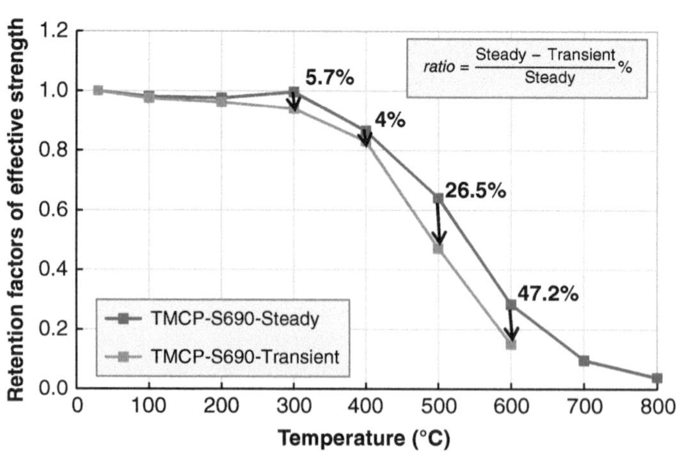

Fig. 6.24 Comparison of retention factors of effective strength of TMCP-S690 between steady-state and transient-state tests.

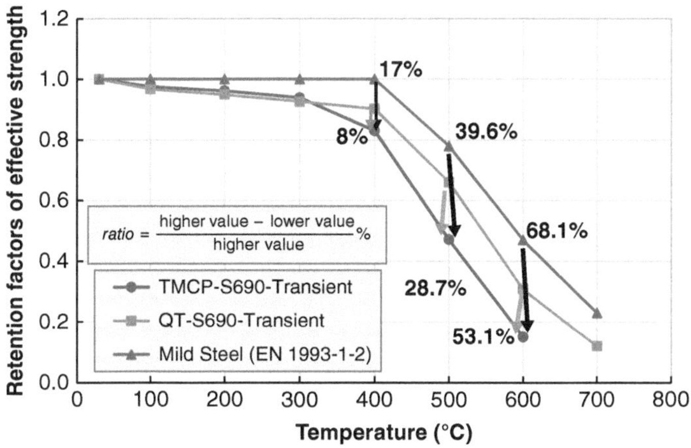

Fig. 6.25 Comparison of retention factors of effective strength between mild steel, TMCP-S690 and QT-S690 based on transient-state results.

higher the temperature is, the larger is the difference. The reason for the difference is due to the microstructural changes. The initial microstructure of HSS is dominated by martensite or bainite structure that is transformed from the ferrite and cementite structures through the heat-treatment process. The high strength is owing to the resulted finer grain sizes. With the temperature increasing, recrystallization happens in HSS where the larger grains grow at the expense of the smaller grains (Mittemeijer, 2010). In other words, the finer martensite or bainite structure gradually approaches a coarser structure with cementite and ferrite which characterizes the microstructure of mild steel. Thus, the mechanical properties of HSS at high temperatures become similar to those of mild steel (Xiong and Liew, 2016a & 2020a). Since the retention factor is calculated as a ratio of the high temperature strength and the normal temperature strength, the HSS with higher normal temperature strength has a smaller retention factor.

Fig. 6.25 also shows that the strength retention factor of TMCP-S690 is smaller than that of QT-S690 by 8% ~ 53.1% when the temperature is higher than 300 °C. Such difference is due to their different thermal creep behaviors. According to abovementioned descriptions on their microstructures under heat-treatment, the microstructure of QT-S690 steel is mainly characterized by the martensite structure whereas it is dominated by bainite structure in TMCP-S690. Also, the heating during test is in fact another round of tempering process. The tempered martensite usually yields relatively finer martensitic laths, whereas the tempered bainitic microstructure shows relatively larger bainitic plates (Wurmbauer et al., 2010), leading the TMCP-S690 steel to have a relatively poorer creep behavior (Abe, 2004; Aghajani et al., 2009). Overall, the effective strength of TMCP-S690 drops more than that of QT-S690 at the elevated temperatures.

6. Stress-strain relationship

Various stress-strain-temperature (σ-ε-T) relationships for NSS are available in the literature. The relationships could be bilinear (Jeane, 1985; Contro et al., 1988), tri-linear (Corraddi et al., 1990), or quadrolinear (Ianizzi et al., 1991). There are also models using an elliptical curve to approximate the transition part between the proportional stress and yield stress (Purkiss, 1988; Rubert and Schaumann, 1988). Besides, the models from Lie and Chabot (1990), Poh (2001) and Eurocode 3 (EN 1993–1–2, 2005) have been widely used in numerical analysis. Their comparison is shown in Fig. 6.26. The Lie and Chabot's

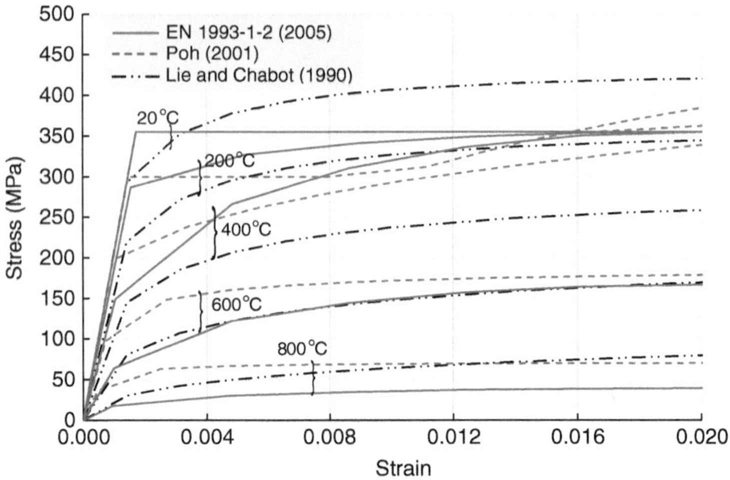

Fig. 6.26 Comparison between stress-strain curves of steel at elevated temperatures.

model is developed based on compression tests on hollow steel tubes, thus, implicitly considered the effects of thermal creep at elevated temperatures. The thermal creep is not included in Poh's model. No strain-hardening is considered in the Eurocode 3 model, and the implementation of it is more complicated than the Lie and Chabot's model since more parameters are involved.

It is recommended that the Eurocode 3 model can be used for HSS at elevated temperatures (Xiong and Liew, 2016a & 2020a). A comparison between the test stress-strain curves of TMCP-S690 steel and the fitted stress-strain curves based on the Eurocode 3 model is shown Fig. 6.27. The fitted curves are obtained by substituting the tested elastic moduli and strengths at elevated temperatures into the Eurocode 3 stress-strain model. Fig. 6.27 shows a reasonable agreement between the test and fitted stress-strain curves.

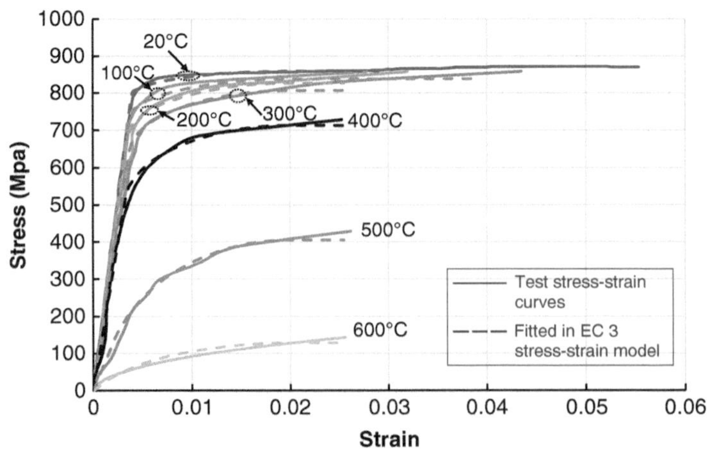

Fig. 6.27 Test and fitted stress-strain curves of TMCP-S690 from transient-state tests.

6.4 Temperature fields

Calculation methods and design charts have been provided to determine the temperature fields of steel and reinforced concrete members in some design codes. For example, Eurocode 3 (EN 1993–1–2, 2005) provides a simplified formula to calculate the temperatures of both unprotected and protected steel members based on section factors and uniform temperature distributions in the steel section and the fire protection. Eurocode 2 proposes design charts to determine the temperature profiles of slabs, beams and columns at elevated temperatures (EN 1992–1–2, 2004). However, for steel-concrete composite members, no relevant methods are provided in Eurocodes. Considering there is thermal contact resistance at the steel-concrete interface, advanced finite element method (FEM) or finite difference method (FDM) can be resorted to for the calculation of the temperature field. FEM is applicable to not only the individual members, but also the subassemblies or entire structures. FDM is limited to the individual members only but the advantage of it is the seamless integration of the calculated temperatures into the fire engineering approaches provided in Section 6.6 for calculating the fire resistance of the steel-concrete composite members, e.g., buckling resistance, N-M interactive curves, etc. To illustrate the implementation of FDM for temperature calculation of composite members, an introduction to the calculation procedure of temperature field of a protected circular CFST column is given hereinafter.

6.4.1 Heat transfer analysis by fdm

Heat transfer in the circular CFST column can be described by the partial differential governing equation. given in Eq. (6.2), it is formulated in a polar coordinate system. T is the temperature and t is the fire exposure time; λ, ρ, c are temperature-dependent thermal conductivity, density and specific heat of material, respectively.

$$\frac{\rho c}{\lambda}\frac{\partial T}{\partial t}=\frac{\partial^2 T}{\partial r^2}+\frac{1}{r}\frac{\partial T}{\partial r} \tag{6.2}$$

Because of the temperature gradient in concrete, the cross-section of the column needs to be discretized into elements. Assuming it is a 1-D heat transfer problem, ring elements are usually adopted for the discretization as shown in Fig. 6.28, where i denotes the ith time step; d_t, d_c, d_s and d_p are the sizes of time, concrete element, steel element, and fire protection element, respectively, herein, the subscript "c", "s" and "p" respectively stand for concrete, steel and fire protection, unless otherwise stated; n_c and n_p are the numbers of elements of concrete and fire protection along the radius of section, and the steel tube is only discretized into one element due to the very large conductivity; T_m^i, T_{m-1}^i, and T_{m+1}^i denote the temperatures at node m, $m+1$, m-1 at the ith time step, respectively, the element temperature is represented by the nodal temperature. Using the governing equation. to solve the temperature at node m ($2 \le m \le n_c$), the partial differential terms in Eq. (6.2) at node m should be replaced with

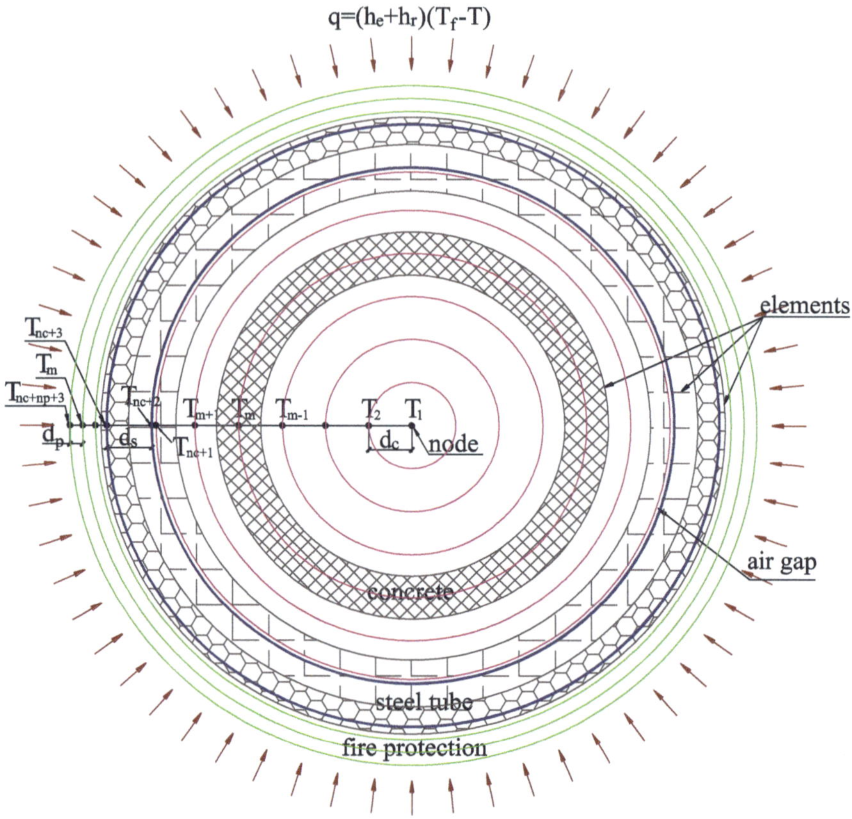

Fig. 6.28 Cross-section discretization of a typical protected circular CFST column.

the finite difference quotients. The replacement can be done by Taylor series expansion with ignoring the high order truncation errors, for example

$$T_{m-1}^{i} = T_{m}^{i} - \Delta r \frac{\partial T_{m}^{i}}{\partial r} + \frac{(\Delta r)^{2}}{2!} \frac{\partial^{2} T_{m}^{i}}{\partial r^{2}} + O\left((\Delta r)^{3}\right) \tag{6.3}$$

$$T_{m+1}^{i} = T_{m}^{i} + \Delta r \frac{\partial T_{m}^{i}}{\partial r} + \frac{(\Delta r)^{2}}{2!} \frac{\partial^{2} T_{m}^{i}}{\partial r^{2}} + O\left((\Delta r)^{3}\right) \tag{6.4}$$

Adding above equations. and ignoring $O((\Delta r)^3)$ gives

$$\frac{\partial^{2} T_{m}^{i}}{\partial r^{2}} = \frac{T_{m-1}^{i} - 2T_{m}^{i} + T_{m+1}^{i}}{(\Delta r)^{2}} \tag{6.5}$$

Making subtraction between Eq. (6.3) and Eq. (6.4) gives

$$\frac{\partial T_{m}^{i}}{\partial r} = \frac{T_{m+1}^{i} - T_{m-1}^{i}}{2\Delta r} \tag{6.6}$$

The ignored terms $O((\Delta x)^3)$ is called local truncation error. This error is only for one time step (i^{th} step). For all time steps in a heating time range [0, t], the number of time steps is $N = t/\Delta t$, and the error will be accumulated. The accumulated error is called global truncation error as given in Eq. (6.7). It is found, the approximation in space domain has consistency of order 2 and is said to be second-order accurate.

$$\text{Global truncation error} \approx O\left((\Delta x)^3\right)\times N \approx O\left((\Delta x)^2\right) \tag{6.7}$$

Similarly, the term $\frac{\partial T}{\partial t}$ in Eq. (6.2) could also be expanded based on Taylor series. Ignoring the local truncation error $O((\Delta t)^2)$ gives

$$\frac{\partial T_m^i}{\partial t} = \frac{T_{m+1}^{i+1} - T_{m-1}^i}{\Delta t} \tag{6.8}$$

Above approximation results in a global truncation error of $O(\Delta t)$. That is to say, the approximation in time domain is first-order accurate. Therefore, the approximation in the time domain is less accurate than that in the space domain. As a result, the subdivision in the time domain should be finer than that in the space domain.

Substituting Eq. (6.5), Eq. (6.6) and Eq. (6.8) into Eq. (6.2) gives the temperature calculation formula at node m.

$$\frac{p_c}{\lambda_c}\frac{T_m^{i+1} - T_m^i}{\Delta t} = \frac{T_{m-1}^i - 2T_m^i + T_{m+1}^i}{(\Delta r)^2} + \frac{1}{r}\frac{T_{m+1}^i - T_{m-1}^i}{2\Delta r} \tag{6.9}$$

It should be mentioned that Eq. (6.9) can also be formulated based on the energy conservation in each element. According to the law of conservation of energy (Incropera et al., 2007), the left-hand side of Eq. (6.9) is in fact the heat energy for temperature increase at Node m and the right-hand side is the heat energy conducted by the adjacent Node $m - 1$ and Node $m + 1$, they should be balanced by each other. By rearranging Eq. (6.9), the nodal temperature T_m^{i+1} can be solved from Eq. (6.10).

$$T_m^{i+1} = T_m^i + \frac{d_t}{p_c A_m}\left[\lambda_c^{\left(\frac{T_{m-1}^i + T_m^i}{2}\right)}\frac{T_{m-1}^i - T_m^i}{d_c}l_{m1} + \lambda_c^{\left(\frac{T_{m+1}^i + T_m^i}{2}\right)}\frac{T_{m+1}^i - T_m^i}{d_c}l_{m2}\right] \tag{6.10}$$

where $l_{m1} = (m - 1.5)d_c$, $l_{m2} = (m - 0.5)d_c$ and $A_m = \frac{1}{2}\left(l_{m2}^2 - l_{m1}^2\right)$. p_c is the thermal capacity of concrete, and $\lambda_c^{\left(\frac{T_{m-1}^i + T_m^i}{2}\right)}$ means that the thermal conductivity of concrete λ_c is a function of the average temperature $\left(\frac{T_{m-1}^i + T_m^i}{2}\right)$ between adjacent nodes (Xiong et al., 2016). Based on the law of conservation of energy, the temperature calculations for the other nodes are given as follows.

$$\text{Node 1}: T_1^{i+1} = T_1^i + \frac{4d_t}{p_c d_c^2}\lambda_c^{\left(\frac{T_2^i + T_1^i}{2}\right)}\left(T_2^i - T_1^i\right) \tag{6.11}$$

$$\text{Node}\left(n_c+1\right): T_{nc+1}^{i+1}=T_{nc+1}^{i}+\frac{d_t}{p_c A_c}\left[\lambda_c^{\left(\frac{T_{nc}^i+T_{nc+1}^i}{2}\right)}\frac{T_{nc}^i-T_{nc+1}^i}{d_c}l_c+h_{tr}\left(T_{nc+2}^i-T_{nc+1}^i\right)l_{cs}\right] \tag{6.12}$$

where $l_c = (n_c - 0.5)d_c$, $l_{cs} = n_c \bullet d_c$, and $A_c = \left(l_{cs}^2 - l_c^2\right)/2$

$$\text{Node}\left(n_c+2\right): T_{nc+2}^{i+1}=T_{nc+2}^{i}+\frac{d_t}{p_s A_s}\left[h_{tr}\left(T_{nc+1}^i-T_{nc+2}^i\right)l_{cs}+\lambda_s^{\left(\frac{T_{nc+3}^i+T_{nc+2}^i}{2}\right)}\frac{T_{nc+3}^i-T_{nc+2}^i}{d_s}l_s\right] \tag{6.13}$$

where $l_{cs} = n_c \bullet d_c$, $l_s = n_c \bullet d_c + 0.5d_s$, and $A_s = \left(l_s^2 - l_{cs}^2\right)/2$

Node (n_c+n_p+3):

$$T_{nc+np+3}^{i+1}=T_{nc+np+3}^{i}+\frac{d_t}{p_p A_{ep}}\left[\lambda_p^{\left(\frac{T_{nc+np+2}^i+T_{nc+np+3}^i}{2}\right)}\frac{T_{nc+np+2}^i-T_{nc+np+3}^i}{d_p}l_{p1}+\left(h_e+h_r\right)\left(T_f^i-T_{nc+np+3}^i\right)l_{p2}\right] \tag{6.14}$$

where $l_{p1} = D/2+\left(n_p-0.5\right)d_p$, $l_{p2} = D/2+n_p \cdot d_p$, $A_{ep} = \left(l_{p2}^2 - l_{p1}^2\right)/2$, and D is the external diameter of the CFST column.

The temperature fields of square CFST columns and CES columns (usually in square section) in fire could be similarly determined, except the governing equation. presented in Eq. (6.15) should be used, the heat transferred in such columns is in fact in a 2-D manner. A typical cross-section discretization for a protected square CFST column is illustrated in Fig. 6.29. More information regarding the temperature calculation of composite members using the finite difference method can be found in the authors' published paper (Xiong et al., 2016).

$$\frac{\rho c}{\lambda}\frac{\partial T}{\partial t}=\frac{\partial^2 T}{\partial x^2}+\frac{\partial^2 T}{\partial y^2} \tag{6.15}$$

6.4.2 Parameters for heat transfer analysis

For above temperature calculation, the coefficient of convection h_e can be taken as 25 W/m²·K for an exposure to ISO-834 fire (EN 1991–1–2, 2002), and the coefficient of thermal radiation h_r can be calculated according to Eq. (6.16) where T_f is the fire temperature and T represents the surface temperature of the column.

$$h_r = \Phi \cdot \varepsilon_m \cdot \varepsilon_f \cdot \sigma\left[\left(T_f+273\right)^2+\left(T+273\right)^2\right]\left(T_f+T+546\right) \tag{6.16}$$

The configuration factor Φ can be taken as 1.0 by ignoring the shadow and position effects (EN 1991–1–2, 2002). The surface emissivity ε_m may be taken as 0.7

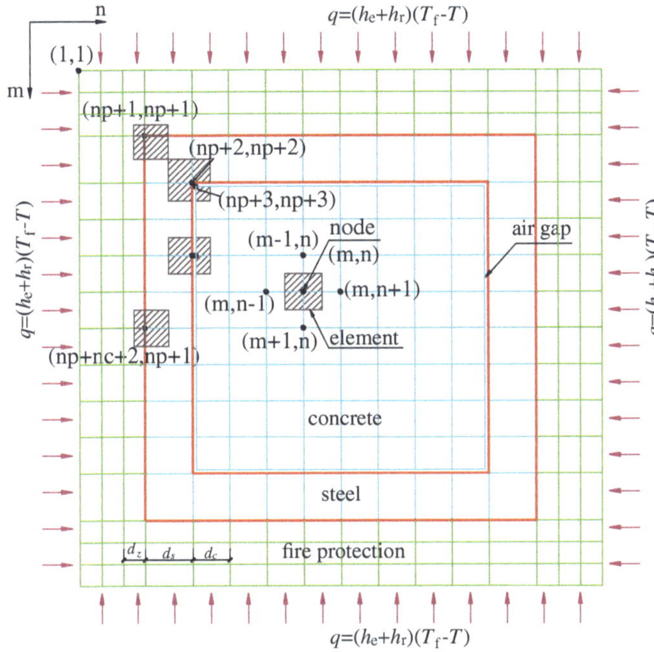

Fig. 6.29 Cross-section discretization of a typical protected square CFST column.

for unprotected columns (EN 1993–1–2, 2005) and 0.8 for the protected ones (EN 1991–1–2, 2002). The emissivity of the fire ε_f is equal to 0.8 (Lu et al., 2011). The Stephan Boltzmann constant σ should be taken as 5.67 × 10–8 W/m^{2}K^4.

Because of the air gap at the steel-concrete interface, a thermal contact resistance h_{tr} should be considered. In the literature, a constant value of 100 W/m^2·K is suggested by CIDECT (Renaud, 2004), whereas Ding and Wang (2008), and Espinos et al. (2010) recommended a value of 200 W/m^2 K. A better agreement between the tested and predicted temperatures has been found for composite columns using high strength materials when the higher value equal to 200 W/m^2·K was used (Xiong and Liew, 2020b & 2021). For CFST columns, the thermal contact resistance between the steel tube and fire protection (sprayed) can be ignored as the fire protection would deform along with the steel tube and the air gap between them is not significant (Xiong and Liew, 2020b).

6.5 Prescriptive methods

6.5.1 Concrete filled steel tubular columns

The outer surface of the concrete filled steel tubular member may be exposed directly to the fire thus it is susceptible to the strength and stiffness degradations under fire conditions. This weakness can be overcome either by applying on the steel tube with a suitable thickness of fire-retardant material, or providing reinforcement in the concrete core so that when the steel tube is sacrificed under fire, the reinforced concrete column can resist the loads under fire conditions.

For a CFST column with normal strength materials (NSS and NSC), a complete thermal-stress analysis as covered in Sections 6.6 and 6.7 may not be necessary if the temperature of the outer steel section does not exceed 350 °C as recommended by EN 1994–1–2 (2005).

For a CFST column with HSC (f_{ck} > 50 N/mm^2) and HSS (f_y > 460 N/mm^2), the fire resistance is deemed to be satisfied provided that the temperature of outer steel tube is less than 300 °C and the load level is lower than 0.65.

In case where the aforementioned minimum temperature is exceeded, passive fire protection may be necessary. Their performance can be assessed in accordance with BS EN 13,381–2 (2014) for vertical screens and BS EN 13,381–6 (2012) for coating and sprayed materials. The thickness of fire protection could be determined in accordance with the supplier's guidance or calculated by the fire engineering approach described in Section 6.6 and advanced calculation models introduced in Section 6.7. Alternatively, reinforcing steel bars may be provided in the concrete core. To determine the standard fire ratings, the minimum reinforcement ratios and concrete cover for CFST columns with NSS and NSC are shown in Table 6.2. For CFST columns with HSC and HSS, the fire engineering approach and advanced calculation models may be used to determine the fire resistance.

Table 6.2 Minimum values to achieve standard fire ratings for CFST column sections.

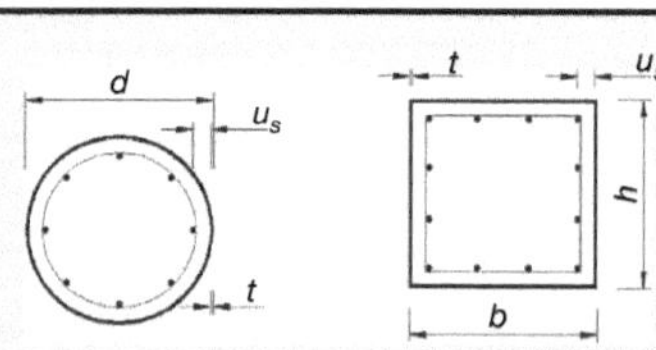		Standard Fire Rating				
		R30	R60	R90	R120	R180
Load Level	Minimum Values					
$\eta_{fi} \le 0.28$	Minimum h, b or d in (mm)	160	200	220	260	400
	Minimum ratio $A_s / (A_c + A_s)$ in (percent)	0	1.5	3.0	6.0	6.0
	Minimum u_s in (mm)	–	30	40	50	60
$\eta_{fi} \le 0.47$	Minimum h, b or d in (mm)	260	260	400	450	500
	Minimum ratio $A_s / (A_c + A_s)$ in (percent)	0	3.0	6.0	6.0	6.0
	Minimum u_s in (mm)	–	30	40	50	60
$\eta_{fi} \le 0.66$	Minimum h, b or d in (mm)	260	450	550	–	–
	Minimum ratio $A_s / (A_c + A_s)$ in (percent)	3.0	6.0	6.0	–	–
	Minimum u_s in (mm)	25	30	40	–	–

Note: A_s, A_c are the sectional area of reinforcements and concrete respectively.

6.5.2 Concrete encased steel columns

The concrete cover in concrete encased composite members serves as a fire protection layer for the steel reinforcements and the encased steel section, therefore, additional fire protection is normally not required. To determine the fire resistance of concrete encased steel composite column with concrete grade up to C90/105, Table 6.3 may be followed provided the following requirements are met:

- The CES column needs to be part of a braced frame and has a length not exceeding 30 times the minimum cross-sectional dimension chosen;
- The steel contribution ratio δ of the CES column is not smaller than 0.3;
- At least 2kg/m^3 of monofilament propylene fibres need to be included in the concrete mix if concrete grade exceeds C55.

6.6 Fire engineering approaches

6.6.1 Design criteria

To protect composite members from fire hazards, both active and passive fire protections could be utilized. In cases where the mechanical resistance of them in fire is also required, the composite members should be designed and constructed in such a way that they maintain their load-bearing function during the complete duration of the fire including decay phase or during a required period of time (EN 1994–1–2, 2005). This can be ensured by Eq. (6.17). If the mechanical resistance in fire is adequate, the fire protection is not required.

$$E_{fi,d,t} \leq R_{fi,d,t} \tag{6.17}$$

$E_{fi,d,\,t}$ is the design effect of actions for the fire situation, including the effects of thermal expansions and deformations. Only the effects of thermal deformations resulting from the thermal gradients across the cross-section of members need to be considered. The effects of axial or in-plain thermal expansions may be neglected (EN 1994–1–2, 2005), which could be investigated by the advanced calculation models introduced in Section 6.7. $R_{fi,d,\,t}$ is the corresponding design resistance in the fire situation.

The design effect of actions in fire situations $E_{fi,d,\,t}$ may be obtained from a structural analysis for normal temperature design as Eq. (6.18). The boundary conditions at supports and ends of the member, as well as the forces and bending moments at such boundaries and ends, may be assumed to remain unchanged throughout the fire exposure (EN 1994–1–2, 2005).

$$E_{fi,d,t} = \eta_{fi} E_d \tag{6.18}$$

where

E_d is the design value of the corresponding force or moment for normal temperature design

η_{fi} is a reduction factor which is taken as $\dfrac{G_k + \varphi_{fi} Q_{k,1}}{\gamma_G G_k + \gamma_{Q,1} Q_{k,1}}$

G_k is the characteristic value of a permanent action

Table 6.3 Minimum values to achieve standard fire ratings for CES column sections.

	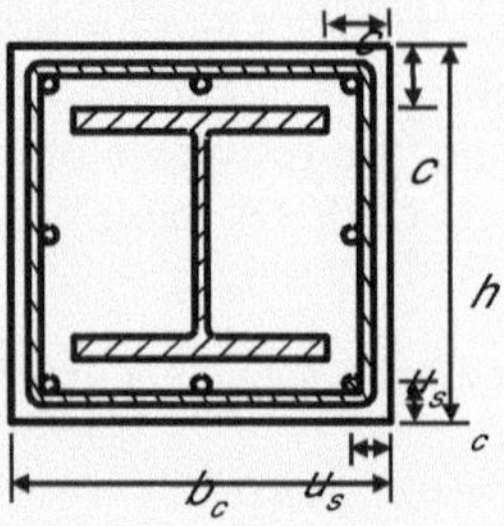	**Standard Fire Resistance**				
		R60	**R90**	**R120**	**R180**	**R240**
	Minimum c in (mm) Minimum u_s in (mm)	50 30	50 30	75 40	75 50	75 50
1	Minimum h_c and b_c in (mm) for load level[a] $\eta_{fi,t}$ ≤ 0.3	See note 2				
1.1	steel contribution ratio[c] $\delta \ge 0.30$	180/190 /190/200[b]	190/200 /200/210	250/260 /260/270	280/300 /300/320	300/320 /330/350
1.2	steel contribution ratio[c] $\delta \ge 0.40$	180/190 /190/200	190/200 /200/210	240/250 /250/260	270/290 /290/310	290/310 /320/340
1.3	steel contribution ratio[c] $\delta \ge 0.50$	180/190 /190/200	190/200 /200/210	230/240 /240/250	260/280 /280/300	280/300 /310/330
2	Minimum h_c and b_c in (mm) for load level[a] $\eta_{fi,t}$ ≤ 0.5	See note 2				
2.1	steel contribution ratio[c] $\delta \ge 0.30$	210/220 /230/250[b]	230/250 /260/280	290/310 /320/340	320/340 /350/380	340/360 /380/430
2.2	steel contribution ratio[c] $\delta \ge 0.40$	210/220 /230/250	230/250 /260/280	280/300 /310/330	310/330 /340/370	330/350 /360/400
2.3	steel contribution ratio[c] $\delta \ge 0.50$	210/220 /230/250	230/250 /260/280	280/300 /310/330	310/330 /340/370	320/340 /350/390
3	Minimum h_c and b_c in (mm) for load level[a] $\eta_{fi,t}$ ≤ 0.7	See note 2				
3.1	steel contribution ratio[c] $\delta \ge 0.30$	280/300 /320/340[b]	320/340 /360/400	400/420 /440/520	480/500 /520/630	530/560 /590/700
3.2	steel contribution ratio[c] $\delta \ge 0.40$	270/290 /310/330	310/330 /350/390	380/400 /420/470	440/460 /480/560	490/510 /540/630
3.3	steel contribution ratio[c] $\delta \ge 0.50$	260/280 /300/320	300/320 /340/380	360/380 /400/450	410/430 /450/530	450/470 /500/590

[a] $\eta_{fi,t}$ is the load level for fire design according to EN1994-1-2;
[b] The minimum cross-sectional dimensions are displayed in the format of A/B/C/D, where A is for concrete grade below C50; B is for concrete grade up to C60; C is for concrete up to C80; and D for concrete grade up to C90. The concrete grade is classified according to EN1992-1-1.
[c] Steel contribution ratio δ is calculated according to EN1994-1-1.

Table 6.4 Recommended values of $\Psi_{1.1}$ and $\Psi_{2.1}$.

Imposed loads in buildings	$\Psi_{1.1}$	$\Psi_{2.1}$
Category A: domestic, residential areas	0.5	0.3
Category B: office areas	0.5	0.3
Category C: congregation areas	0.7	0.6
Category D: shopping areas	0.7	0.6
Category E: storage areas	0.9	0.8
Category F: traffic areas, vehicle weight ≤ 30kN	0.7	0.6
Category G: traffic areas, vehicle weight ≤ 160 kN	0.5	0.3
Category H: roofs	0	0
Wind loads on buildings	0.2	0

$Q_{k,1}$ is the characteristic value of the leading variable action 1
γ_G is the partial factor for permanent action, = 1.35
$\gamma_{Q,1}$ is the partial factor for variable action 1, = 1.50
φ_{fi} is the combination factor for fire situation given either by $\Psi_{1.1}$ (frequent value) or $\Psi_{2.1}$ (quasi-permanent value). The use of $\Psi_{2.1}$ is recommended by EN 1991–1–1 (2002) and EN 1991–1–2 (2002) which is given in Table 6.4.

The reduction factor η_{fi} reflects the load level in fire situations. It changes with the load ratio $Q_{k,1}/G_k$. As a simplification, the recommended value of $\eta_{fi} = 0.65$ may be used, except for imposed loads in Category E introduced in Table 6.4 where the recommended value is 0.7.

6.6.2 Simple calculation model

A simple calculation model (SCM) has been provided in Eurocode 4 (EN 1994–1–2) for resistant design of steel-concrete composite columns. It is only applicable to individual members under pure compression, and the calculation procedure is similar to the buckling resistance design at normal temperature, except the temperature dependent mechanical properties should be taken into account. The buckling resistance of a composite column in fire situation based on SCM can be calculated as:

$$N_{\mathrm{fi,Rd}} = \chi_{\mathrm{fi}} N_{\mathrm{fi,pl,Rd}} \tag{6.19}$$

$$N_{\mathrm{fi,pl,Rd}} = \sum_j \frac{A_{\mathrm{a},\theta} f_{\mathrm{ay},\theta}}{\gamma_{\mathrm{M,fi,a}}} + \sum_k \frac{A_{\mathrm{c},\theta} f_{\mathrm{c},\theta}}{\gamma_{\mathrm{M,fi,c}}} + \sum_m \frac{A_{\mathrm{s},\theta} f_{\mathrm{sy},\theta}}{\gamma_{\mathrm{M,fi,s}}} \tag{6.20}$$

To implement SCM, the cross-section of the composite column should be subdivided into various parts where $A_{\mathrm{a},\theta}$, $A_{\mathrm{c},\theta}$, and $A_{\mathrm{s},\theta}$ are the areas of elements of steel profile, concrete, and reinforcing rebars at temperature θ, respectively; $f_{\mathrm{a},\theta}$, $f_{\mathrm{c},\theta}$, and $f_{\mathrm{s},\theta}$ are strengths of them. $\gamma_{\mathrm{M,fi,a}}$, $\gamma_{\mathrm{M,fi,c}}$, and $\gamma_{\mathrm{M,fi,s}}$ are partial factors of said materials at fire situation, taking as 1.0 for all. The buckling reduction factor χ_{fi} is calculated as

$$\chi_{fi} = \frac{1}{\Phi_{fi} + \sqrt{\Phi_{fi}^2 - \bar{\lambda}_{fi}^2}} \tag{6.21}$$

$$\Phi_{fi} = 0.5\left(1 + \eta + \bar{\lambda}_{fi}^2\right) \tag{6.22}$$

$$\eta = \alpha_{fi}\left(\bar{\lambda}_{fi} - 0.2\right) \tag{6.23}$$

α_{fi} is the imperfection factor corresponding to a relevant buckling curve selected for the fire resistant design. For composite columns such as CFST and CES columns with normal strength materials (NSS and NSC), the buckling curve is recommended as "c" (EN 1994–1–2, 2005). For CFST columns with HSC and UHSC, the buckling curve "c" can still be used (Xiong and Liew, 2020b) except HSS is used in which case the buckling curve "d" should be adopted (Xiong and Liew, 2021). For CES columns using high strength materials, the selection of buckling curve needs further investigation.

$\bar{\lambda}_{fi}$ is the non-dimensional slenderness ratio which can be calculated according to Eq. (6.24). $N_{fi,pl,Rk}$ is the characteristic value of $N_{fi,pl,Rd}$ calculated in Eq. (6.20), and $N_{fi,cr}$ is the Euler buckling load of column under fire calculated according to Eq. (6.25)

$$\bar{\lambda}_{fi} = \sqrt{\frac{N_{fi,pl,Rk}}{N_{fi,cr}}} \tag{6.24}$$

$$N_{fi,cr} = \frac{\pi^2 (EI)_{fi,eff}}{L_{fi,eff}^2} \tag{6.25}$$

$(EI)_{fi,eff}$ is the effective flexural stiffness calculated from Eq. (6.26) where $E_{a,\theta}$ and $E_{s,\theta}$ are the Young's modulus of steel profile and reinforcing rebars at temperature θ, respectively and $E_{c,\theta}$ is the tangent modulus of concrete; $\varphi_{a,\theta}$, $\varphi_{c,\theta}$, and $\varphi_{s,\theta}$ are reduction factors considering the effects of thermal stresses, they can be taken as 1.0, 0.8, and 1.0 for steel profile, concrete, and reinforcing rebars, respectively (EN 1994–1–2, 2005; Wang, 2002).

$$(EI)_{fi,eff} = \sum_j \left(\varphi_{a,\theta} E_{a,\theta} I_{a,\theta}\right) + \sum_k \left(\varphi_{c,\theta} E_{c,\theta} I_{c,\theta}\right) + \sum_m \left(\varphi_{s,\theta} E_{s,\theta} I_{s,\theta}\right) \tag{6.26}$$

The column buckling length under fire $L_{fi,eff}$ depends on the fire situations and structural behavior. For a column partially heated on the middle height with the two ends insulated in a standard fire test, the buckling length of column under fire is different from that at normal temperature. It is related with the boundary conditions, support height, flexural stiffness of the fire exposed and unexposed parts of column, and the length of column unexposed to fire (Xiong and Yan, 2016). Previous studies on CFST columns indicate that, irrespective of the strengths of the steel and concrete, the normal temperature buckling length can be used for fire resistant design of the protected columns, but the fire exposure time-dependent buckling length needs to be used for the unprotected columns. An exception is that the normal temperature buckling length can be adopted for both protected and unprotected columns provided they are pinned-pinned at both ends. For a column in a braced frame, the buckling length under fire may be determined according to EN 1994–1–2 (2005), for example, the buckling length of a column on an intermediate storey is given by 0.5 L and that of a column on the top floor can be taken as 0.7 L, where L is the actual length of column. For a column on the lowest floor, the buckling length may vary from 0.7 L to 0.5 L depending on the rotation rigidity of the column base. For a column in unbraced frame, the buckling length of the column under fire is related with the lateral stiffness of the frame. In such cases, the advanced calculation models introduced in Section 6.7 may be used to determine the fire resistance of the columns.

6.6.3 N-M interaction model

SCM is only applicable to the axially loaded columns with additional moments resulted from the initial imperfections, e.g. out-of-straightness and residual stresses etc. The bending moment distribution induced from the initial imperfection is shown in Fig. 6.30(a). SCM is not applicable to beam-columns subjected to combined axial force and bending moments, such bending moments are induced by transverse loads, end moments, axial load eccentricities, and thermal bowing under non-uniform heating, etc. The bending moment distributions of beam-columns are shown in Fig. 6.30(b) to (d). The American Code ASCE/SFPE 29–05 (2007) is also limited to the axially loaded but CFST columns. It is not applicable for the beam-columns, Thus, it is not applicable for composite columns with moment distributions shown in Fig. 6.30(b) to (d).

Considering a simplified N-M interaction curve has been adopted by Eurocode 4 for normal temperature design of composite beam-columns (EN 1994–1–1, 2004), such curve can also be extended for fire resistant design of them. This maintains a good consistency with the normal temperature design (Xiong and Liew, 2018). Based on that, the shape of N-M interaction curves at elevated temperatures is similar to that at normal temperature as shown in Fig. 4.5, except the internal resistance at Points A, B, C, and D should be calculated based on the temperature-dependent mechanical properties of the constituent materials of the composite column. It should be mentioned that the internal resistance at said points are derived based on plastic analysis

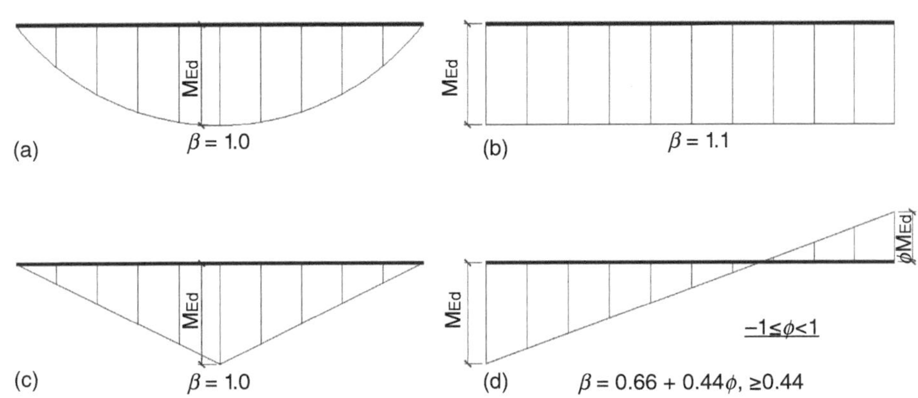

Fig. 6.30 Moment diagrams of beam-columns.

that requires the cross-section of the steel profile (e.g., steel tubes in CFST columns) to be at least compact. The section can be similarly classified at both normal temperature and fire situations, but a reduction factor of 0.85 should be applied under fire to the coefficient $\varepsilon = \sqrt{235/f_y}$ as a consideration for the effect of increasing temperature (EN 1993–1–2, 2005). Therefore, the section classification is more stringent for fire resistant design. The internal resistance at Points A, B, C, and D for CFST columns is introduced hereinafter. The N-M interaction curve and the corresponding stress distributions are illustrated in Fig. 6.31. It should be mentioned that the determination of such curves at elevated temperatures is irrespective of the strengths of materials. Said internal resistance and N-M interaction curves can be similarly determined for CES columns.

1. Point A (0, $N_{fi,pl,Rd}$):
   Full cross-section is under uniform compression. No bending moment is resulted from the compressive stresses on the cross-section. The plastic axial resistance $N_{fi,pl,Rd}$ is calculated according to Eq. (6.27). It is proposed that the increase of concrete compressive strength and decrease of steel yield strength caused by the confinement from steel tube are not taken into account, irrespective of the shape of the cross-section. This assumption is based on the fact that the steel tube is softened under fire and cannot provide sufficient confinement to the core concrete.

$$N_{fi,pl.Rd} = \sum_j \left(A_{a,\theta,j} f_{ay,\theta,j} / \gamma_{M,fi,a} \right) + \sum_k \left(A_{c,\theta,k} f_{ck,\theta,k} / \gamma_{M,fi,c} \right) \tag{6.27}$$

2. Point B ($M_{fi,pl,Rd}$, 0):
   The cross-section is subject to pure bending without axial compression. The height of neutral axis can be calculated based on a force equilibrium where the tensile resistance of the steel section within the height of $2h_n$ is equal to the compression resistance of the concrete portion in the compression zone (out of the height of $2h_n$). The plastic moment resistance $M_{fi,pl,Rd}$ can be calculated by taking the moment about either line on the cross-section paral-

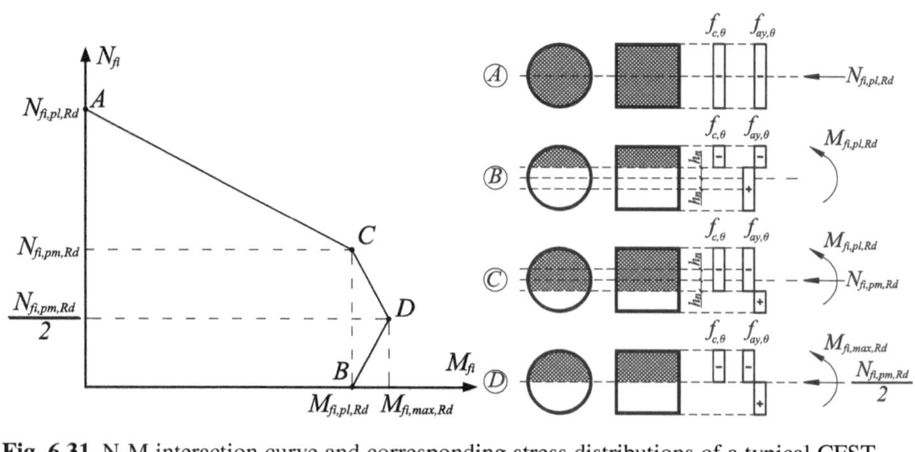

Fig. 6.31 N-M interaction curve and corresponding stress distributions of a typical CFST column under fire.

lel to the neutral axis. Herein, the centerline of the cross-section is referenced, and $M_{fi,pl,Rd}$ is determined according to Eq. (6.28). The subscripts p and q represent the numbers of steel and concrete elements within the height of $2h_n$; $W_{a,n,p}$ and $W_{c,n,q}$ are their plastic modulus bending about said centerline.

$$M_{fi,pl,Rd} = \left(\sum_j W_{a,j} f_{ay,\theta,j} - \sum_p W_{a,n,p} f_{ay,\theta,p}\right) / \gamma_{M,fi,a} + \left(\sum_k W_{c,k} f_{ck,\theta,k} - \sum_q W_{c,n,q} f_{ck,\theta,q}\right) / 2\gamma_{M,fi,c} \tag{6.28}$$

3. Point C ($M_{fi,pl,Rd}$, $N_{fi,pm,Rd}$):
The location of neutral axial is the same as that at Point B. The cross-section is under partial compression, and the axial force is resultant from the compressive stresses. The axial force is equal to the compression resistance of concrete in the compression zone plus the compression resistance of steel section within the height of $2h_n$. It is mentioned in Item 2) that the compression resistance of the steel section within the height of $2h_n$ equals the compression resistance of concrete located out of the height of $2h_n$. Thus, the axial force is actually equal to the full cross-sectional compression resistance of concrete and determined from Eq. (6.29).

$$N_{fi,pm,Rd} = \sum_k \left(A_{c,\theta,k} f_{ck,\theta,k} / \gamma_{M,fi,c}\right) \tag{6.29}$$

4. Point D ($M_{fi,max,Rd}$, $N_{fi,pm,Rd}/2$):
The maximum plastic moment resistance $M_{fi,max,Rd}$ is calculated as,

$$M_{fi,max,Rd} = \sum_j W_{a,j} f_{ay,\theta,j} / \gamma_{M,fi,a} + \sum_k W_{c,k} f_{ck,\theta,k} / 2\gamma_{M,fi,c} \tag{6.30}$$

which is based on $h_n = 0$. The difference between Point D and Point B decreases with the increase of steel amount. An increase in the yield strength of steel significantly increases the sectional resistance at both Points B and D. It is expected that an increase in concrete strength will not have a significant effect on the flexural strength at Point B because steel yielding governs the strength, but they tend to increase the distance between Point B and Point D along the axis of N_{fi} in the interaction curve shown in Fig. 6.31.

Due to the deteriorated elastic moduli of steel and concrete under fire, the lateral deflection should be larger than its counterpart at normal temperature. Therefore, the second-order moment due to member curvature effect at fire situation is higher. The second-order effect can also be considered by using an equivalent design moment like normal temperature design, which is calculated through multiplying the greatest first-order design moments within the column length by an amplification factor k as given in Eq. (4.21), except that the temperature-dependent material properties should be used in the calculation. It should be mentioned that there are two first-order bending moments for a beam-column, i.e., one is caused by the initial imperfection, and the other is taken from global analysis and may be an "independent" moment or induced by the load eccentricity. The equivalent design moment from the two first-order bending moments are illustrated in Fig. 4.7 and calculated according to Eq. (4.22).

For fire resistant design of composite columns such as CFST and CES columns with normal strength materials (NSS and NSC), the member initial bow imperfection e_0 can be taken as $L/150$ corresponding to a buckling curve "c" (EN 1994–1–1, 2004; EN 1994–1–2, 2005), L is the actual length of column. For CFST columns with HSC and UHSC, the initial bow imperfection of $L/150$ can still be used (Xiong and Liew 2020b) except HSS is used where an initial bow imperfection of $L/100$ is recommended (Xiong and Liew, 2021; EN 1994–1–1, 2004). For CES columns using high strength materials, the value of initial bow imperfection needs further investigation.

After the equivalent design moment is obtained, the following condition should be satisfied for the stability check within the column length for a composite beam-column under fire.

$$\frac{M_{fi,Ed}}{\mu_{fi,d} M_{fi,pl,Rd}} \le \alpha_{fi,M} \tag{6.31}$$

$M_{fi,Ed}$ is the equivalent design moment in fire. For normal temperature design according to EN 1994–1–1 (2004), the coefficient α_M is generally taken as 0.9 for steel grades between S235 and S355 inclusive and as 0.8 for steel grades S420 and S460, considering the premature crush of concrete before the steel profile fully yields. The higher the steel grade is, the more likely the concrete is prematurely crushed. However, in the fire situation, such premature crush of concrete may not be considered, attributed to the fact that the concrete generally loses part of its strength under fire and shows better ductility than it does at normal temperature. Thus, it is less prone to be crushed before the steel profile yields. In this regard, it is recommended the coefficient $\alpha_{fi,M}$ for fire resistant design could be taken as 1.0, irrespective of the material strength.

The N-M interaction curve contracts with the increase of fire exposure time, whereas the external bending moment increases with time owing to the second-order effect. The fire resistance of the composite column is determined when the increased bending moment meets the contracted N-M interaction curve.

6.6.4 Differences between scm and nmim

Both simple calculation model (SCM) and N-M interaction model (NMIM) are applicable for axially loaded columns with the additional moments resulted from the initial imperfections. Both of them consider the second-order effect through an amplification factor. The difference is that the SCM implicitly considers the amplification factor through an imperfection factor that is used to reduce the inherent cross-sectional compressive resistance of the column. The relationship between the imperfection factor under fire and the initial bow imperfection of column is formulated in Eq. (6.31) according to the previous study (Xiong and Liew, 2018).

$$\alpha_{fi} = \frac{c}{r^2\left(\bar{\lambda}_{fi} - 0.2\right)} e_0 \tag{6.32}$$

where c is the distance from the centroid of the cross-section to the most stressed fiber of the section, and r is the sectional radius of gyration, they are constant values. As the imperfection factor under fire is constant, the magnitude of the initial bow imperfection is actually variable according to Eq. (6.32) because the non-dimensional slenderness $\bar{\lambda}_{fi}$ varies with the temperature. However, the NMIM considers explicitly the amplification factor with a constant initial bow imperfection.

Previous studies (Xiong and Liew, 2020b & 2021) have demonstrated that at low to medium load levels, the difference between the predictions on fire resistance time by SCM and MNIN is not noticeable, but at high load levels (e.g. load levels higher than 0.65), their difference is quite large. The reason is because of the different considerations for initial bow imperfection in SCM and NMIM as discussed above. Through a comparison with test results, SCM is found not applicable to the highly loaded columns with either normal or high strength materials (Xiong and Liew, 2020b & 2021). Fortunately, the load level η_{fi} for fire resistant design (a ratio between unfactored loads and factored loads as given in Eq. (6.18)) is quite low since it considers only the permanent and leading variable loads. The maximum load level is 0.74 when the leading variable load vanishes. Because the load case to achieve the maximum load level is rare, a load level of 0.65 is recommended for practical design by Eurocode 4 (EN 1994–1–2, 2005). In this regard, SCM is still applicable to fire resistant design of composite columns. However, NMIM is recommended considering its continuity and consistency in predictions.

6.7 Advanced calculation models

Advanced calculation models are not only applicable to individual members, but also subassemblies and entire structures. Compared with the fire engineering approaches introduced in Section 6.6, the advanced calculation models give an improved approximation of the actual structural performance in fire situations with vivid demonstrations on the developments of deformations, stresses and failure modes. The advanced calculation models may be accomplished in commercial finite element software, such as ABAQUS, ANASYS, etc.

The implementation of advanced calculation models usually comprises of separate calculation models for the determination of temperature field within composite members (heat transfer analysis model) and mechanical behavior of the composite structure or of any part of it with the imported temperature fields (thermal-stress analysis model). It may be used in association with any time-temperature heating curve in the heat transfer analysis, and the combined effects of mechanical actions, geometric imperfections and thermal actions can be considered in the thermal-stress analysis (EN 1994–1–2, 2005).

It is highly important to establish the validity for the advanced calculation models. They can be verified through comparisons with relevant test results, in terms of deformations, temperatures, fire resistance times, etc. Sensitivity analysis is recommended for checking the effects of some critical parameters such as member length, section size, load level, boundary condition, mesh size, etc.

Special considerations for high strength materials

7

Outline

7.1 High tensile steel section ($f_y > 460$ N/mm^2)

High strength steel sections are available in hot-rolled or plated section. Mega shape hot-rolled sections with nominal yield strength of 460 MPa are available and typically used for top-down construction. High strength flat steel plates with nominal yield strength up to 690 MPa, as shown in Table 2.3 in Chapter 2, are increasingly used in practice as they can be manufactured to form thicker plates. Thicker steel plates

Design of Steel-Concrete Composite Structures Using High Strength Materials. DOI: 10.1016/C2019-0-05474-X can be welded to form complex column shapes which are extremely difficult to be manufactured into hot-rolled profiled shapes. Steel plated sections are typically shop fabricated under QC environment and then transported to the construction site. Steel plated box section can act as permanent formwork before casting concrete and resist construction loads. At the service stage, steel sections act as reinforcements to contribute to the compression resistance of composite columns as well as provide confinement of inner concrete to enhance its compressive strength, leading to a more ductile failure mode.

The steel specifications in different national standards vary from each other as the chemical composition and mechanical properties may be different. However, they are generally comparable in term of tensile strength to yield strength ratio, elongation at failure and ultimate tensile strength as shown in Table 2.4. The strength of steel is categorized generally in terms of nominal thickness and manufacturing methods. Mechanical properties for hot rolled and hollow section structural steel are given in EN 1993–1–1. In Australian standard AS 4100, both cold formed and hot rolled structural steel grades are provided. In American standard AISC 360–16, yield strength and ultimate strength for steel plates and hollow structural steel are given.

High strength steel tends to have relatively poor ductility performance comparing to mild steel. EN 1994–1–1 and AISC 360–16 set a limit of steel yield strength when designing composite members. EN 1994–1–1 specifies that all the rules in EN 1994–1–1 apply to the structural steel of nominal yield strength not more than 460 MPa. In AISC 360–16, the yield strength of structural steel should not exceed 525 MPa. Extensive research has been carried out on CFST columns using high strength steel and ultra-high strength steel with good structural performance. EN 1993–1–12 provides the additional design values for steel grades up to S700 with nominal yield strength of 700 MPa. This paths the way to support future use of higher strength steel for composite construction.

7.1.1 Hot forming

For high tensile steel in the quenched and tempered condition produced in according with EN 10025–6 (2004) and steel in the thermo-mechanically controlled condition in accordance with EN 10149–2 (1996), the hot forming is only permitted up to the stress relief annealing temperature. If higher temperatures are used, an additional quenching and tempering operation shall be required in which case the manufacturer shall be consulted.

7.1.2 Cold forming

For high tensile steel in the quenched and tempered condition in accordance with EN 10025–6 (2004), the minimum inside bend radii for cold forming without cracks induced should be conformed to the values in Table 7.1.

Table 7.1 Minimum bend radii for cold forming of quenched and tempered steels.

Steel class	Minimum inside bend radii (mm)	
	Axis of bend in transverse direction	**Axis of bend in longitudinal direction**
S500Q/QL/QL1	3t	4t
S550Q/QL/QL1	3t	4t
S620Q/QL/QL1	3t	4t
S690Q/QL/QL1	3t	4t

Note: The values are applicable for bend angles ≤ 90° and plate thickness **t** ≤ 16 **mm**.

Table 7.2 Minimum bend radii for cold forming of thermo-mechanically controlled steels.

Steel class	Minimum inside bend radii for nominal thickness in mm		
	$t \leq 3$	$3 < t \leq 6$	$t > 6$
S500MC	1.0t	1.5t	2.0t
S550MC	1.0t	1.5t	2.0t
S600MC	1.0t	1.5t	2.0t
S650MC	1.5t	2.0t	2.5t
S700MC	1.5t	2.0t	2.5t

Note: The values are applicable for bend angles ≤ 90°.

For high tensile steel in the thermo-mechanically controlled condition in accordance with EN 10149–2 (1996), the minimum inside bend radii for cold forming is given in Table 7.2.

7.1.3 Cutting

High tensile steel plates can be cold sheared. The maximum thickness of shearing will depend on the power available in the shear machine and the material used in the shear blades. The maximum thickness of shearing is generally smaller by 30% ~ 40% relative to mild steels, which indicates slower shearing rate. The quality of the sheared edge is influenced by the machine setup and therefore the cutting blades should be well maintained.

The high tensile steels can also be cut by oxy-fuel gas flame, abrasive water jet, and plasma techniques. Water jet and plasma cutting result in less or almost no heat around the cut edges relative to flame cutting. However, care should be taken as cutting underwater could result in a high hardness edge owing to the quenching effect.

Hardness of free cut edges should be checked after cutting. For high tensile steels as concerned by EN 10025–6 (2004) and EN 10149–2 (1994), the permitted maximum hardness (HV 10) at the free cut edge is 450 (EN 1090–2, 2008). Hardness testing with a load of HV10 shall be performed in accordance with EN 1043–1 (2011) or EN ISO 6507–1 (2005). The hardness testing is carried out on prepared four samples of steel, and four hardness tests should be done for each sample. Preheating of material may be necessary in order to achieve the required hardness of free cut edges.

The cut edges should be free from any sharp notches. If necessary, the free edge surfaces shall be smoothed by grinding or machining in which case the minimum depth of grinding or machining shall be 0.5 mm.

7.1.4 Bolt holes

The execution of bolt holes for high tensile steels may be done by process such as drilling, punching, laser, plasma or other thermal cutting. The local hardness and quality of cut edges around a finished hole should fulfill the requirements as for cutting.

7.1.5 Welding

With increasing product thickness and increasing strength level, cold cracking (hydrogen cracking) can occur. The cold cracking is caused by following factors in combination:

- Amount of diffusible hydrogen in the weld metal
- Brittle structure of the heat affected zone (HAZ)
- Significant tensile stress concentrations in the welded joint

Welding procedures for avoiding the hydrogen induced cold cracking may be determined in accordance with Method A and Method B in EN 1011–2 (2001), Japanese code JIS B 8285 (2010) or U.S code AWS D1.1 (2010). The most effective way of avoiding the cold cracking is to reduce the hydrogen input from the welding consumables to the weld metal. Thus, low hydrogen consumable shall be selected. In addition, it is also important to slow down the cooling rate at the heat affect zone, by control of weld run dimensions in relation to metal thickness, or by applying preheat and controlling interpass temperature, or if necessary, by post-heat on completion of welding which typically a maintenance of the preheat temperature. Stress concentrations can be reduced when the cooling rate is slowed down.

Preheat can be used not only for high tensile steels but also for thicker mild steels. The necessity and requirements for preheat during welding should be consulted with steel manufacturers. In case where it is absent, Method B of EN 1011–2 (2001) may be adopted. Method B is based on extensive experience and data which is mainly, but not exclusively, for low alloy high tensile steels. However, it applies only to normal fabrication restrain conditions. Higher restraint situations such as cruciform welding of tubular joints may need higher preheat temperature or other precautions to prevent hydrogen cracking. In addition, it only refers to welding of parent metal at temperatures above 0 °C.

Preheat should be extended to a zone of width of at least 4 times the thickness of the plate per side on both sides of the weld seam. For thickness greater than 25 mm, 100 mm adjacent to the seam on both sides is adequate.

Single layer fillet welds generally have a lower internal stress than butt welds. The preheat temperatures determined for single layer fillet welds therefore can be approximately 60 °C lower than butt welds. Multi-layer fillet welds and butt welds have similar stress conditions. Therefore, the same preheat temperature shall be used to avoid cold cracks.

In cases where adequate preheat is impracticable, it is advisable to use austenitic or Ni-based welding consumables. It is then possible to avoid the use of preheat because of the comparatively low internal stress level of the welded joints and the better solubility of the hydrogen in austenitic weld metal.

Post-heat may be necessary when there is an increased risk of cold cracking, such as submerged arc weld for high tensile steels and a thickness greater than 30 mm. The post-heat can be implemented by means of soaking, such as $2h/250\ ^\circ C$, immediately after the welding.

7.1.6 Hot-Dip galvanization

Hot-dip galvanization is a coating process where steel elements are submerged in a bath of molten zinc or zinc alloy at a temperature of approximately 450 °C, and withdrawn when the metallurgical reaction developing the coating is complete. Before galvanization, the steel elements should be in general pickled in either hydrochloric or sulphuric acid to remove impurities, such as oil/grease, paint, rust or mill scales.

During pickling, hydrogen ions are released and penetrate the grain boundaries of steel. Hydrogen ions are unstable and combine in voids of steel matrix to form stable hydrogen molecules on the grain boundaries. The hydrogen molecules create pressure from inside voids. As a result, microscope surface cracks are formed, and the steel elements may experience premature failure when subjected to tensile loads. The hydrogen induced cracking is called hydrogen embrittlement.

Mild steels are generally not embrittled by the absorption of hydrogen during pickling, owing to better ductility. However, if steels with yield strength greater than 650 N/mm^2 (including high strength bolts of Grade 10.9) or harder than approximately 340 HV, care should be taken to minimize the effect of hydrogen embrittlement.

The likelihood of hydrogen embrittlement is increased by excessive pickling temperature, prolonged pickling time, and poor inhibition of the pickling acid. To remove the potential for hydrogen embrittlement, heating to 150 °C after pickling and before galvanizing will result in expulsion of hydrogen from the grain boundaries of steel.

Another way is to use mechanical cleaning, such as shot or sand blasting, to remove the impurities instead of pickling. The abrasive blast cleaning does not generate hydrogen while it is removing the impurities. Nevertheless, a flash pickling after abrasive blast cleaning is needed to remove any final traces of blast media before hot-dip galvanization.

Test for the likelihood of hydrogen embrittlement for galvanized high tensile steels can be referred to ASTM A 143/A 143M (2003).

7.1.7 Inspection of welds

The inspection and testing procedures for welds on high tensile steels are similar to that on mild steels, except special attention should be paid on the hydrogen induced cracks as mentioned in Section 7.1.5. Visual examination and non-destructive testing (NDT) methods, such as radiographic or ultrasonic inspection, may be adopted. The visual examination is used to detect surface cracks whereas the NDT testing is employed to detect both surface and internal discontinuities. Personnel performing the visual examination and NDT testing shall have documented training and qualifications according to EN ISO 9712 (2012).

Due to the risk of delayed cracking of high tensile steel welds, a period of at least 48 h is generally required before the inspection. The period shall be stated in the inspection records. For welds with heat-treatment to reduce the hydrogen content, the inspection may be carried out immediately after the heat-treatment.

For direct visual examination, the access shall be sufficient to place the eye within 600 mm of the examined weld at an angle not less than 30° relative to background plane. An additional light source may be necessary to increase the contrast and relief between imperfections and the background. The visual examination shall be done in accordance with EN ISO 17637 (2011).

In case where the radiographic inspection is chosen, Class B techniques, being more sensitive to cracks compared with Class A, should be used for high tensile steel welds. The technical requirements for Class B inspection are given in EN ISO 17636 (Part 1 and Part 2, 2013).

When the ultrasonic inspection is adopted, special attention should be paid for high tensile steel plates made from thermo-mechanically controlled process (TMCP) as discussed in DNV-OS-F101 (2012). General requirements for ultrasonic inspection shall conform to EN ISO 17640 (2010).

7.2 High strength concrete (f_{ck} > 50 N/mm^2)

Detailed strengths for different concrete grades can be found in American, Australian and European codes. However, ACI 318–14 specifies that the concrete compressive strength depends on the proportioning of concrete mixtures and testing and acceptance of concrete. Therefore, only the limits of strength are provided for the application of different structural components and frames. EN 1994–1–1 covers the design of composite structures with both normal-weight concrete and light-weight concrete. Strength classes should be greater than C20/25 but lower than C50/60. Also, AS 3600 and AISC 360–16 set similar limitations for the concrete grade. However, apart from the lightweight concrete and normal strength concrete, high strength concrete with the compressive strengths over 50 MPa has been utilized in concrete filled steel tubular members and concrete encased steel composite columns. The readers are referred to Chapter 1 for the projects adopting high strength concrete in composite construction. To ensure the construction quality, the water to cement ratio should be strictly controlled as it is difficult to ensure that the inner steel tube is fully filled with concrete if internal diaphragms and stiffeners are provided to connect steel beams to concrete filled tube. One potential method to fix this problem is to use self-consolidating concrete (SCC). SCC is a type of flowable concrete which can flow into two steel plates without additional vibration. Other types of concrete have been investigated such as eco-oriented cement concrete typically involving the use of fly ashes or ground granulated blast furnace slag to replace cement which is more environmentally friendly.

7.2.1 Cement

High strength concrete can be produced with conventional Portland cement combined with fly ash or ground granulated blast furnace and silica fume. However, high early strength cements should be generally avoided since the rapid rise in temperature due to hydration can be accompanied by early age autogenous shrinkage and cracking. To obtain higher strength mixtures while maintaining good workability, it is necessary to carefully evaluate the cement composition and fineness and its compatibility with the chemical admixtures. Experience has shown that low-C_3A cements generally produce concrete with improved rheology.

7.2.2 Coarse aggregate

Aggregate plays an important role on the strength of concrete. The low water to cement ratio used in high strength concrete causes densification in both the matrix and interfacial transition zone, and the aggregate may become the weak link in the development of the mechanical strength. Extreme care is necessary, therefore, in the selection of aggregate to be used in the high strength concrete. The higher the targeted compressive strength, the smaller the maximum size of coarse aggregate should be adopted. Up to 70 N/mm^2, compressive strength can be produced with a good coarse aggregate of a maximum size ranging from 20 to 28 mm. Crushed rock aggregates which are not too angular and elongated should preferably be used.

7.2.3 Fine aggregate

The particle size distribution of fine aggregate that meets the Eurocode specifications (EN 932, EN 933, EN 1097, EN 1744, etc.) is adequate for the high strength concrete mixtures. Fine sands should not be used, particularly those with high absorption.

7.2.4 Supplementary cementitious materials

Using supplementary cementitious materials, such as blast furnace slag, fly ash and natural pozzolans, not only reduces the production cost of concrete, but also addresses the slump loss problem. Generally, silica fume is necessary to produce the high strength concrete.

7.2.5 Superplasticizer

Superplasticizer should be used to achieve maximum water reduction. The compatibility between cement and chemical admixtures and the optimum dosage of an admixture or combination of admixtures should be determined by laboratory experiments.

7.2.6 Mix proportion design

The basic proportioning of high strength concrete mixture follows the same method as for normal strength concrete, with the objective of producing a cohesive mix while minimizing the void content. This can be done by theoretical calculations or subjective laboratory trials. The basic strength to water/cement ratio relationships used for producing normal strength concrete are equally valid when applied to high strength concrete, except that the target water/cementitious materials ratio can be in the range 0.18–0.3.

The mechanical properties of high-performance concrete (HPC) containing fly ash supplied by a company whose main business is power generation using coal-fired power plants have been investigated. High performance concretes with GGBFS, silica fume and their combination have also been investigated for comparison. The mix proportion of the HPCs are given in Table 7.3. The compressive strength of the HPC was characterized according to the relevant ASTM standards. The results for the compressive strengths are summarized in Table 7.4. The compressive strength of the cubes increased with curing from 7 days to 28 days. At both 7 and 28 days, the compressive strengths of the 20% fly ash and 40% fly ash replacement reduced compared to that of the control specimen with only cement. The compressive strength for 50% GGBFS replacement was comparable to the control while the compressive strength for 70% GGBFS replacement reduced by almost 25% at 28 days compared to the control. High performance concrete mix-

Table 7.3 Mix proportions of high performance concrete with fly ash, GGBFS or silica fume.

Mix Type	Cement	Fly Ash	GGBFS	Silica Fume	Water	Fine aggregates	Coarse aggregates	Super plasticizer (% of binder by mass)	Slump (mm)
Control	485	0	0	0	145	690	1150	1.25	70
SF-7.5	449	0	0	36	145	690	1150	1.35	65
SF-15	412	0	0	73	145	690	1150	1.75	75
GGBFS-50	243	0	243	0	145	690	1150	1.75	80
GGBFS-70	145	0	340		145	690	1150	3.00	70
FA-20	388	97	0	0	145	690	1150	1.75	85
FA-40	291	194	0	0	145	690	1150	1.75	85
SF-7.5 & GGBFS-50	206	243	0	36	145	690	1150	2.50	85
SF-7.5 & FA-40	255	194	0	36	145	690	1150	2.25	80

Table 7.4 Compressive strength of high performance concrete with fly ash and GGBFS or silica fume.

Mix Type	Compressive strength (MPa)	
	7-day	**28-day**
Control	66.3	81.4
SF-7.5	72.7	95.5
SF-15	65.5	98.3
GGBFS-50	63.1	80.3
GGBFS-70	40.9	60.5
FA-20	53.8	64.7
FA-40	35.8	47.2
SF-7.5 & GGBFS-50	59.3	86.1
SF-7.5 & FA-40	38.1	61.9

tures with silica fume were higher than those of the control. Specifically, mixtures with 7.5% and 15% silica fume replacement for cement showed 17% and 21% higher compressive strengths compared to the control at 28 days. Combinations of 7.5% silica fume and 50% GGBFS resulted in similar compressive strengths compared to the control while those with 7.5% silica fume and 40% fly ash showed around 24% lower compressive strengths compared to the control. In this case, more studies are necessary to verify the performance of fly ash-silica fume mixtures and whether higher amount of silica fume or lower amounts of fly ash can achieve compressive strengths similar to the control. Based on the promising mechanical properties of mixtures with 50% GGBFS and 7.5% silica fume and reduced environmental impact of mixtures with GGBFS, it would be feasible to replace the cement with GGBFS for such HPC applications.

7.2.7 Quality control

When superplasticizer is used, concrete tends to lose workability rapidly. High strength concrete containing such materials must therefore be transported, placed, and finished before they lose their workability. Many modern superplasticizers can retain reasonable workability for a period of about 100 min, but care is still needed, particularly on projects where ready-mixed concrete delivery trucks require long journey times. Often, to avoid drastic reduction in slump and resultant difficulty in placing, only part of the superplasticizer is mixed during batching with the balance being added on site prior to pouring.

The same production and quality control techniques for normal strength concrete should also be applied to high strength concrete. In fact, the importance of strict control over material quality as well as over the production and execution processes cannot be over-emphasized for high strength concrete. In general, production control should include not only correct batching and mixing of ingredients, but also regular inspection and checking of the production equipment, e.g. the weighing and gauging equipment, mixers and control apparatus. With ready-mixed concrete supply, this control should extend to transport and delivery conditions as well.

The main activities for controlling quality on site are placing, compaction, curing and surface finishing. Site experience indicates that more compaction is normally needed for high strength concrete with high workability than for normal strength concrete of similar slump. As the loss in workability is more rapid, prompt finishing also becomes essential. To avoid plastic shrinkage, the finished concrete surface needs to be covered rapidly with water-retaining curing agents.

7.2.8 Casting of concrete in steel tubes

In general, the concrete casting in hollow steel tubes can be conducted by means of free-falling, tremie tube, flexible hose, or pumping. The free-falling casting is limited by a maximum height to eliminate segregation, whereas the flexible hose, connected to a hopper at top of column, can effectively reduce air entrapment in concrete compared with the free-falling method. The concrete casting by free-falling and flexible hose can be employed for CFST columns without internal diaphragms which are free of obstructions for casting.

Instead of the hopper and flexible hose, the concrete can be placed in tremie pipe and then cast into the column tubes. Meanwhile, poker vibrator is attached to the tremie pipe to compact the concrete. Casting by tremie pipe is shown in Fig. 7.1 (a). The tremie should be fabricated of heavy-gage steel pipe to withstand all anticipated handling stresses and should have a diameter large enough to prevent aggregate-induced blockages. Pipes with diameter of 200 to 300 mm are generally recommended (ACI 304R-00, 2003). A stable platform should be provided to support the tremie during the placement of concrete. Tremie pipes should be embedded in the fresh concrete with a depth of 1.0 to 1.5 m. The embedment depths depend on placement rates and setting time of the concrete. All vertical movements of the tremie pipes should be done slowly and carefully to prevent loss of seal.

Self-compacting concrete may be placed into the steel tube using high pressure pump as shown in Fig. 7.1 (b) & (c). An opening with doubler plate is created on the surface of the steel tube, to which the concrete is being pumped into the steel tube, which acts as the permanent formwork. The most important achievement of concrete filled steel tubular construction is the saving of construction time due to the significantly improved speed of the construction. The pumping is carried out from bottom to the top of the column. The location of pumping inlet should be at about 300 mm from the floor level. The maximum height of concrete pumping depends on the capacity of the pumping machine but should not exceed 60 m. For composite columns with internal diaphragm plates, the pumping rate should not exceed 1m/min to avoid the air entrapment.

7.3 Ultra high-performance concrete (f_{ck} > 120 MPa)

The American Concrete Institute (ACI) Committee 239 emerging technology report on Ultra-High Performance Concrete (ACI 239R-18) defines UHPC as “concrete that

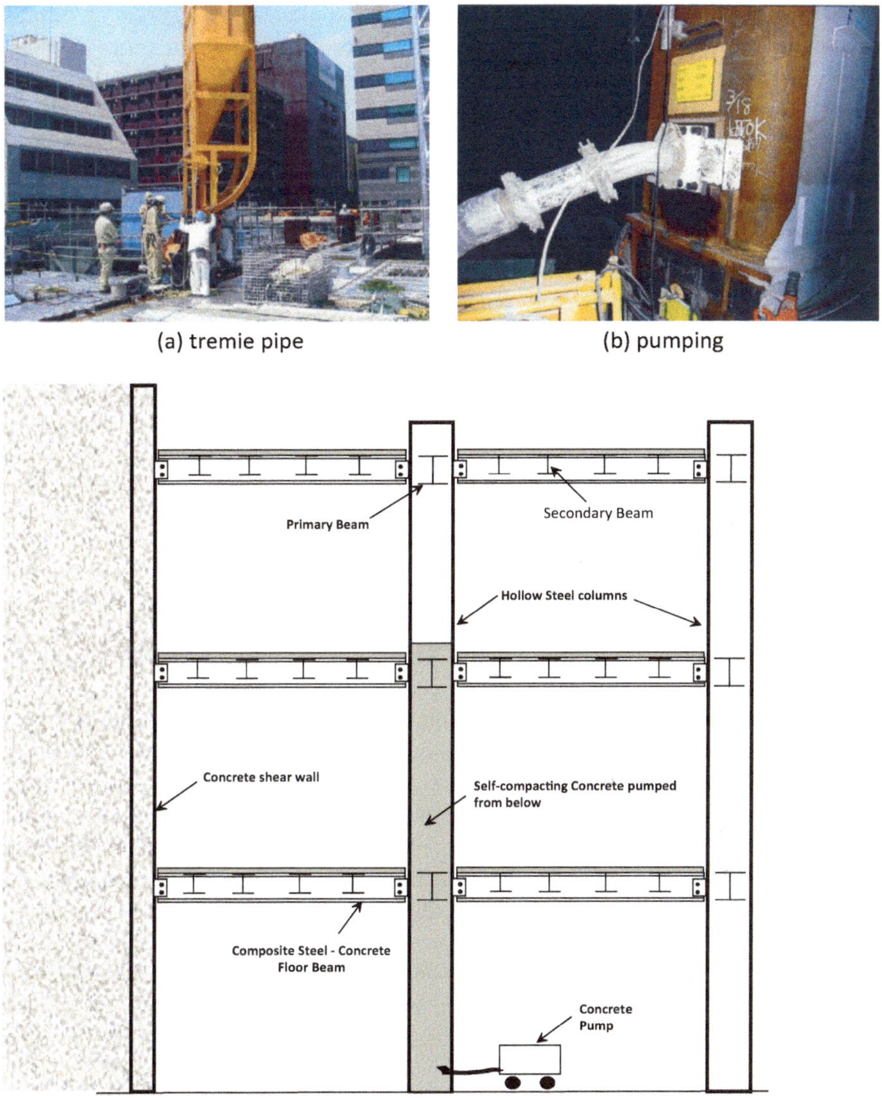

Fig. 7.1 Concrete casting methods for CFST columns.

has a specified compressive strength of at least 150 MPa and complies with specified durability, ductility and toughness requirements using reproducible test methods; fibers are generally included to achieve specified requirements". On the other hand, the American Society of Testing and Materials (ASTM C1856/C1856M – 17) defines UHPC as "a cementitious mixture that has a specified compressive strength of at least 120 MPa [17 000 psi], generally containing fibers and has other properties measured by standard test methods that comply with specified durability, ductility and toughness requirements". Thus, depending on the applications, UHPCs of different strengths can be adopted. UHPCs are widely used for different applications such as in bridges, facades and infrastructure requiring impact resistance including security and blast resistance. UHPCs generally incorporate the following components: cement, silica fume, sand, quartz powder or silica flour, high-range water reducing admixture, water and fibers. More details of the individual components are provided below.

7.3.1 Cement

UHPC often eliminates larger aggregates and incorporates only fine aggregates. Due to the smaller size of the aggregates, the aggregate surface to be enveloped with cementitious paste is higher than in normal concrete. This leads to incorporation of a high paste volume in UHPC to achieve good flowability. The cement content in UHPC often ranges from 700–1400 kg/m^3. However, due to the low w/c ratio, the cement cannot be completely hydrated and significant amount of cement in the UHPC remains unhydrated. Therefore, this cement can be replaced partially by supplementary cementitious materials such as fly ash or ground granulated blast furnace slag.

7.3.2 Silica fume

Silica fume is a main constituent of a typical UHPC mixture and plays an important role in improving the rheological and mechanical properties of UHPC. Silica fume plays three prominent roles in UHPC, firstly, it has a filling effect which improves the particle packing density. Secondly, it has a lubricating effect enhancing the rheological properties due to its spherical nature. Thirdly, it exhibits pozzolanic reactivity leading to the production of additional calcium silicate hydrates thus contributing to enhanced mechanical properties. Different studies use a range of dosages of silica fume in UHPC from 10% to 30% by mass of the UHPC mixture. However, due to its high specific surface area, silica fume increases the water demand in UHPC. Therefore, it is necessary to use silica fume in conjunction with high range water reducing admixtures. Higher dosages of silica fume require higher contents of super plasticizer and result in more sticky mixture. Further, high dosage of silica fume can also result in higher autogenous shrinkage at early ages which can in turn provoke microcracking affecting the long-term durability of structures. Therefore, the silica fume content in UHPC needs to be optimized.

7.3.3 Supplementary cementitious materials

Using supplementary cementitious materials (SCMs), such as blast furnace slag, fly ash and natural pozzolans, not only reduces the production cost of concrete, but also has technical advantages. Use of SCMs can reduce the heat of hydration as well as enhance the durability of the UHPC. Specifically, due to its reaction with cement hydration products such as calcium hydroxide, the SCMs can enhance the durability. The SCMs can also improve the particle packing efficiency of UHPC. However, incorporation of SCMs can affect the mechanical properties of UHPC and therefore, it is recommended to use optimal amount of SCMs which do not compromise the mechanical properties of the UHPC

7.3.4 Coarse aggregates

Typically, UHPC mixtures do not include coarse aggregates. Given the superior performance of UHPC, aggregates may become the weakest phase and the strength of the concrete may not be a function of the water-cement ratio but rather the aggregate type and quantity. Therefore, coarse aggregates are typically eliminated in UHPC unless for some special mixtures which may incorporate granite, bauxite or iron ore coarse aggregates.

7.3.5 Fine aggregates

The matrix of UHPC is generally optimized for particle packing. In other words, the grading of the fine aggregates is specially chosen to ensure that densest packing of the sand in conjunction with the cement and the silica fume. Specifically, the particle size and mechanical properties of the constituents are chosen so that the number of contact points between the particles are increased and level of stress transferred between particles through the paste is reduced thus reducing the possibility of microcrack formation along with increased mechanical properties.

7.3.6 Quartz powder

Quartz powder is commonly used in the manufacture of UHPC. Quartz is generally considered to be chemically inert at ambient temperatures and is used as a micro filler in UHPC. Quartz powder has several effects in UHPC. Firstly, the fine size of the quartz powder enhances the particle packing in UHPC. Secondly, while quartz powder is generally considered to have no chemical reaction, it is known to have physical effect and accelerate the hydration reaction of cement. The fine size of the quartz powder is known to provide more nucleation sites for calcium silicate hydrate growth through cement hydration.

7.3.7 Superplasticizer

Superplasticizer should be used to achieve maximum water reduction. The compatibility between cement and chemical admixtures and the optimum dosage of an admixture or combination of admixtures should be determined by laboratory experiments.

7.3.8 Fibers

The matrix of UHPC is very brittle so the ductility enhancement in UHPC is achieved through the incorporation of fibers. Specifically, the fibers are used to obtain the elastic-plastic or strain-hardening behavior in tension. UHPCs often contain up to 1 – 3% by volume of steel or polyethylene (PE) fibers and the amount of fibers is controlled by the fiber aspect ratio, shape and its impact on the workability of the UHPC.

7.3.9 Mix proportion design

The mix design of UHPC considers several factors including particle size distribution, cement content, supplementary cementitious material content, water/cementitious materials ratio, dosage of superplasticizer, and fiber content to satisfy the requirements of UHPC in both fresh and hardened states. The particle size of the different constituents such as cement, supplementary cementitious materials, aggregates and quartz powder should be selected in such a way that the grading curve for the mixture of granular materials matches with the optimum grading curves provided by the packing models. The values of the mix design parameters can be selected within their ranges as reported in literature, for example, the water to total binder ratio 0.15–0.24, cement and silica fume content 700–1400 kg/m^3, silica fume to cement ratio 0.10–0.35, quartz and sand 1000–1400 kg/m^3 and steel fiber content 77–250 kg/m^3.

The authors have investigated UHPC mixtures containing different types of Class F fly ash, amounts of fly ash, along with comparison with another commonly used mineral admixture ground granulated blast furnace slag (GGBFS), effect of aggregate size on the mechanical properties of concrete, i.e. compressive strength, modulus of elasticity, and flexural strength. The mix proportions are presented in Table 7.5 and results are summarized in Table 7.6 and Fig. 7.2.

Results indicate that the compressive strength of the UHPC decreased with the increase in fly ash and GGBFS contents at various ages from 7 to 56 days as expected (Table 7.6). However, UHPC with compressive strength greater than 120 MPa can be produced even with 30% fly ash or 41% GGBFS

The flexural tensile strength of the UHPC with 30% GGBFS is comparable to that of the control concrete. With the increase in the GGBFS content to 41%, however, the flexural tensile strength concrete is increased by about 20% compared to the control concrete (Table 7.6). The enhancement in the flexural strength may be due to the improvement in bond between the fibers and matrix. For the UHPCs with fly ash, the flexural tensile strength increased with an increase in fly ash content to 15% and showed strain hardening behavior similar to the control (Fig. 7.2). However, further increase in fly ash content to 30% results in decreased flexural tensile strength (Fig. 7.2), indicating an optimal fly ash content exists for the flexural tensile strength. The higher tensile strengths of the UHPCs with 15% fly ashes may be explained by the improvement of the interfacial bond between the UHPC matrix and fibers due to reduced $Ca(OH)_2$ and increased C-S-H by pozzolanic reactivity of the fly ashes in addition to the effect of silica fume. In addition, the replacement of cement by fly ashes can lower the rate of cement hydration and pozzolanic reaction of silica fume at early stage, which may lead to more homogeneous microstructure and refined pore structure both in the matrix and at fiber-matrix interface. These may also contribute to the improved flexural tensile strength. When the fly ash is increased to 30% as cement replacement, on the other hand, more residual fly ash is likely present in the system, which reduces the bond between the fibers and the matrix. Further research is needed to evaluate the fiber-matrix bond in combination with the optimization of the GGBFS or fly ash content in UHPC. Given the results, there is a huge potential for producing UHPCs by replacing part of the Portland cement with GGBFS or fly ash.

Table 7.5 Mix proportions of UHPCs.

Mix ID	FA*/GGBFS, % by mass of CM	Fly ash type	Mix proportion, kg/m³								Flow (mm)
			Water	Cement	FA/ GGBFS	Silica fume	Sand	SP[a]	SRA[b]	Fibers[c]	
Control-FA	0	–	202	1021	–	102	1040	15.0	10.2	77.0	167
FA1-15%	15% fly ash	FA1	202	853	169	102	988	14.0	10.2	77.0	166
FA1-30%	30% fly ash	FA1	201	680	335	102	955	13.0	10.2	77.0	170
FA2-15%	15% fly ash	FA2	202	853	169	102	988	15.0	10.2	77.0	169
Control-GGBFS	0	–	204	1030	–	103	1050	15.5	10.2	77.7	150
GGBFS-30%	30% GGBFS	–	201	680	335	102	1035	11.5	10.2	76.6	160
GGBFS-41%	41% GGBFS	–	201	557	456	101	1033	10.6	10.1	76.5	200
Control-140	0		193	1011	0	179	1011	15.0	–	78.0	138
FA–15–140	15% fly ash		187	805	171	171	1011	14.0	–	78.0	147

[a]ADVA 181N, W.R. Grace (Singapore) Pte. Ltd
[b]Eclipse Floor, W.R. Grace (Singapore) Pte. Ltd
[c]Straight high carbon steel wire fibers, Dramix@, Bekaert, Belgium Pte, Ltd of length 13 mm, diameter 0.16 mm, aspect ratio of 81, modulus of elasticity of 200 GPa, tensile strength of 2500 MPa.

Table 7.6 Density and mechanical properties of UHPCs.

Mix ID	FA*/ GGBFS,% by mass of CM	Fly ash type	Demold density, kg/m³	Compressive strength, MPa Elastic modulus, GPa			Flexural strength, MPa	
				7-day	28-day	56-day	28-day	28-day
Control-FA	0	–	2414	129.1	142.3	147.2	40.8	15.4
FA1-15%	15% fly ash	FA1	2373	120.1	134.6	141.8	42.6	19.3
FA1-30%	30% fly ash	FA1	2375	110.5	127.3	133.6	42.8	12.8
FA2-15%	15% fly ash	FA2	2388	126.8	137.1	142.2	43.3	19.2
Control-GGBFS	0	–	2456	136.1	137.7	157.5	–	14.8
GG-BFS-30%	30% GGBFS	–	2409	109.2	129.0	138.2	–	14.3
GG-BFS-41%	41% GGBFS	–	2413	108.1	123.0	125.6	44.1	17.6
Control-140	0	–	2417	–	137.3*	–	43.7	17.0
FA–15–140	15% fly ash	FA	2371	–	141.3*	–	44.3	14.3

*The 28-day compressive strengths for Control-140 and FA-15–140 were determined using cylinder specimens while those of the others were determined using 100 mm cubes.

Additional research shows that UHPC with 28-day compressive strength of >140 MPa can be produced with 15% fly ash as cement replacement (Table 7.6).

7.3.10 Durability of UHPC

In general, the durability of UHPC is better than that of conventional concrete. Due to its dense matrix and reduced porosity compared to normal concrete, UHPC demonstrates improved resistance against many harmful liquids and gases, chloride attack, carbonation, alkali-silica reaction etc. The water permeability coefficient of UHPC is also one to two order of magnitude lower than of normal concrete. Steel reinforcement embedded in UHPC also exhibits a reduced corrosion risk compared to normal concrete. Tests carried out on UHPC exposed to mid-tide marine exposure site in the Maine state in United States of America, also did not show any observable degradation in the mechanical or durability properties after exposure for 5 and 15 years, indicating the excellent durability properties of UHPC (Thomas et al. 2012). UHPC beams installed in the aggressive environment in the water-cooling tower of a nuclear power plant in France also exhibited excellent durability after 10 years indicating that UHPC have excellent durability performance (ACI 239R-18).

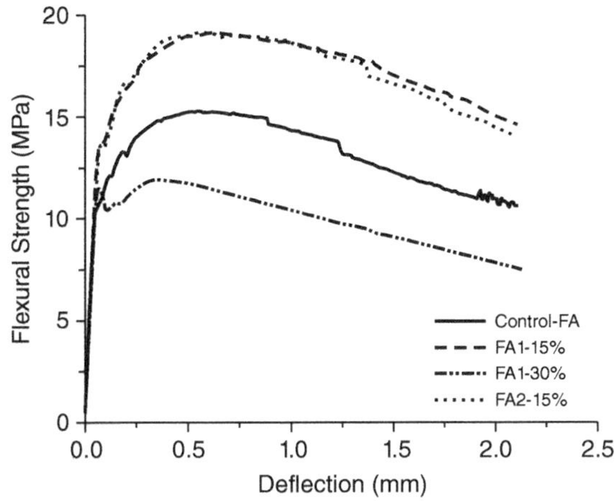

Fig. 7.2 Flexural stress-deflection curves of the concretes with 15% fly ashes (with different fineness's and chemical compositions) and 30% fly ash in comparison to that of the control concrete without fly ash.

7.3.11 Creep and shrinkage of UHPC

Creep and shrinkage occur over the service life of a concrete structure and are dependent on the age and environmental conditions to which the concrete structures are exposed. Creep and shrinkage can result in deformations which if excessive can result in cracking of concrete or excessive element deformations. Several factors are known to influence the creep and shrinkage behavior including the age of loading, the temperature, humidity, degree of hydration etc.

Similar to normal concrete, UHPC also experiences volume changes with sustained load. The creep is known to depend on the curing regime since many researchers use early age thermal treatment for UHPC. It was found that early-age creep coefficients of UHPC were within the range of normal concrete. This is because at early ages, the UHPCs are still gaining strength and not completed self-desiccated. Further, the early age was found to be almost twice the creep as mature age and structural effects of early age creep of UHPC could be mitigated during design (Haber et al. 2018).

The long-term creep coefficients (ratio of ultimate creep strain to elastic strain) of UHPC were found to vary between 0.31 (for steam cured) to 0.8 (for ambient curing). In comparison, the creep coefficients for normal concrete are normally in the range of 1.5 to 3.0 (Graybeal 2006). Further, the specific creep value (that is the creep coefficient divided by the elastic modulus of the concrete), ranged from 5.7 (for steam curing) – 21.1 (for ambient curing) με/MPa compared to the specific creep values of 35 to 145 με/MPa experienced by the normal concrete (Graybeal 2006). Therefore, the long-term creep of UHPC may be less than normal concrete and maybe directly attributed to the higher strength and modulus of elasticity of the UHPC mixtures.

Volume changes due to the loss of moisture from capillary pores is referred to as shrinkage. When the volumetric contraction occurs due to the internal consumption of water during hydration of cement, it is referred to as autogenous shrinkage. The American Concrete Institute (ACI) defines autogenous shrinkage as "change in volume produced by continued hydration of cement, exclusive of the effects of applied load and change in either thermal condition or moisture content" (ACI 231R-10, 2010). It can also be looked as the deformation of cement paste in a closed system (Mehta & Monteiro, 2006). While present in all types of concrete, autogenous shrinkage is most significant in high performance concrete or UHPC with low w/c ratio and high cement content. Further, UHPC usually contains silica fume, which accelerates the water consumption at early ages due to accelerated cement hydration and pozzolanic reactions making autogenous shrinkage more critical in such systems. Particularly, autogenous shrinkage occurs within the first 24h and even before 12h in a stage where the tensile strength of concrete is too low to resist the shrinkage cracking resulting in crack propagation by shrinkage stresses, therefore caution should be taken to minimize restraint due to formwork and existing structural elements. While the early age shrinkage may be higher, the long-term shrinkage of UHPC is commonly within the range associated with normal concrete. The long-term shrinkage may be due to the low w/c ratio (low water content and low permeability) of UHPC. One way to overcome the high autogenous shrinkage in UHPC is by incorporation of fibers. The fibers help to bridge the shrinkage cracks and redistribute the stresses thus lowering the shrinkage in UHPC. From the point of view of a practicing engineer, a freshly poured UHPC mixture is expected to show height reduction and this contraction can be overcome by over-pouring the element. Therefore, such shrinkage may be considered by engineers during the design and construction process.

Joints in composite construction

8

Outline

8.1 General

Joints are classified as either rigid or simple in traditional design. In fact, the joints could be classified into nominally pinned, semi-rigid and rigid in accordance with EN 1993–1–8, 2005. The semi-rigid and rigid joints are capable of resisting bending moments. Joint behavior should be considered in structural analysis based on each type of joint, for example, the design moment-rotation characteristics of semi-rigid joints should be used in structural analysis to determine internal forces for individual members.

As recommended by EN 1993–1–12 (2007), the rules for semi-rigid connections are not applicable for steel sections with grade higher than S460, and the connection should be either rigid or pinned. The moment resistance of the connection made of high strength steel with steel grade greater than S460 should be determined based on elastic distribution of forces over the components of the connection rather than based on full plasticity for connections with steel grades not greater than S460.

The partial factors γ_M for joint design in accordance with EN 1993–1–8 (2005) are given in Table 8.1, regardless of the steel grade.

Design of Steel-Concrete Composite Structures Using High Strength Materials. DOI: 10.1016/C2019-0-05474-X

Table 8.1 Partial factors for design using Eurocode 3 and Eurocode 4.

Applications	γ_M	Values
Resistance of cross-sections	γ_{M0}	1.0
Resistance of members to instability assessed by member checks	γ_{M1}	1.0
Resistance of cross-sections in tension to fracture	γ_{M2}	1.25
Resistance of bolts Resistance of welds Resistance of plates in bearing	γ_{M2}	1.25
Slip resistance		
• at ultimate limit state	γ_{M3}	1.25
• at serviceability limit state	$\gamma_{M3,ser}$	1.1
Preload of high strength bolts	γ_{M7}	1.1

8.2 Column splices

8.2.1 Bolted splice joints

For open steel sections encased by concrete in CES columns, a non-bearing splice joint may be used as shown in Fig. 8.1 (a). This splice is detailed with a physical gap remained between the upper and lower steel sections, therefore, the forces are transferred through the bolts and cover plates at flanges and web, rather than the direct bearing between the ends of the steel sections. The bearing splice joint is not recommended mainly due to two facts; one is the difficulties in fabrication and installation to flatten and tighten the steel sections at ends, and the other is the complexity in load transferring. For example, when the splice joint is subject to bending the tension force

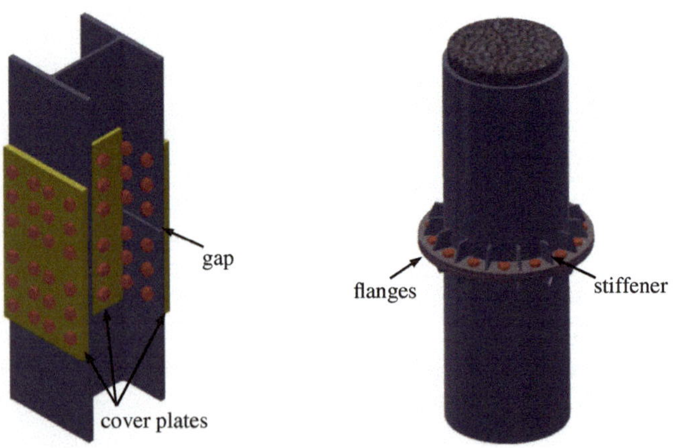

(a) Steel section splicing in CES column (b) Flange splicing in CFST column

Fig. 8.1 Bolted splices for steel sections in CES and CFST columns.

is resisted by bolts and cover plates, but the compression force is resisted by direct bearing. This causes difficulties in determining the flexural stiffness of the splice joint. Nevertheless, the bearing splice joint can be used in axially loaded columns or columns with small bending moments where no tension force is induced on the cross-section.

The non-bearing splice joint can also be used for tube splicing in CFST columns but the blind bolts are needed. Alternatively, the flange splicing can be adopted as shown in Fig. 8.1(b), in which case additional stiffeners are usually required to strength the flange plate to achieve full rigidity to ensure uniform load transfer from one steel tube to the other through the flange plates.

The splice joints may have limited moment resistance and, thus, they should be placed near to the contraflexure point of moment of the column.

8.2.2 Welded splice joints

The welded splice joints with either full penetration or partial penetration butt welds may be adopted for steel sections in composite columns as shown in Fig. 8.2(a) and (b). Peripheral on-site welds are required. Guide plates are provided to connect and align the column segments by bolts before welding is applied at site, as shown in Fig. 8.2(c). These guide plates and bolts will be removed after the column is fully welded and full continuity is achieved.

It is not necessary for splice joints with the full penetration butt welds to check the joint resistance against the design loads, unless an undermatched electrode is used for the welds. In such cases, the welded joint resistance should be determined by using the strength of electrode rather than that of the steel column section. For splices with the partial penetration butt welds, the splice joint resistance should be calculated on the basis of method for fillet welds with an effective throat thickness equal to the penetration depth, a, as shown in Fig. 8.2.

Due to the limitation of transportation, composite columns are practically delivered to site in lengths of about 12 m and welded on site to form a continuous column. The splices should be located as close as possible to the contraflexure point of the column to avoid large moment, and they are often practically located at 1.2 to 1.5 m above the floor level to facilitate on-site positioning and welding.

When the column splice is placed at the junction with the change of steel section thickness, a gradual change of plate thickness is preferred to avoid stress concentration at the welded joint as shown in Fig. 8.3.

Fig. 8.4 shows the typical splice joint details with the change of column cross sections located at the beam to column joints. The slope for the change of section is practically not greater than 1:6 to provide a smoother flow of stress in the column.

For boxed column fabricated by four steel plates welded at the corners, full penetration butt weld formed continuously along the column height is generally required, and a backing plate is placed at the back of the joint near the weld root to shield molten weld metal. Both single-V and single-bevel butt welds shown in Fig. 8.5 may be provided.

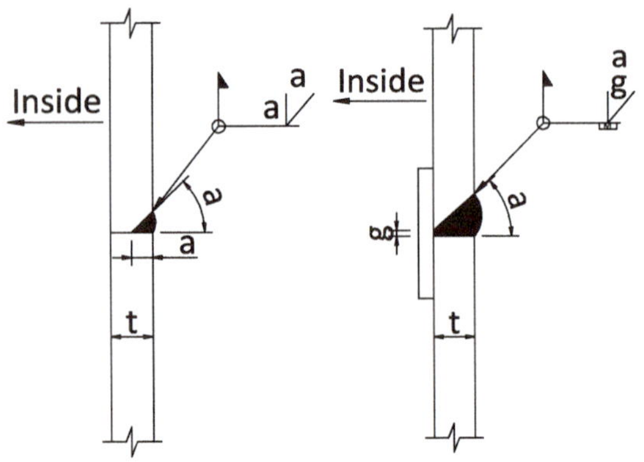

(a) Partial penetration

(b) Full penetration

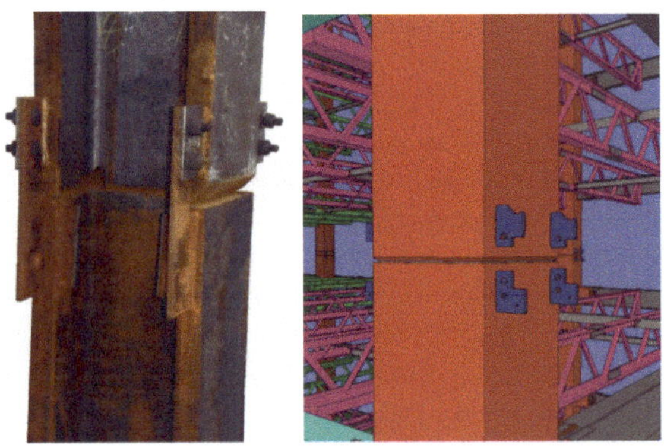

(c) Temporary guided plates to hold the two columns in position before welding

Fig. 8.2 Typical partial and full penetration butt welds for composite column splices.

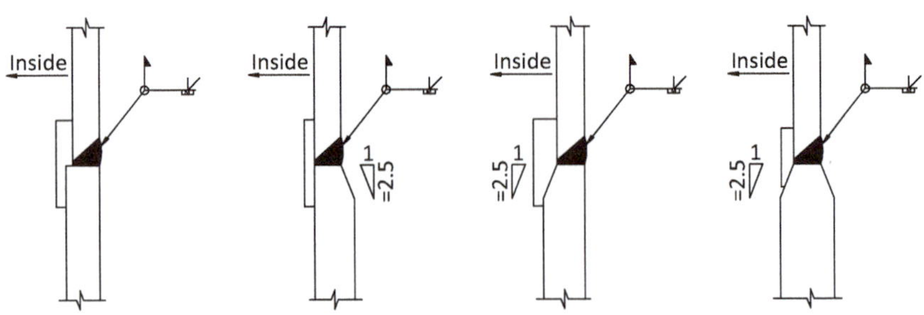

Fig. 8.3 Splicing details for change of plate thickness in composite columns.

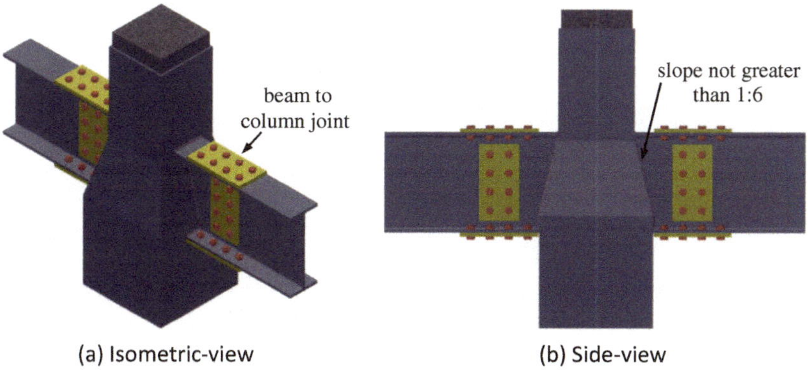

Fig. 8.4 Splicing details for change of cross-sectional sizes of CFST column.

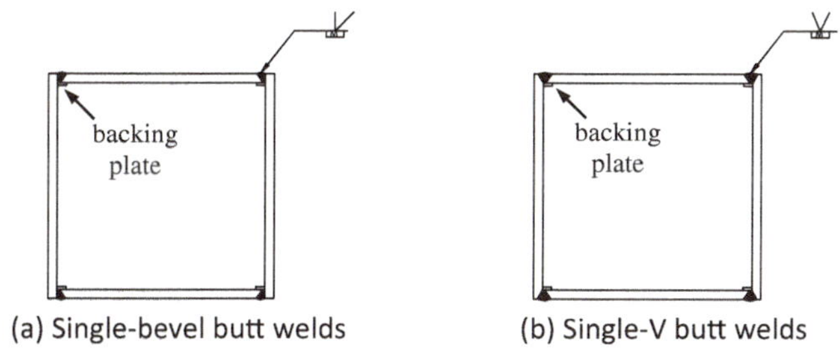

Fig. 8.5 Full penetration welds for welded sections.

8.3 Steel beam to composite column joints

8.3.1 Simple connections

8.3.1.1 Fin plate connection

Fin plates may be used for beam to CFST column connection. The fin plate is a length of plate welded to the CFST column in workshop to which the supported beam web is bolted on site to form a simple connection as shown in Fig. 8.6. The flanges of beam are not connected to the column.

The fin plate connections should have sufficient rotational capacity which may be developed from the bearing deformation of the bolts against the fin plate and/or the beam web, and from the shear deformation of the bolts. Practical recommendations

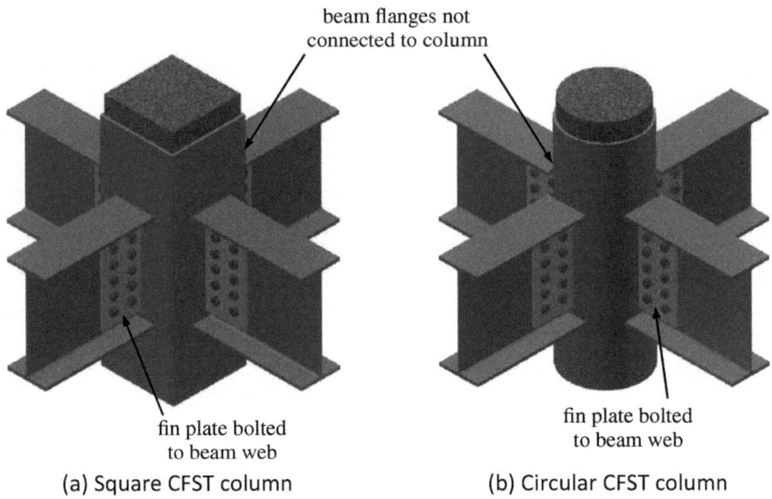

Fig. 8.6 Simple connection between steel beam and CFST column with fin plates.

are given below in order to ensure that the beam is properly restrained and possesses sufficient rotational capacity:

- The fin plate is positioned as close as possible to the top flange of beam in order to provide positional restraint to the beam;
- The fin plate depth is at least 0.6 times the supported beam depth in order to provide the beam with sufficient torsional restraint;
- For robustness design, all bolt end and edge distances on the fin plate and the beam web should be at least 2 times the diameter of bolt hole. Full strength fillet welds are recommended for the fin plate;
- At least two M20 G8.8 bolts and the fin plate thickness of at least 6 mm should be adopted in order to achieve minimum tie force for structural integrity.

The supported beam and column should be checked for shear and bearing. The design procedure can be referred to SCI Publication P358 (2014) and SN017a-EN-EU NCCI (2005). With a single vertical line of bolts, the connection shear resistance will be in the range of 25% to 50% of the beam shear resistance. Using two vertical lines of bolts increases the resistance, but as the eccentricity of the load also increases, the benefit does not double and the best that can be achieved is around 75% of the beam shear resistance.

8.3.1.2 Through fin plate connection

The simple connection between beam to CFST column could also be achieved by through fin plate. Such fin plate is inserted through the steel tube and welded on it in workshop to which the supported beam web is bolted on site as shown in Fig. 8.7. Opening should be cut on the steel tube for the passage of fin plate and all-round fillet

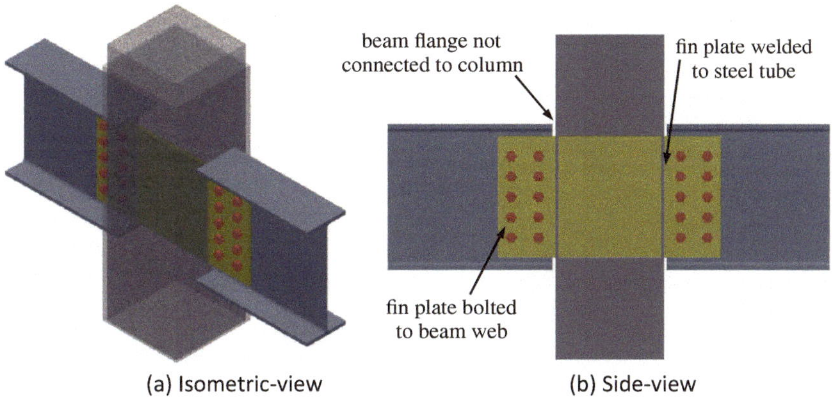

Fig. 8.7 Simple connection between steel beam and CFST column with through fin plate.

welds may be used to connect it to the steel tube. The design procedure is similar to that of the welded fin plate except it is not necessary to check the column tube against the local shear and punching shear. However, the local bearing capacity of the concrete under the fin plate in column should be checked in accordance with EN 1994–1–1, Clause 6.7.4 (6) (2004).

8.3.1.3 T-Stub

T-stub may be used for simple connection between steel beam and concrete encased steel (CES) column as shown in Fig. 8.8. The T-stub is either welded to the flange of the encased steel section through a fin plate or precast in concrete through bolts in the workshop, the web of the T-stub is bolted to the web of the supported beam. The flanges of the supported beams are not connected to the column. The fin plate and bolts are designed against combined shear force and bending moment that is introduced by the eccentric shear force. The web of T-stub can be designed as fin plate as shown in Fig. 8.6 and Fig. 8.7.

8.3.1.4 Supporting bracket

Steel bracket (or steel corbel) is sometimes used as simple connection between steel beam and CFST column as shown in Fig. 8.9. The design procedure requires a check for the bracket subjected to combined shear and bending moment. In addition, the beam web at the end should be checked for local bearing in accordance with EN 1993–1–1 (2005). Vertical stiffeners may be added to prevent the local bearing failure. Full strength welds are recommended for the bracket connected to the column tube, and the internal diaphragm plate is required to prevent local tearing failure of the steel tube. Bolts with slot holes may be used to accommodate the movement of steel beam or site tolerance for installation.

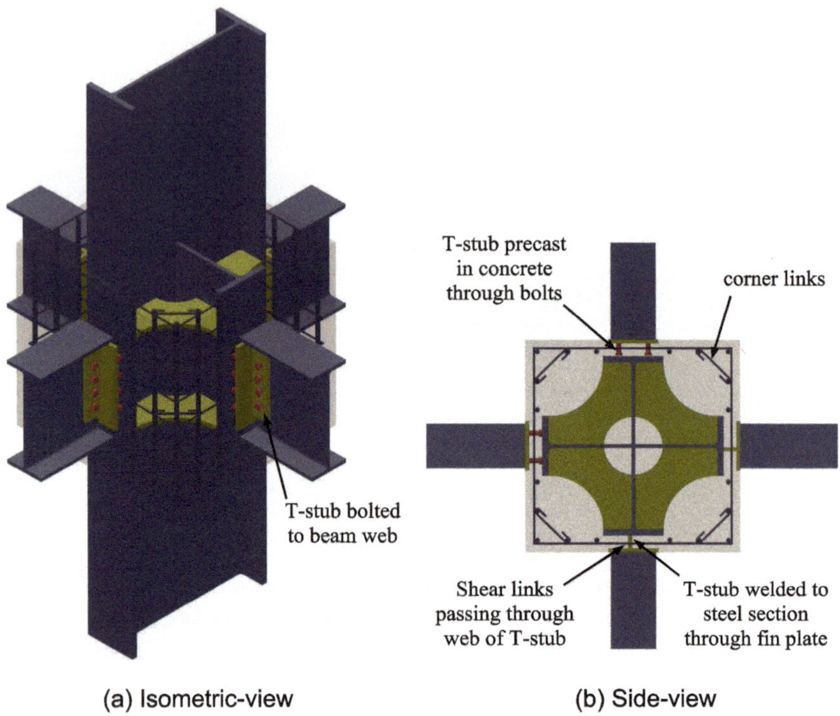

Fig. 8.8 Simple connection between steel beam and CES column with T-stub.

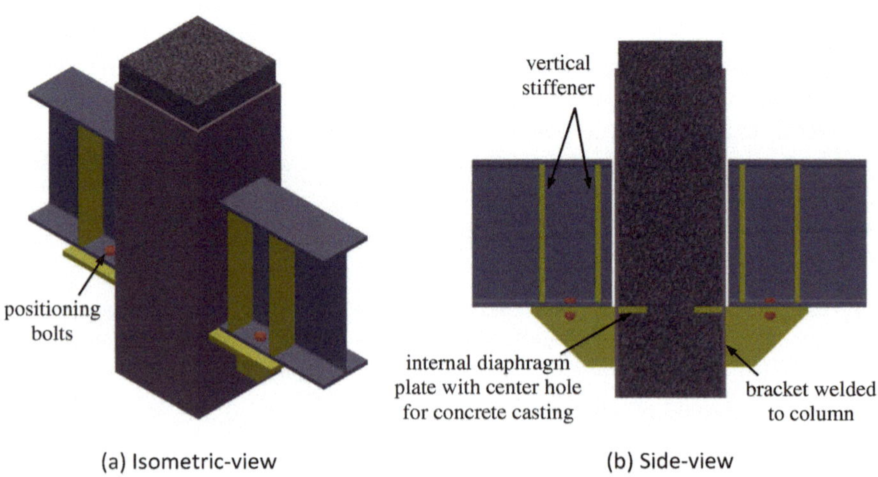

Fig. 8.9 Simple connection between steel beam and CFST column with steel bracket.

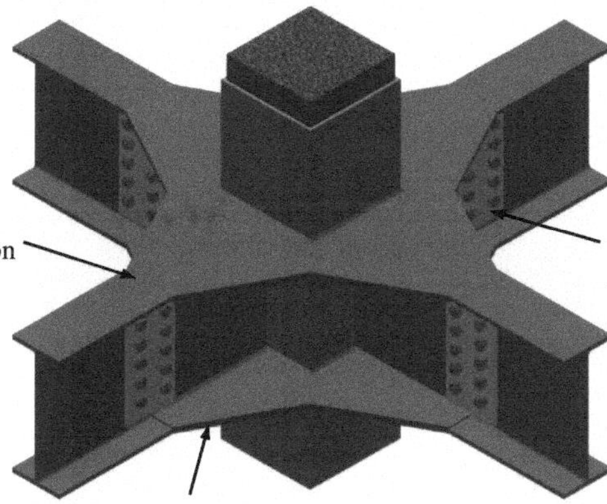

Fig. 8.10 Moment connection between steel beam and CFST column with external diaphragm plates.

8.3.2 Moment connections

8.3.2.1 External diaphragm plate

Typical moment connection between steel beam and CFST column with external diaphragm plates is shown in Fig. 8.10. The external diaphragm plates are welded to the flanges of the floor beam to transfer bending moment, whereas the beam web is generally connected by high strength friction bolts to transfer the shear force.

The external diaphragm plates are pre-welded to the CFST column tube in the workshop. The dimensions of them depends generally on the sizes of the floor beams. Such connection is strong and ductile, and the failure of it occurs at the beam to diaphragm plate junction, which is out of the joint area. The disadvantage is the higher cost since it involves large amounts of steel plates and welding work as compared with the other types of connection. The following requirements should be observed when the external diaphragm plates are used for the beam to CFST column connection:

- Matching electrode is recommended for all welds
- The diaphragm plates should have a minimum yield strength higher than or equivalent to that of the floor beams
- The thickness of diaphragm plate should be at least 6 mm to 10 mm larger than that of connected beam flange in case of misalignment during installation
- The minimum width of the diaphragm plate cmin shown in Fig. 8.11 should be at least $\sqrt{2}/2b_f$ where b_f is the width of the floor beam flange

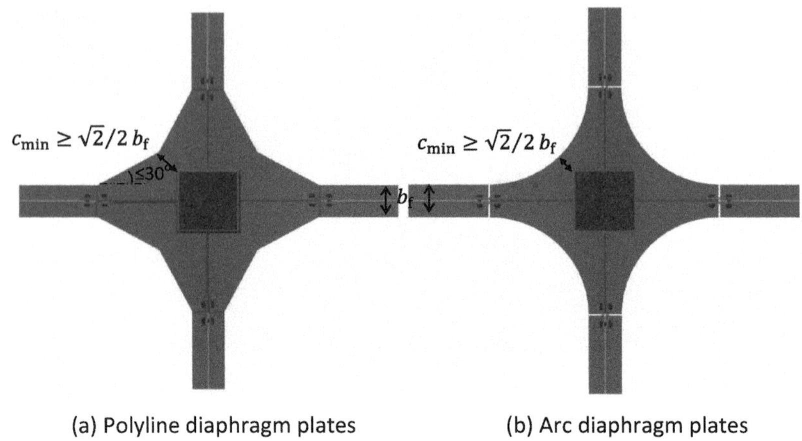

Fig. 8.11 Joint detailing for CFST columns with external diaphragm plates.

When the external diaphragm plates are used for edge or corner columns, the detailing shown in Fig. 8.12 may be adopted. The diaphragm plates should be continuous around the column.

8.3.2.2 Internal diaphragm plate

Internal diaphragm plates shown in Fig. 8.13 may be used for moment transferring from the beam to a CFST column. An outstanding beam segment with flanges aligning to the internal diaphragm plates is welded in workshop, and then connected with the floor beam on site. The internal diaphragm plate is sometime preferred by architects due to aesthetic reason, but special attention should be paid to the casting of concrete since the internal diaphragm plate may obstruct the flowing of concrete. Air voids may form either beneath or above the diaphragm plate when the concrete is infilled from top or bottom of the column. Therefore, the access holes should be

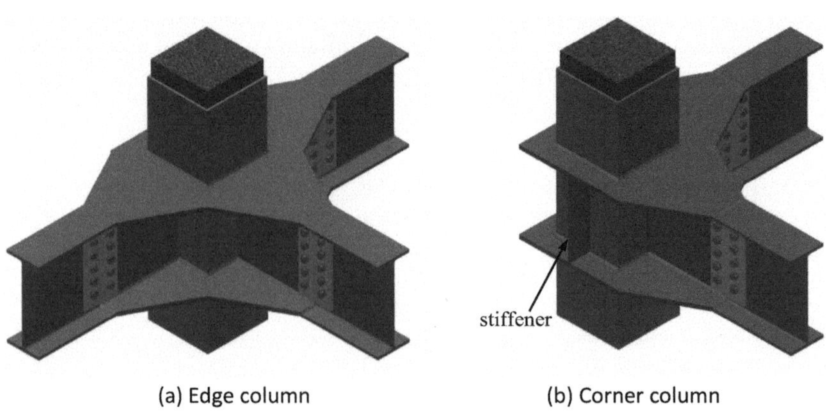

Fig. 8.12 Detailing for edge and corner CFST columns with external diaphragm plates.

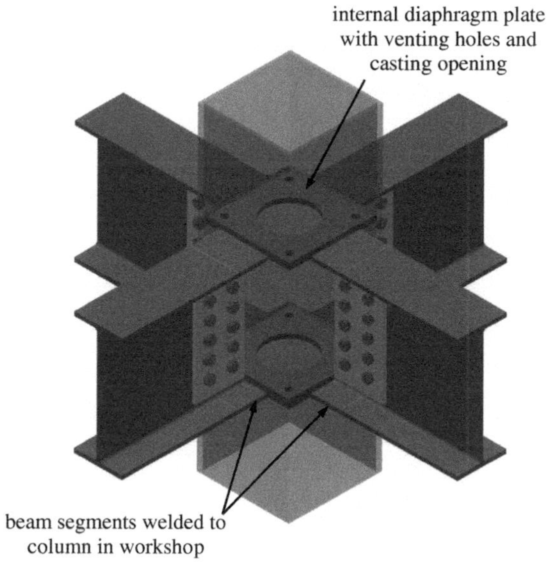

Fig. 8.13 Moment connection between steel beams and CFST column with internal diaphragm plates.

provided on the internal diaphragm plates to allow the ease of flowing of concrete as show in Fig. 8.14. The opening could be either rectangular or circular. The side length or diameter of the opening should be at least 100 mm and the area of the opening should be greater than 15% the cross-sectional area of the concrete core. In addition, four venting holes should be provided at the corners with a hole diameter of 30 mm but not less than the thickness of the internal diaphragm plate.

The internal diaphragm plates can also be used for CES columns to transfer bending moments from floor beams which are directly welded to the encased steel section

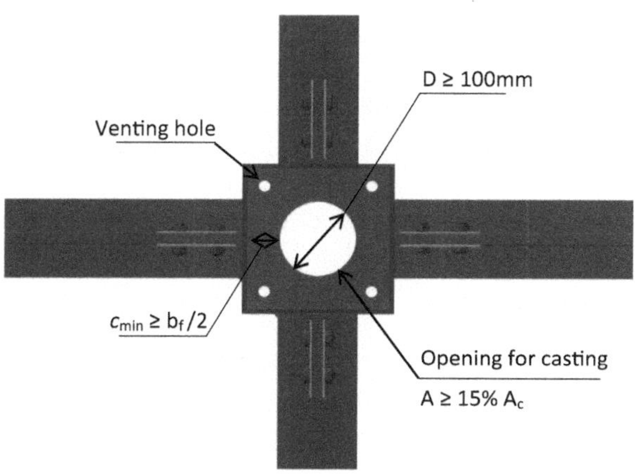

Fig. 8.14 Detailing for internal diaphragm plate.

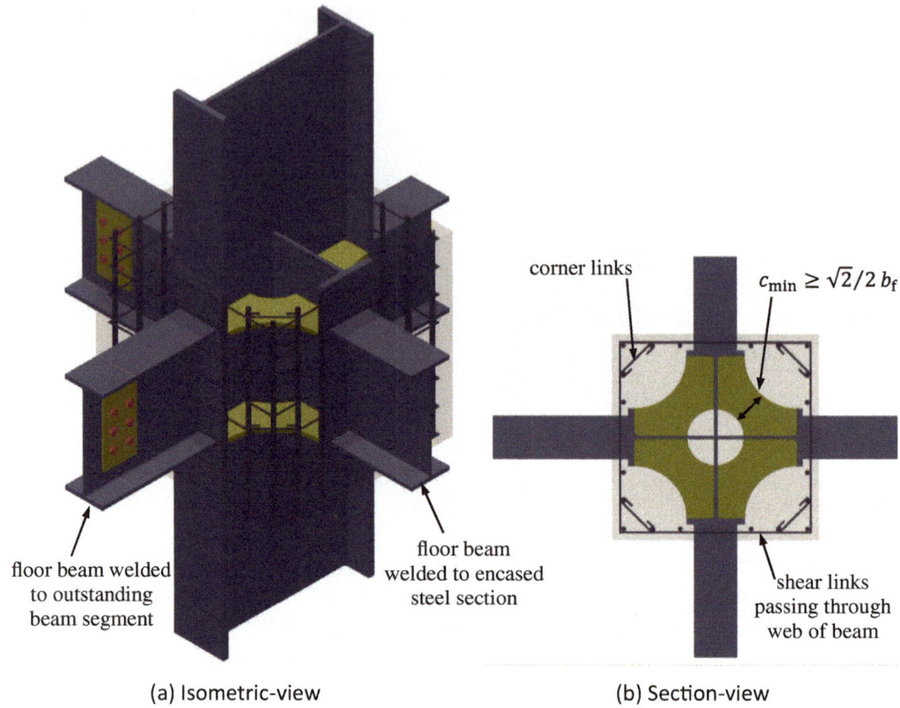

Fig. 8.15 Moment connection between steel beam and CES column with internal diaphragm plates.

or connected to the prefabricated outstanding beam segment. The joint detailing is shown in Fig. 8.15. The shear links of column are welded or pass through the web of beam.

The internal diaphragm plates are generally welded at the same level of the flanges of the beam. The welding work should be carefully carried out to prevent distortion and micro-cracking of the steel tube or the encased steel section at the junction area with the diaphragm plate since the column tube or encased steel section receives welding heat from both inner and outer sides of it. Electroslag or electrogas welds are usually used for internal diaphragm plates in square or rectangular CFST columns. However, for circular CFST columns, it is difficult to perform such welding. External diaphragm plates may be used for smaller size columns whereas internal diaphragm plate may be based for large size tubes in which case the common shielded metal arc welding may be done directly inside the steel tube.

Design checks to the internal diaphragm plates are not necessary when the following requirements are satisfied

- Matched or overmatched electrodes are used for all welds.
- The diaphragm plates should have a minimum yield strength higher than or equivalent to that of the floor beams;

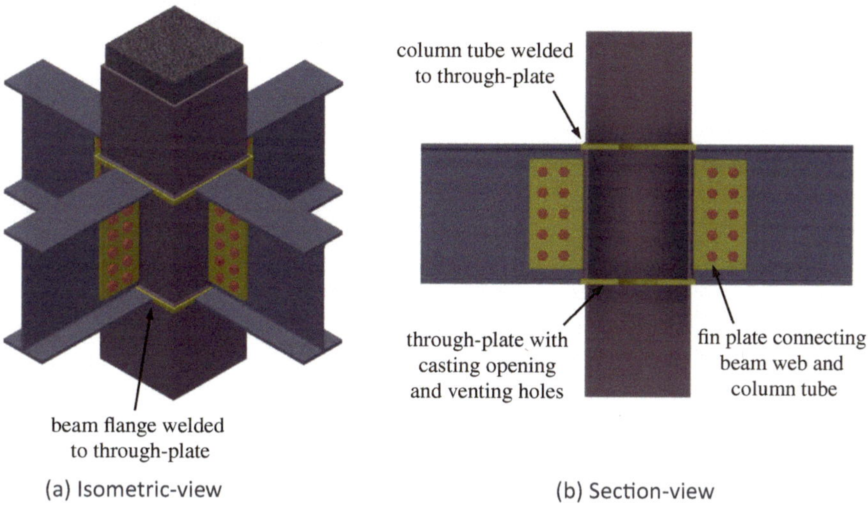

Fig. 8.16 Moment connection between steel beam and CFST column with through-plate.

- The thickness of diaphragm plate should be at least 3 mm to 5 mm larger than that of the flange of the floor beam to avoid misalignment;
- The width of internal diaphragm c_{min} should be at least equal to $b_f/2$ for CFST column as shown in Fig. 8.14 and $\sqrt{2}/2b_f$ for CES column as shown in Fig. 8.15, where b_f is the width of the flange of floor beam.

8.3.2.3 Other types of moment connections

Through-plate connection shown in Fig. 8.16 is a combination of connections with external and internal diaphragm plates. The upper and lower column tubes are discontinuously welded to the through-plates, and practically, butt welds are required to achieve full-strength design. Since the butt welds could be subjected to large tensile force due to the bending moment transferred by the column tubes, the through-plate should be checked regarding the through-thickness tearing failure, this could be referred to EN 1993–1–10 (2005) for mild steel plates and EN1993-1–12 (2007) for high strength steel plates. In practice, the thickness of through-plate is 6 mm ~ 10 mm larger than that of the flanges of the floor beam. Opening and venting holes should also be provided on the through-plate for concrete casting.

Fig. 8.17 shows a type of moment connection with through-beam which continuously passes through the column. The flanges of the through-beam in column should be cut for concrete casting. In order to enhance the bond resistance between the embedded flanges and the concrete core, shear studs may be welded on said flanges.

8.3.2.4 Detailing for connections with unequal beams

When the connected beams have unequal depths, they may share the external or internal diaphragm plates with a tapered outstanding beam as shown in Fig. 8.18. The

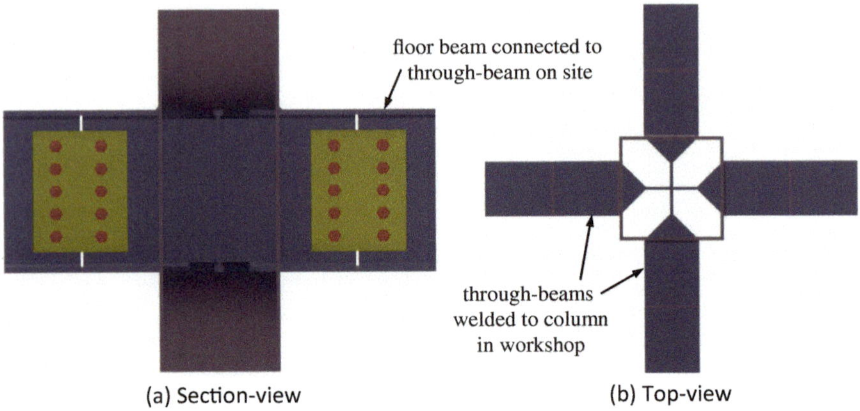

Fig. 8.17 Moment connection between steel beam and CFST column with through-beam.

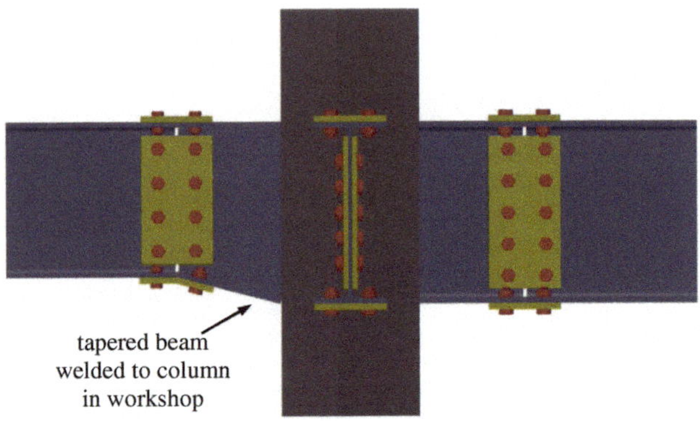

Fig. 8.18 CFST column connected with beams with unequal depths using tapered beam.

slope should be made so that the force at the bottom flange can be smoothly transferred into the column. In practice, the slope should be smaller than 1:3.

When the difference of depths of the unequal beams is so large that it is difficult to use the tapered beam, additional external or internal diaphragm plates should be welded as shown in Fig. 8.19.

8.4 Reinforced concrete beam to composite column joints

8.4.1 Simple connections

Fig. 8.20 shows the use of corbel for simple connection between reinforced concrete (RC) beam and CFST column. If more than one corbel is used, the top plates of the corbels may be continuously connected around the column tube to reduce the local

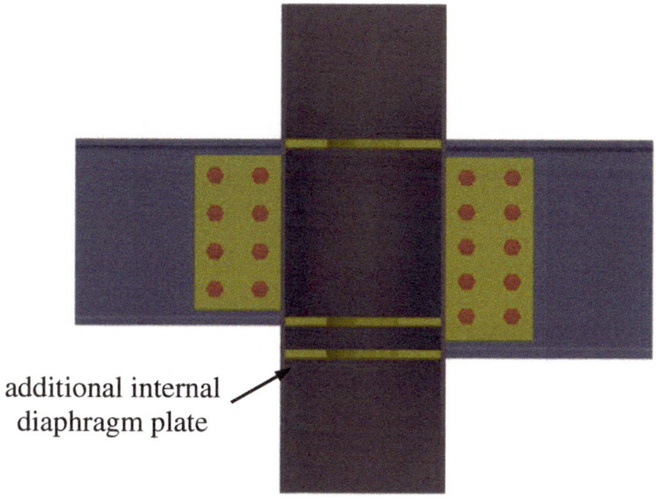

Fig. 8.19 CFST column connected with beams with unequal depths using additional diaphragm plates.

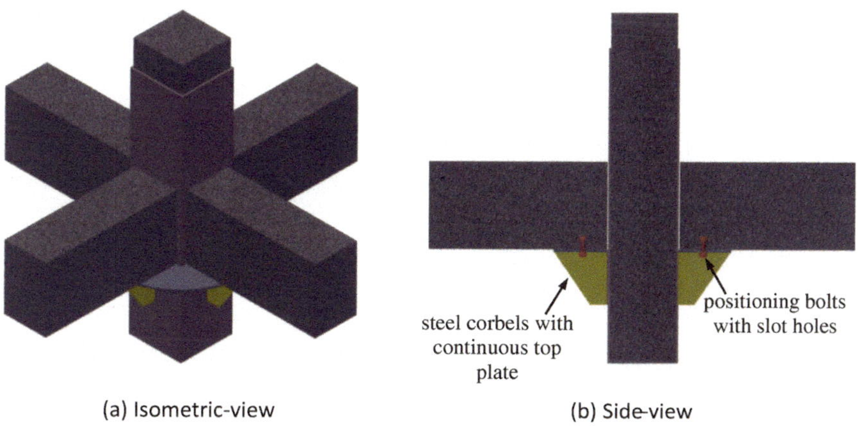

Fig. 8.20 Simple connection between RC beam and CFST column with external corbel.

stress concentration and distortions caused by the force transmitted from the corbels. Installation bolts with slot holes may be used for positioning the RC beam onto the steel corbel meanwhile facilitating the movement of beam end.

If the head room is limited, the steel corbels may be embedded in the RC beam as shown in Fig. 8.21 (a). In cases where the reaction force from the beam is large, the corbel could be continuously passing through the column as shown in Fig. 8.21 (b). Proper checks should be conducted on the steel corbel to ensure that it can resist the design shear force and nominal moment due to load eccentricity. Sufficient shear studs should be provide to ensure the steel corbel is properly connected to the RC beam to prevent longitudinal slippage due to horizontal load acting on the structure.

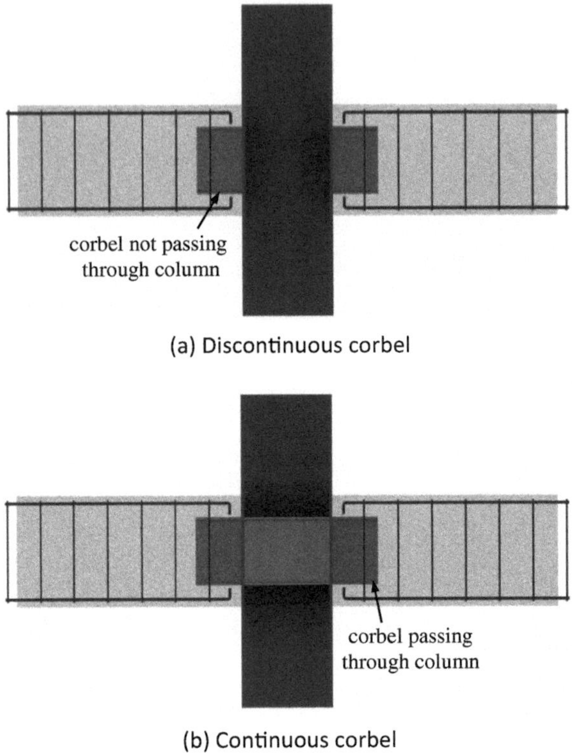

Fig. 8.21 Simple connection between RC beam to CFST column using embedded corbels.

8.4.2 Moment connections

8.4.2.1 Through reinforcements

To transfer bending moment of RC beam to the CFST column, the longitudinal reinforcements may pass continuously through the column tube as shown in Fig. 8.22.

Holes or slot holes may be cut for individual reinforcement or bundle of reinforcements. Since the column tube is impaired by the cut holes, strengthening of column tube will be necessary which can be achieved by welding a casing plate as shown in Fig. 8.22. The casing plate can be welded on either outer surface or inner surface of the column tube. The thickness of the casing plate should be determined so that the impaired column tube can be counterbalanced by the casing plates.

8.4.2.2 Ring corbel with welded reinforcements

In case where it is not practical to cut holes for through reinforcements, the ring corbel similar to the external diaphragm plate may be used. The longitudinal reinforcements are anchored to the ring corbel at top and bottom by groove welds as shown in Fig. 8.23. The effective throat thickness and length of the groove welds should be determined based on the tensile forces transmitted from the longitudinal

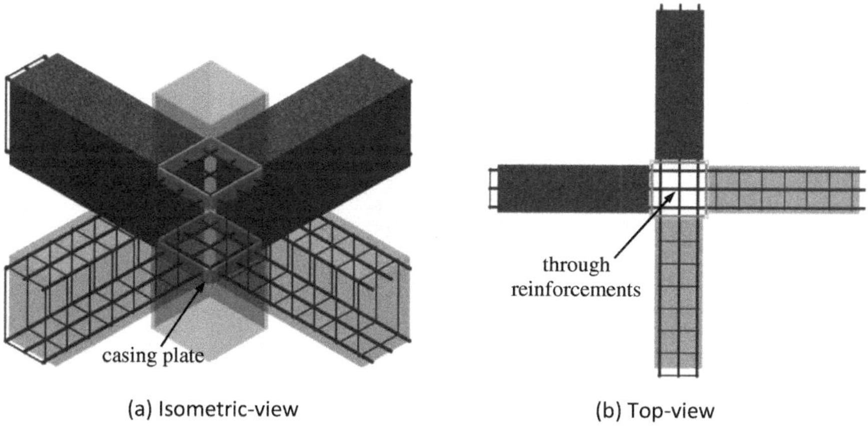

Fig. 8.22 Moment connection between RC beam and CFST column with through reinforcements.

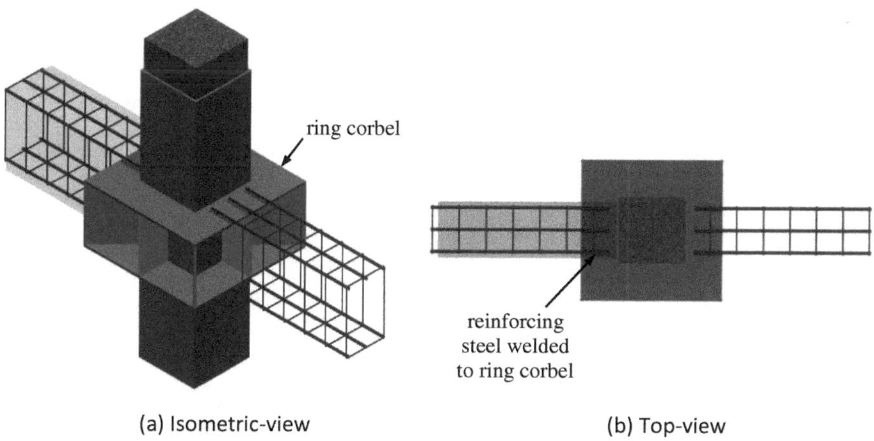

Fig. 8.23 Moment connection between RC beam and CFST column with ring corbel.

reinforcements. The weldability of reinforcements may be checked according to EN 10080 (2005). All welding work could be carried out conforming to ISO 17660–1 (2006).

8.5 Column base joints

8.5.1 Simple connections

CFST column bases with simple connection to foundation may be used for portal frames with bracing members or multi/high-rise buildings with multi-storey

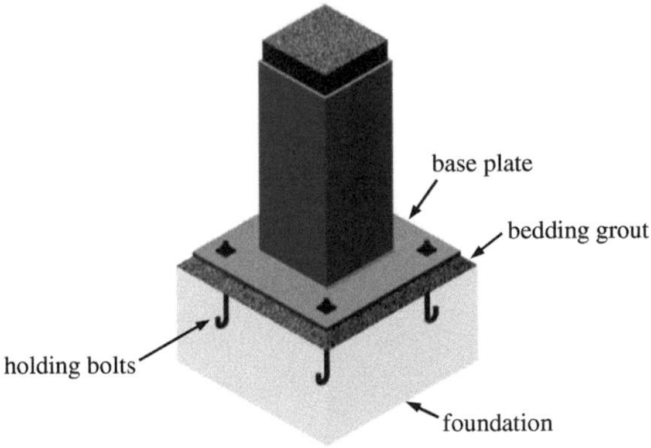

Fig. 8.24 Typical CFST column base with simple connection to the foundation.

basement. Typical column base with simple connection is shown in Fig. 8.24. Design checks for such base connection could be referred to SCI Publication P358 (2014).

The column tubes are practically connected to the base plate by single-V butt welds. Base plates are usually flame cut or sawn from steel plates with the same strength to the column tubes. In practice, the base plate is set to be at least 100 mm larger all round than the column tube, with a thickness equal to or greater than that of the column tube.

Venting holes with a diameter of 50 mm should be provided to release the trapped air and also for inspection during casting. The hole should be provided near the center of the base plate to ensure the grout reaches the center. In cases where the venting holes are also used for casting of grout, the diameter of such holes should be increased to 100 mm.

The bedding grout shall be of non-shrinkage and at least equal in strength to that of the foundation concrete. The bedding grout is normally used for purposes of base plate leveling and corrosion prevention for the holding-down bolts. A bedding spacing of 25 mm to 50 mm is generally adopted, which gives reasonable access for thoroughly filling the space under the base plate. The bedding grout may be fine concrete with a maximum aggregate of 10 mm. The usual mix is 1:1.25:2 with a water-cement ratio of between 0.4 and 0.45 (Davison and Owen, 2012).

The holding-down bolts are mainly used for proper installation, they should be sufficiently robust to withstand loads experienced during erection resulting from the wind loads and the lack of verticality, etc. The use of holding down bolts could be referred to BS 7419 (2012) which covers bolts with square head and neck, and bolts with hexagon head and round neck. The embedded length of the bolts in concrete is usually in the range of 16 to 18 times bolt diameter. The length for threads should be at least 100 mm plus the bolt diameter. The threaded portion of the bolts should be protected during the concrete casting. In practice, Class 8.8 bolts are mostly used. M24 bolts are recommended for base plates with thickness up to 50 mm, whereas increasing to M36 for plates over 50 mm thick. Generally, minimum four holding-down bolts are used and positioned at the four corners of the base plate. Clearance

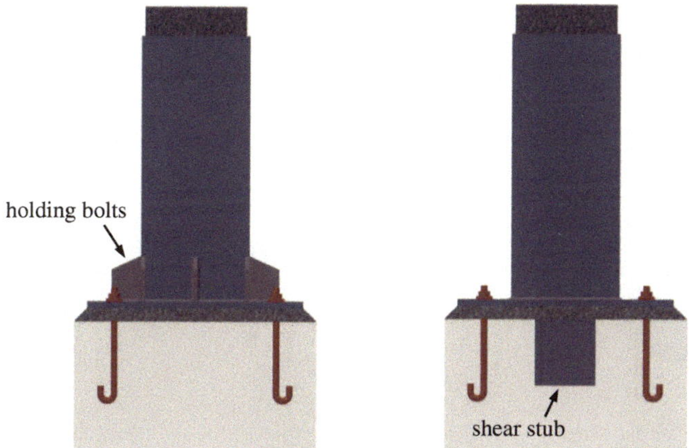

(a) Connection with stiffeners (b) Connection with shear stub

Fig. 8.25 Reinforcements to column base connection.

holes in the base plate should be 6 mm greater than the bolt diameter to allow for adjustment. For base plates thicker than 60 mm, the clearance holes may be increased accordingly. The use of washers may be referred to ISO 7091 (2000). Alternatively, the washers may be cut from plates.

In cases where the axial load is large and it is not practical to use a very thick base plate to develop a uniformly distributed pressure under the base plate, stiffener plates may be welded to increase the bearing area as shown in Fig. 8.25(a). In cases where the transfer of high shear force to the column base is needed, a shear stub (shear key) welded to the underside of the base plate may be necessary as shown in Fig. 8.25(b).

8.5.2 Moment connections

8.5.2.1 Exposed column bases

Typical exposed column bases for CFST column are shown in Fig. 8.26. When necessary, double base plates may be employed to reduce the force acting on the plate stiffeners, and to reduce the number of stiffeners required.

The exposed column base can be designed based on small eccentricity, moderate eccentricity and large eccentricity of the axial load as shown in Fig. 8.27. Elastic distribution of bearing pressure under base plate is conservatively assumed. For the calculations, the dimensions of base plate should be firstly assumed and then justified so that the compression resistance of foundation concrete, tensile resistance of bolts and the yield strength of base plate and stiffeners are not exceeded.

The exposed column base details for CES columns are shown in Fig. 8.28. Bedding grout with a depth of 50 mm is needed and the holding-down bolts are required for installation and positioning. Shear force is resisted by friction between the base plate and the bedding grout, whereas the bending moment is resisted by the longitudinal

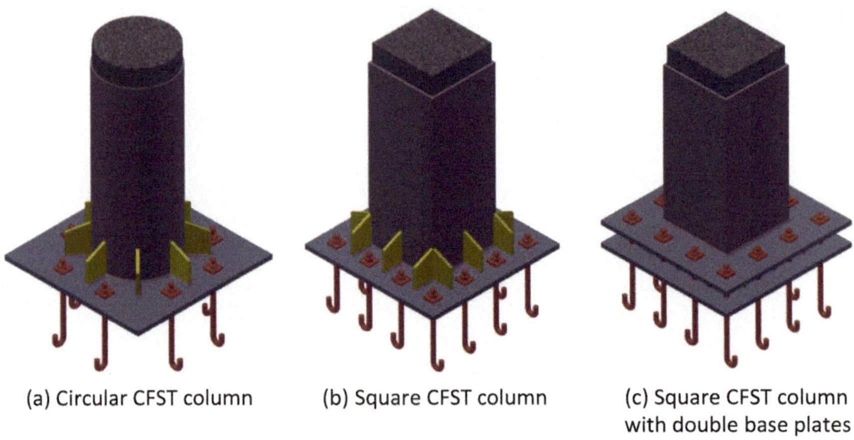

Fig. 8.26 Typical exposed column bases for CFST columns.

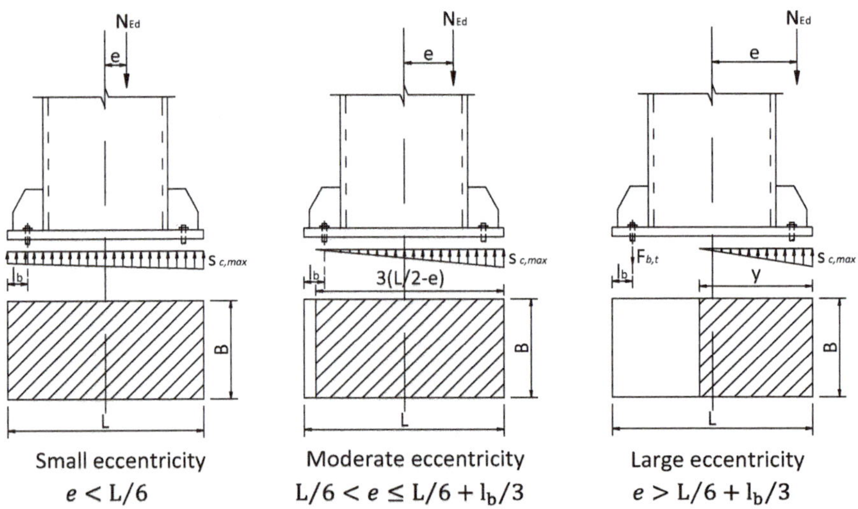

Fig. 8.27 Stress distributions under column base.

reinforcements anchored directly into the concrete foundation. In cases where said friction is inadequate to resist the shear, a shear key welded to the underside of the base plate may be used. Stiffener plates are used to spread the compression force of the encased steel section to foundation, and the shear studs are usually provided onto the encased steel section near the column base to improve the bonding strength at steel-concrete interface, this is because the shear force at column base is generally quite large. The weld collars of shear studs should comply with the requirements of ISO 13918 (2017). The overall height of a stud should not be less than $3d$, d is the bolt diameter. The center to center spacing of studs in the direction of the shear force

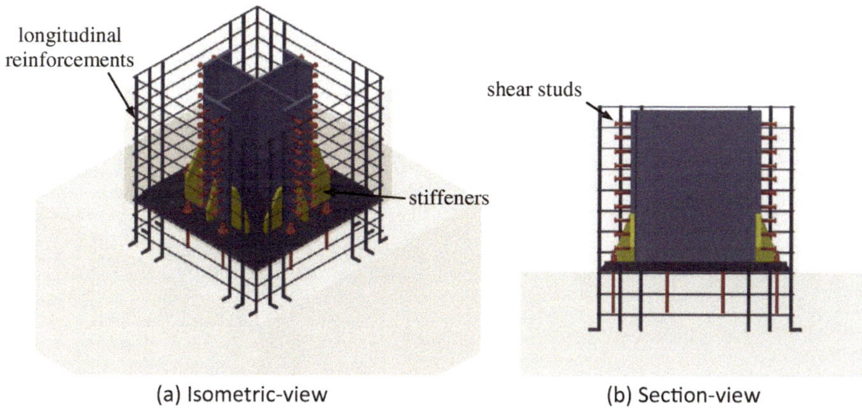

Fig. 8.28 Typical exposed column base for CES columns.

should not be less than 5*d*; whereas the spacing in the direction transverse to the shear force should not be less than 2.5*d*. The diameter of a welded stud should be not greater than 2.5 times the thickness of flange of the steel section.

8.5.2.2 Embedded column bases

The column base of a CFST column can be embedded into the concrete foundation as shown in Fig. 8.29. Similarly, the holding-down bolts are required for installation and positioning. In practice, the embedded height of column above the foundation should be greater than 3 times the column section height.

The design axial load is directly transferred into foundation through base plate, in cases where the bearing area under base plate is not sufficient, stiffener plates may be used as shown in Fig. 8.28. The design moment and shear force are resisted by foundation concrete in bearing. In practice, shear studs are welded onto the column tube to

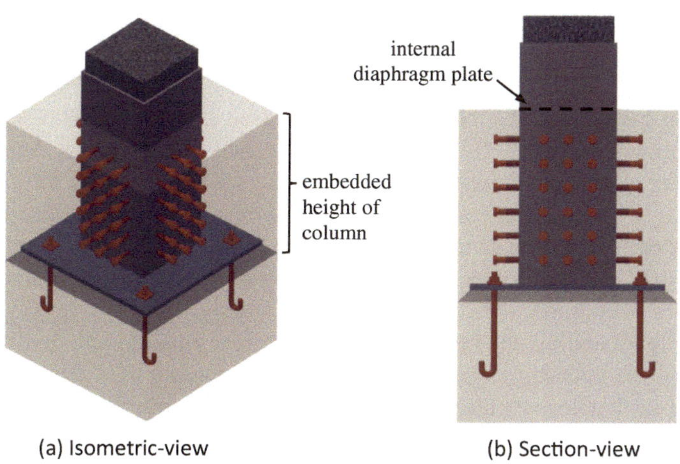

Fig. 8.29 Typical embedded column base for CFST column.

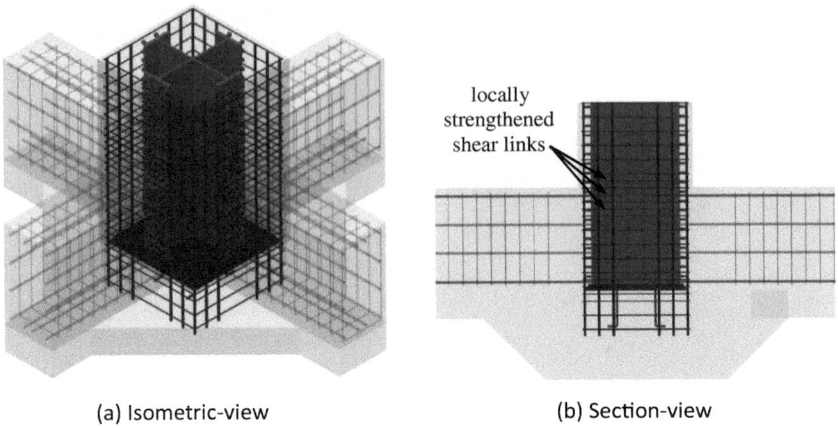

Fig. 8.30 Typical embedded column base for CES column.

improve bonding strength, and also the flexural stiffness and energy dissipation capacity of the column base. The details of shear studs in terms of bolt diameter, height, and spacing etc. are similar to those for CES columns introduced in Section 8.5.2.1.

In practice, an internal diaphragm plate shown in Fig. 8.29 is welded inside column tube at location where the column is not embedded. This is to prevent crush of concrete due to the very large shear force. Opening is provided on the diaphragm plate for concrete casting.

A typical embedded column base for CES column is shown in Fig. 8.30. Holding-down bolts are required, and the embedded height of column should be not less than 3 times the column section height. The longitudinal reinforcements of column are anchored into the foundation. To prevent crush of concrete mentioned above, the shear links of column can be strengthened by reducing spacing at the location of column not embedded.

8.5.2.3 Concrete encased column bases

The column base of a CFST column can be encased and reinforced outside the foundation as shown in Fig. 8.31. The concrete encased column bases can be used for shallow foundations or splices between concrete columns and CFST columns. The encased height should be at least 3 times the column cross-sectional height. In case of local crushing of concrete at top of the encased height, the top shear reinforcements should be strengthened, for example, the spacing of top 3 shear reinforcements is set to be 30 mm ~ 50 mm.

The design bending moment is resisted by the longitudinal reinforcements at each side of the encasing concrete. Design for the longitudinal reinforcements may be referred to that for reinforced concrete columns as provided in Eurocode 2 (EN 1992–1–1, 2004). The design shear force is resisted by the shear reinforcements which can also be designed in accordance with the Eurocode 2.

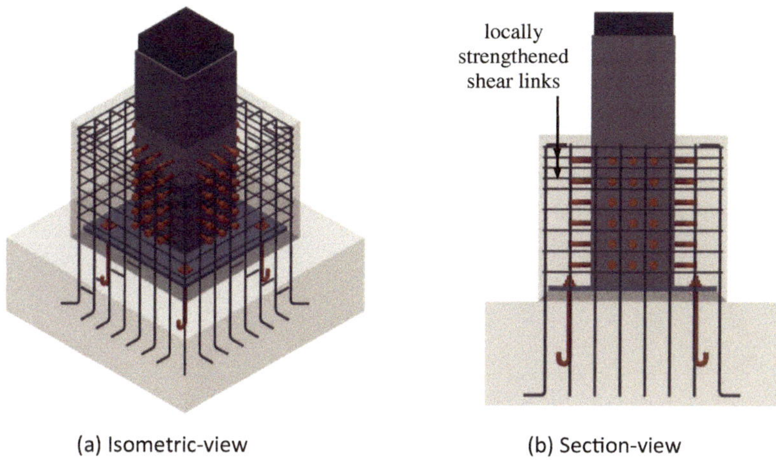

Fig. 8.31 Typical concrete encased column base for CFST column.

Alternatively, the columns at the basement level could be fully encased with concrete to resist high axial compression force and accident loads such as car collision (with car park at the basement level), and to protect the steel tube against fire. This is shown in Fig. 8.32. The disadvantage of providing full concrete encasement is the reduction of usable floor area.

Liew (2019) has written a design guidebook for buildable steel connections preferred by the steel fabricators. This work is based on feedback from the steel fabrica-

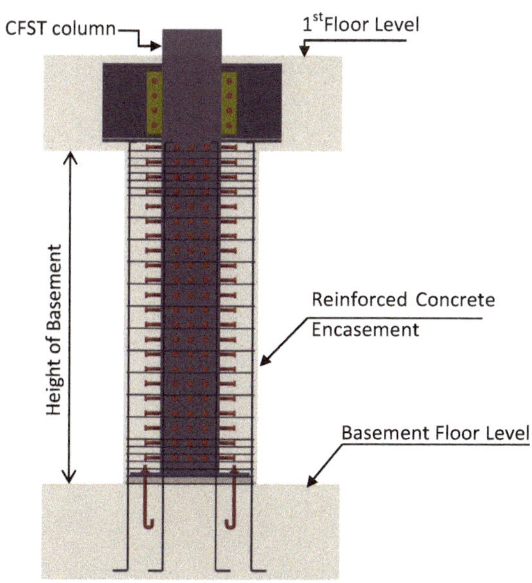

Fig. 8.32 Concrete encased CFST column at basement.

tors and consultants who select 44 typical steel connections that are perceived to be more buildable. Full work examples are provided and detailed design calculations are carried out based on EN1993-1–8, The book can be downloaded from: https://ssss.org.sg/~ssssorgs/images/stories/docs/Design_guide_for_buildable_steel_connections_Final_Version_20190327.pdf

Readers may find this book useful for working out some of the detailed design of the connections highlighted in this chapter.

Design flowcharts

A

This appendix provides flowcharts for the design of concrete filled tubular steel tubular members and concrete encased steel composite members at ultimate limit state. Each step of the flowcharts is linked to corresponding subsections and tables in the book.

Design of Steel-Concrete Composite Structures Using High Strength Materials. DOI: 10.1016/C2019-0-05474-X

The excel spreadsheets in Appendix C are programmed based on the design flow-charts below:

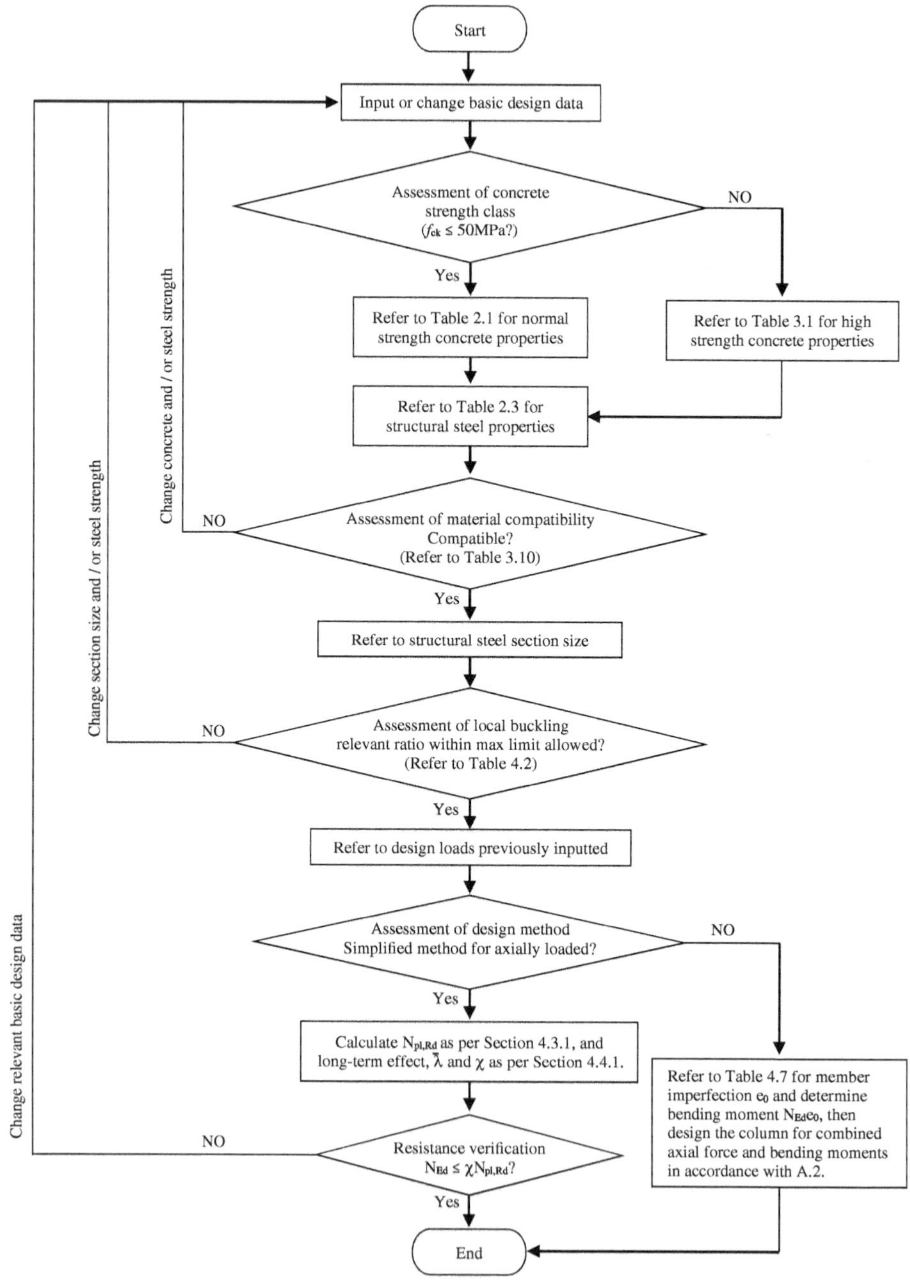

Fig. A.1 Concrete filled steel tubular column subject to axial compression

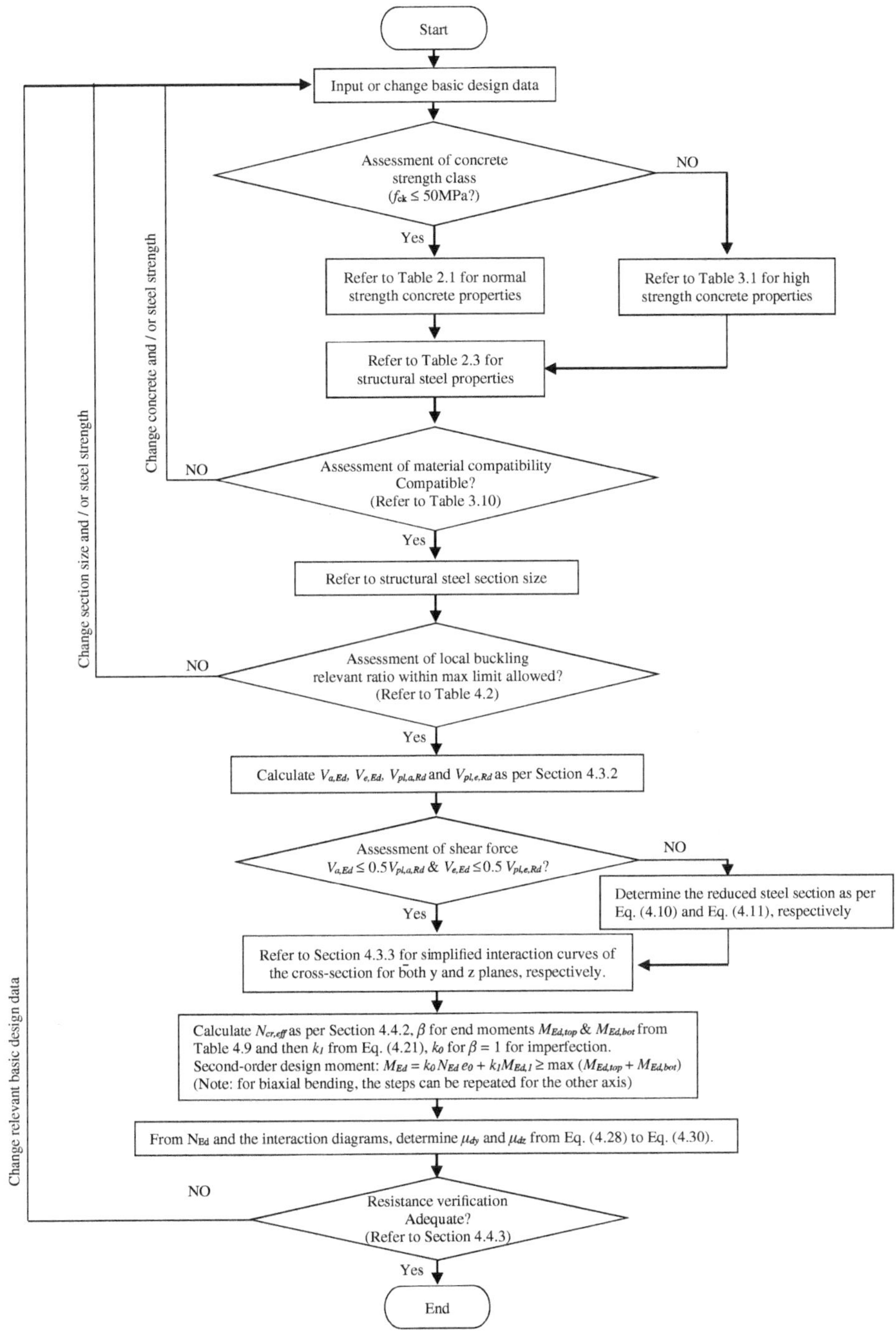

Fig. A.2 Concrete filled steel tubular column subject to combined axial compression and bending moments

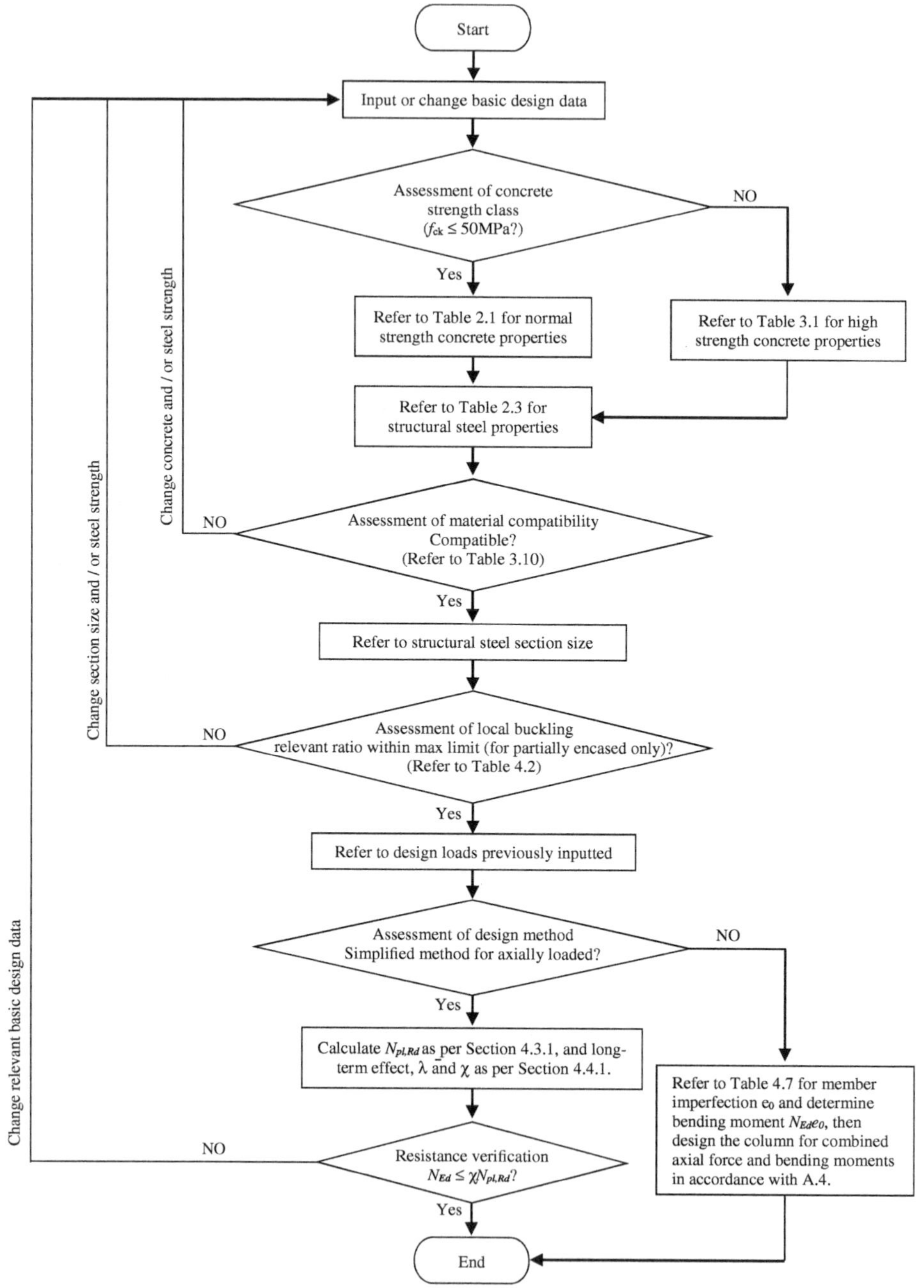

Fig. A.3 Concrete encased steel column subject to axial compression

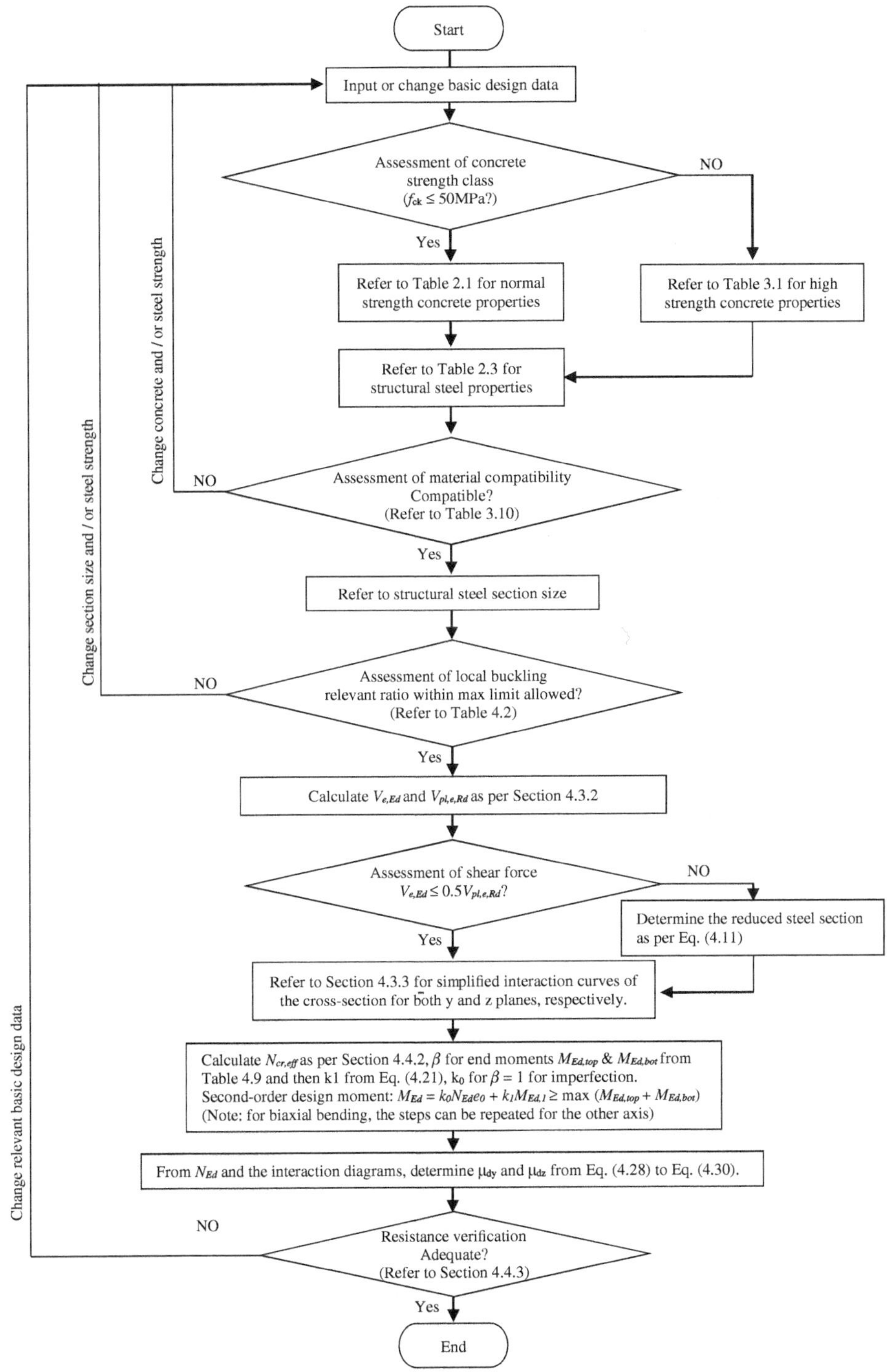

Fig. A.4 Concrete encased steel column subject to combined axial compression and bending moments

Work Examples and Comparison Studies

B

Four examples are provided to demonstrate the design principles highlighted in Chapter 3 and 4. The design of composite columns using high strength materials are also demonstrated. Finally, comparison studies are carried out to illustrate the advantage of using high strength steel and high strength concrete in enhancing the design resistance of composite columns.

B.1 Circular concrete infilled tube subject to compression

B.1.1 General

In Example 1, the axial buckling resistance of a concrete filled steel tubular (CFST) column subject to pure compression is determined. The dimensions of the CFST column are shown in Fig. B.1 and Fig. B.2.

The column lengths and design loads are given as:

Column system length	$L = 4000$ mm
Effective length	$L_{eff} = 4000$ mm
Total design axial load	$N_{Ed} = 11{,}000$ kN
Design axial load that is permanent	$N_{G,Ed} = 4500$ kN

To evaluate and compare their resistance, the following steel and concrete material grades are used:

a) CHS 508 × 12.5 - S355 steel + C40/50 concrete
b) CHS 508 × 12.5 - S355 steel + C90/105 concrete
c) CHS 508 × 12.5 - S460 steel + C40/50 concrete
d) CHS 406 × 12 - S460 steel + C90/105 concrete

B.1.2 CHS 508 x 12.5 S355 steel tube infilled with C40/50 concrete

- Material

Concrete	C40/50, $f_{ck} = 40$ N/mm^2
Steel tube	Grade S355, $f_y = 355$ N/mm^2

- Design strengths and modulus of material

Refer to Table 2.1 and Table 2.3 for the characteristic strength of concrete and steel,

Design of Steel-Concrete Composite Structures Using High Strength Materials. DOI: 10.1016/C2019-0-05474-X

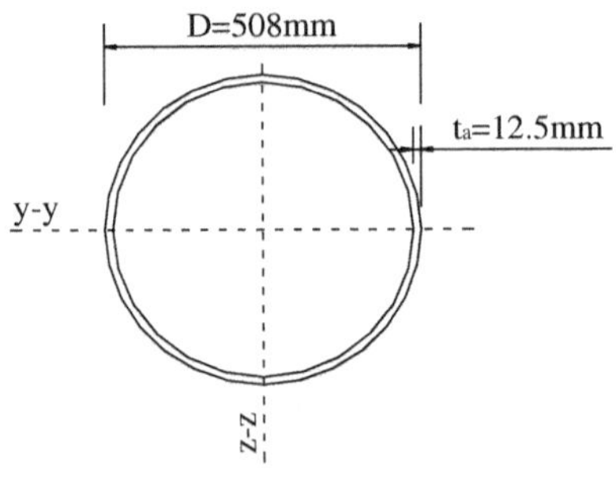

Fig. B.1 Cross-sectional dimensions of CFST column in Example 1.

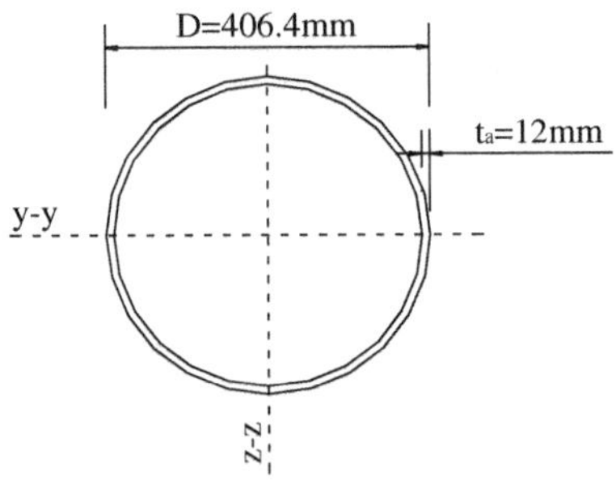

Fig. B.2 Reduced cross-sectional size.

and Table 4.1 for the partial factors, the design strengths are determined as:

$$f_{yd} = f_y/\gamma_a = 355/1.0 = 355\text{N/mm}^2$$
$$E_a = 210\,\text{GPa}$$
$$f_{cd} = f_{ck}/\gamma_c = 40/1.5 = 26.7\text{N/mm}^2$$
$$f_{cm} = f_{ck} + 8 = 40 + 8 = 48\text{N/mm}^2$$
$$E_{cm} = 22(f_{cm}/10)^{0.3} = 22(48/10)^{0.3} = 35.2\,\text{GPa}$$

- Cross sectional areas

$$A_a = (\pi/4)\left[D^2 - (D-2t)^2\right] = (\pi/4)\left[508^2 - (508 - 2\times 12.5)^2\right] = 19458\,\text{mm}^2$$
$$A_c = (\pi/4)(D-2t)^2 = (\pi/4)(508 - 2\times 12.5)^2 = 183225\,\text{mm}^2$$

Unless otherwise stated, the subscript "a" stands for the steel section, and the "c" stands for the concrete section.

- Second moment of areas

$$I_a = (\pi/64)\left[D^4 - (D-2t)^4\right] = (\pi/64)\left[508^4 - (508 - 2\times 12.5)^4\right] \times 10^{-4}$$
$$= 59755\,\text{cm}^4$$
$$I_c = (\pi/64)(D-2t)^4 = (\pi/64)(508 - 2\times 12.5)^4 \times 10^{-4} = 267152\,\text{cm}^4$$

- Check for local buckling (refer to Table 4.2)

$$D/t_a = 508/12.5 = 40.6 < 90(235/f_y) = 90(235/355) = 59.6$$

Resistance against local buckling is adequate!

- Long-term effect

The long-term effect could be evaluated in accordance with EN 1992–1–1: 2004. Herein, the simplified method given in guidebook by Liew and Xiong (2015) "Design Guide for Concrete Filled Tubular Members with High Strength Materials – An Extension of Eurocode 4 Method to C90/105 Concrete and S550 Steel" is referred to.

The age of concrete at the moment considered t is conservatively taken as infinity. For the age of concrete on first loading by effects of creep, although EN 1994–1–1 (2004) recommends $t_0 = 1$ day, it is actually the judgment of designer to determine t_0 since it makes quite a difference whether this age is assumed to be 1 day or 1 month. Herein, t_0 is assumed as 14 days, as said, it could be different. The relative humidity RH for infilled concrete is taken as 50%.

Perimeter of concrete section: $u = \pi(D - 2t_a)$
$= \pi(508 - 2 \times 12.5) = 1517\,\text{mm}$

Notional size of concrete section: $h_0 = 2\text{A}_\text{c}/u$
$= 2 \times 183225/1517.4 = 241.5\,\text{mm}$

Coefficient: $\alpha_1 = (35/f_{cm})^{0.7} = (35/48)^{0.7} = 0.80$

Coefficient: $\alpha_2 = (35/f_{cm})^{0.2} = (35/48)^{0.2} = 0.94$

Coefficient: $\alpha_3 = (35/f_{cm})^{0.5} = (35/48)^{0.5} = 0.85$

$$\text{Factor: } \varphi_{\text{RH}} = \left(1 + \frac{1 - RH/100}{0.1\sqrt[3]{h_0}}\alpha_1\right)\alpha_2 = \left(1 + \frac{1 - 50/100}{0.1\sqrt[3]{241.5}} \times 0.80\right) \times 0.94 = 1.54$$

Factor: $\beta(f_{cm}) = 16.8/\sqrt{f_{cm}} = 16.8/\sqrt{48} = 2.42$

Factor: $\beta(t_0) = 1/\left(0.1 + t_0^{0.2}\right) = 1/\left(0.1 + 14^{0.2}\right) = 0.56$

Factor: $\varphi_0 = \varphi_{\text{RH}}\beta(f_{cm})\beta(t_0) = 1.54 \times 2.42 \times 0.56 = 2.08$

$$\text{Factor: } \beta_H = 1.5\left[1 + (0.012RH)^{18}\right]h_0 + 250\alpha_3 = 1.5\left[1 + (0.012 \times 50)^{18}\right] \times 241.5 + 250 \times 0.85 = 576$$

$$\text{Factor: } \beta_c(t, t_0) = \left(\frac{t - t_0}{\beta_H + t - t_0}\right)^{0.3} = \left(\frac{\infty - 14}{576 + \infty - 14}\right)^{0.3} = 1.0$$

Creep coefficient: $\varphi_t = \varphi_0\beta_c(t, t_0) = 2.08 \times 1.0 = 2.08$

- Elastic modulus of concrete considering long-term effect (refer to Eq. (4.18))

Concrete is sensitive to long-term deformations due to creep and shrinkage. To allow for this, the flexural stiffness of concrete section is reduced.

$$E_{c,eff} = \frac{E_{cm}}{1 + \left(N_{G,Ed}/N_{Ed}\right)\varphi_t} = \frac{35.2}{1 + (4500/11000) \times 2.08} = 19\,\text{GPa}$$

- Effective flexural stiffness of cross-section

$$\begin{aligned}(EI)_{eff} &= E_a I_a + 0.6 E_{c,eff} I_c \\ &= \left(210 \times 10^3 \times 59755 \times 10^4 + 0.6 \times 19 \times 10^3 \times 267152 \times 10^4\right) \times 10^{-3} \\ &= 1.56 \times 10^{11}\,\text{kN} \cdot \text{mm}^2\end{aligned}$$

- Elastic critical Euler buckling resistance

$$N_{cr} = \frac{\pi^2 (EI)_{eff}}{L_{eff}^2} = \frac{\pi^2 \times 1.56 \times 10^{11}}{4000^2} = 96209\,\text{kN}$$

- Characteristic plastic resistance of cross-section

$$N_{pl,Rk} = A_a f_y + A_c f_{ck} = (19458 \times 355 + 183225 \times 40) \times 10^{-3} = 14250\,\text{kN}$$

- Relative slenderness ratio

$$\bar{\lambda} = \sqrt{\frac{N_{pl,Rk}}{N_{cr}}} = \sqrt{\frac{14250}{96209}} = 0.385 < 0.5$$

- Confinement coefficients with load eccentricity $e = 0$

Since the slenderness $\bar{\lambda}$ is less than 0.5 and the load eccentricity is equal to 0 (axially loaded column), the confinement effect may be considered for the circular CFST column.

$$\eta_a = \eta_{a0} = \min\left[0.25(3 + 2\bar{\lambda}),\, 1.0\right] = 0.942$$

$$\eta_c = \eta_{c0} = \max\left[4.9 - 18.5\bar{\lambda} + 17\bar{\lambda}^2,\, 0\right] = 0.298$$

- Design plastic resistance of cross-section considering the confinement effect (refer to Eq. (4.4))

$$\begin{aligned}N_{pl,Rd} &= \eta_a A_a f_{yd} + A_c f_{cd}\left(1 + \eta_c \frac{t}{D}\frac{f_y}{f_{ck}}\right) \\ &= \left[0.942 \times 19458 \times 355 + 183225 \times 26.7\left(1 + 0.298\frac{12.5}{508}\frac{355}{40}\right)\right] \times 10^{-3} \\ &= 11727\,\text{kN}\end{aligned}$$

It is worthwhile to note that the yield strength of steel is reduced ($\eta_a < 0$) and strength of concrete increases ($1 + \eta_c \frac{t}{D} \frac{f_y}{f_{ck}} > 0$) with the consideration of confinement effect.

- Steel contribution ratio

$$\delta = A_a f_{yd}/N_{pl,Rd} = \left(19458 \times 10^{-3} \times 355\right)/11727 = 0.59 < 0.9$$

- Imperfection factor

Refer to Table 4.7, the buckling curve is taken as "a". Thus the imperfection factor α is determined as 0.21 according to Table 4.7.

- Buckling reduction factor

$$\Phi = 0.5[1 + \alpha(\overline{\lambda} - 0.2) + \overline{\lambda}^2] = 0.5\left[1 + 0.21 \times (0.385 - 0.2) + 0.385^2\right] = 0.593$$

$$\chi = \frac{1}{\Phi + \sqrt{\Phi^2 - \overline{\lambda}^2}} = \frac{1}{0.593 + \sqrt{0.593^2 - 0.385^2}} = 0.957$$

- Buckling resistance

According to Section 4.4.1, the axial buckling resistance is checked as:

$$N_{b,Rd} = \chi N_{pl,Rd} = 0.957 \times 11727 = 11223\,\text{kN} > N_{Ed} = 10000\,\text{kN}$$

Buckling resistance is adequate!

In this section, the normal strength concrete (NSC) C40/50 is replaced by high strength concrete (HSC) C90/105. The steel grade is not changed.

- Design strength

Refer to Table 3.1, Eq. (3.1) and Eq. (3.2), the effective compressive strength and modulus of elasticity of the HSC are calculated as:

$$f_{ck} = 72\,\text{N/mm}^2$$
$$E_{cm} = 41.1\,\text{GPa}$$
$$f_{cd} = f_{ck}/\gamma_c = 72/1.5 = 48\,\text{N/mm}^2$$
$$f_{cm} = f_{ck} + 8 = 72 + 8 = 80\,\text{N/mm}^2$$

- Creep coefficient could be similarly determined as $\phi_t = 1.29$.
- Elastic modulus of concrete considering long-term effect

$$E_{c,eff} = E_{cm}\frac{1}{1 + \left(N_{G,Ed}/N_{Ed}\right)\varphi_t} = \frac{41.1}{1 + (4500/11000) \times 1.29} = 26.9\,\text{GPa}$$

- Effective flexural stiffness of cross-section

$$\begin{aligned}(EI)_{eff} &= E_a I_a + 0.6 E_{c,eff} I_c \\ &= \left(210 \times 10^3 \times 59755 \times 10^4 + 0.6 \times 26.9 \times 10^3 \times 267152 \times 10^4\right) \times 10^{-3} \\ &= 1.69 \times 10^{11}\,\text{kN} \cdot \text{mm}^2\end{aligned}$$

- Confinement effect and design plastic resistance of cross-section

$$N_{cr} = \frac{\pi^2 (EI)_{eff}}{L_{eff}^2} = \frac{\pi^2 \times 1.69 \times 10^{11}}{4000^2} = 104010 \text{ kN}$$

$$N_{pl,Rk} = A_a f_y + A_c f_{ck} = (19458 \times 355 + 183225 \times 72) \times 10^{-3} = 20112 \text{ kN}$$

$$\bar{\lambda} = \sqrt{\frac{N_{pl,Rk}}{N_{cr}}} = \sqrt{\frac{20112}{104010}} = 0.44 < 0.5$$

$$\eta_a = \eta_{a0} = \min\left[0.25(3 + 2\bar{\lambda}), 1.0\right] = 0.970$$

$$\eta_c = \eta_{c0} = \max\left[4.9 - 18.5\bar{\lambda} + 17\bar{\lambda}^2, 0\right] = 0.052$$

$$N_{pl,Rd} = \eta_a A_a f_{yd} + A_c f_{cd}\left(1 + \eta_c \frac{t}{D}\frac{f_y}{f_{ck}}\right)$$

$$= \left[0.970 \times 19458 \times 355 + 183225 \times 48\left(1 + 0.052\frac{12.5}{508}\frac{355}{72}\right)\right] \times 10^{-3}$$

$$= 15562 \text{ kN}$$

- Steel contribution ratio

$$\delta = A_a f_{yd}/N_{pl,Rd} = \left(19458 \times 10^{-3} \times 355\right)/15562 = 0.444 < 0.9$$

- Buckling resistance

$$\text{Buckling curve } = \text{“a”},\ \alpha = 0.21$$

$$\Phi = 0.5[1 + \alpha(\bar{\lambda} - 0.2) + \bar{\lambda}^2] = 0.5[1 + 0.21 \times (0.44 - 0.2) + 0.44^2] = 0.622$$

$$\chi = \frac{1}{\Phi + \sqrt{\Phi^2 - \bar{\lambda}^2}} = \frac{1}{0.622 + \sqrt{0.622^2 - 0.44^2}} = 0.942$$

Thus, the axial buckling resistance is:

$$N_{b,Rd} = \chi N_{pl,Rd} = 0.942 \times 15562 = 14659 \text{ kN}$$

Compared with that using NSC C40/50, the buckling resistance is increased by:

$$\frac{14659 - 11223}{11223} \times 100\% = 30.6\%$$

By replacing C40/50 normal strength concrete with C90/105 high strength concrete, the axial buckling resistance of the CFST column is improved by 31%.

B.1.3 CHS 508 x 12.5 – S460 steel tube infilled with C40/50 concrete

In this section, the mild steel S355 is replaced by S460, and normal concrete grade C40/50 is adopted.

- Confinement effect and design plastic resistance of cross-section

$$N_{pl,Rk} = A_a f_y + A_c f_{ck} = (19458 \times 460 + 183225 \times 40) \times 10^{-3} = 16297\,\text{kN}$$

$$\overline{\lambda} = \sqrt{\frac{N_{pl,Rk}}{N_{cr}}} = \sqrt{\frac{16297}{96209}} = 0.412 < 0.5$$

$$\eta_a = \eta_{a0} = \min\left[0.25(3 + 2\overline{\lambda}),\, 1.0\right] = 0.956$$

$$\eta_c = \eta_{c0} = \max\left[4.9 - 18.5\overline{\lambda} + 17\overline{\lambda}^2,\, 0\right] = 0.166$$

$$N_{pl,Rd} = \eta_a A_a f_{yd} + A_c f_{cd}\left(1 + \eta_c \frac{t}{D}\frac{f_y}{f_{ck}}\right)$$

$$= \left[0.956 \times 19458 \times 355 + 183225 \times 26.7\left(1 + 0.166\frac{12.5}{508}\frac{355}{40}\right)\right] \times 10^{-3}$$

$$= 13687\,\text{kN}$$

- Steel contribution ratio

$$\delta = A_a f_{yd}/N_{pl,Rd} = \left(19458 \times 10^{-3} \times 460\right)/13687 = 0.655$$

- Buckling resistance

$$\text{Buckling curve} = \text{“a”},\ \alpha = 0.21$$

$$\Phi = 0.5[1 + \alpha(\overline{\lambda} - 0.2) + \overline{\lambda}^2] = 0.5[1 + 0.21 \times (0.412 - 0.2) + 0.412^2] = 0.607$$

$$\chi = \frac{1}{\Phi + \sqrt{\Phi^2 - \overline{\lambda}^2}} = \frac{1}{0.607 + \sqrt{0.607^2 - 0.412^2}} = 0.95$$

$$N_{b,Rd} = \chi N_{pl,Rd} = 0.95 \times 13687 = 13003\,\text{kN}$$

Compared with that with the mild steel S355, the axial buckling resistance by using S460 is increased by:

$$\frac{13003 - 11223}{11223} \times 100\% = 15.9\%$$

By replacing S355 steel tube with S460 grade, the axial buckling resistance of the CFST column is improved by 15.9%.

B.1.4 CHS 406 x 12 - S460 steel tube infilled C90/105 concrete

In this section, the normal strength materials are replaced by higher strength steel S460 and high strength concrete C90/105. The compatibility between steel grade and concrete class should be evaluated in accordance with Table 3.10 of Chapter 3. The purpose of using higher strength materials is to reduce the cross-sectional size of the original CFST column but the axial buckling resistance remains the same.

- CHS 406 × 12 with S460 steel is tried
- Section properties

$$A_a = (\pi/4)\left[D^2 - (D-2t)^2\right] = (\pi/4)\left[406.4^2 - (406.4 - 2\times 12)^2\right] = 14900\,\text{mm}^2$$

$$A_c = (\pi/4)(D-2t)^2 = (\pi/4)(406.4 - 2\times 12)^2 = 114800\,\text{mm}^2$$

$$\begin{aligned} I_a &= (\pi/64)\left[D^4 - (D-2t)^4\right] \\ &= (\pi/64)\left[406.4^4 - (406.4 - 2\times 12)^4\right] \times 10^{-4} = 28940\,\text{cm}^4 \end{aligned}$$

$$I_c = (\pi/64)(D-2t)^4 = (\pi/64)(406.4 - 2\times 12)^4 \times 10^{-4} = 104961\,\text{cm}^4$$

- Effective flexural stiffness of cross-section

Creep coefficient is similarly determined as $\phi_t = 1.32$

$$E_{c,eff} = \frac{E_{cm}}{1 + \left(N_{G,Ed}/N_{Ed}\right)\varphi_t} = \frac{41.1}{1 + (4500/11000)\times 1.32} = 26.7\,\text{GPa}$$

$$\begin{aligned} (EI)_{eff} &= E_a I_a + 0.6 E_{c,eff} I_c \\ &= \left(210\times 10^3 \times 28940 \times 10^4 + 0.6 \times 26.7 \times 10^3 \times 104961 \times 10^4\right) \times 10^{-3} \\ &= 7.76 \times 10^{10}\,\text{kN}\cdot\text{mm}^2 \end{aligned}$$

- Confinement effect and design plastic resistance of cross-section

$$N_{cr} = \frac{\pi^2 (EI)_{eff}}{L_{eff}^2} = \frac{\pi^2 \times 7.76 \times 10^{10}}{4000^2} = 47849\,\text{kN}$$

$$N_{pl,Rk} = A_a f_y + A_c f_{ck} = (14900 \times 460 + 114800 \times 72) \times 10^{-3} = 15121\,\text{kN}$$

$$\bar{\lambda} = \sqrt{\frac{N_{pl,Rk}}{N_{cr}}} = \sqrt{\frac{15121}{47849}} = 0.562 > 0.5$$

Since the relative slenderness ratio is higher than 0.5, the confinement effect is not taken into account, thus

$$N_{pl,Rd} = A_a f_{yd} + A_c f_{cd} = (14900 \times 460 + 114800 \times 48) \times 10^{-3} = 12365\,\text{kN}$$

- Steel contribution ratio

$$\delta = A_a f_{yd}/N_{pl,Rd} = \left(14900 \times 10^{-3} \times 460\right)/12365 = 0.554 < 0.9$$

- Buckling resistance

Buckling curve “a” is used, imperfection factor $\alpha = 0.21$

$$\Phi = 0.5[1+\alpha(\bar{\lambda}-0.2)+\bar{\lambda}^2] = 0.5[1+0.21 \times (0.562 - 0.2) + 0.562^2] = 0.696$$

$$\chi = \frac{1}{\Phi + \sqrt{\Phi^2 - \bar{\lambda}^2}} = \frac{1}{0.696 + \sqrt{0.696^2 - 0.562^2}} = 0.904$$

The axial buckling resistance is:

$$N_{b,Rd} = \chi N_{pl,Rd} = 0.904 \times 12365 = 11178\,\text{kN}$$

Thus, the buckling resistance is almost the same as the CHS 508 × 12.5 - S355 steel with C40/50 concrete infilled as calculated in Section B.1.2.

- Reductions of sectional and surface area

The column section area is thus reduced by

$$\frac{\Delta A_f}{A_f} = \frac{\pi/4 \times (508^2 - 406.4^2)}{\pi/4 \times 508^2} = 36\%$$

The surface area of the column is reduced by

$$\frac{\Delta A_s}{A_s} = \frac{\pi \times (508 - 406.4)}{\pi \times 508} = 20\%$$

With the reduced surface area, the cost of fire protection material may be reduced since the labor cost for applying the fire protection material is based on the surface area. In addition, welding work and labor cost will be reduced as the less construction materials are needed for smaller column size.

B.1.5 Summary

For CFST columns subject to axial compression force only (mainly used in braced frames with simple construction), the use of high strength concrete will benefit more than the use of higher grade steels, compared with the increase of cost.

With the use of high strength materials, the column size is reduced. As a result, the fabrication cost of column and labor cost for applying fire protection are reduced, and the usable floor area is increased.

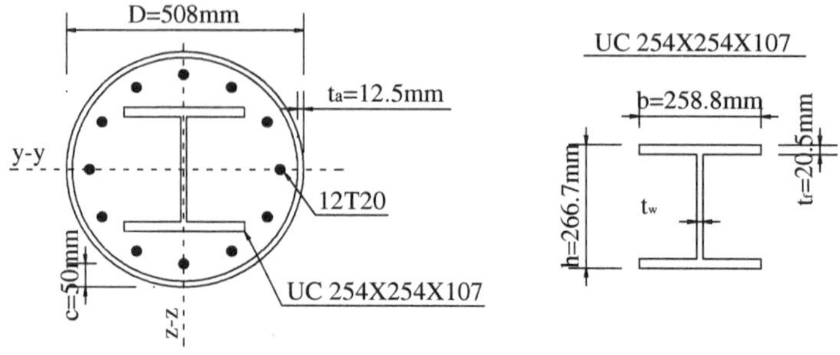

Fig. B.3 Cross-sectional dimensions of CFST column in Example 2.

B.2 Concrete filled steel tube with a UC steel section subject to compression and uniaxial bending

B.2.1 General

In Example 2, the design resistance of a circular concrete filled steel tubular member with encased reinforcements and UC steel section is checked against combined compression and uniaxial bending moment about the major axis. The dimensions of the CFST column are shown in Fig. B.3.

The column lengths and design loads are given as:

Column system length		$L = 4000$ mm
Effective length		$L_{\text{eff}} = 4000$ mm
Total design axial load		$N_{\text{Ed}} = 10{,}000$ kN
Design axial load that is permanent		$N_{\text{G,Ed}} = 4000$ kN
Design moment at bottom around y-y axis		$M_{\text{b,y}} = 700$ kN.m
Design moment at top around y-y axis		$M_{\text{t,y}} = -500$ kN.m

To evaluate the resistance, the following steel, concrete and reinforcing steel are used:

a) CHS 508 × 12.5 and UC 254 × 254 × 107 - S355 steel sections with C50/60 concrete and G460 reinforcements

b) CHS 508 × 12.5 and UC 254 × 254 × 107 - S355 steel sections with C90/105 concrete and G460 reinforcements

c) CHS 508 × 12.5 and UC 254 × 254 × 107 - S500 steel sections with C50/60 concrete and G460 reinforcements

B.2.2 CHS 508 x 12.5 and UC 254 x 254 x 107 - S355 steel sections with C50/60 concrete and G460 reinforcements

- Material

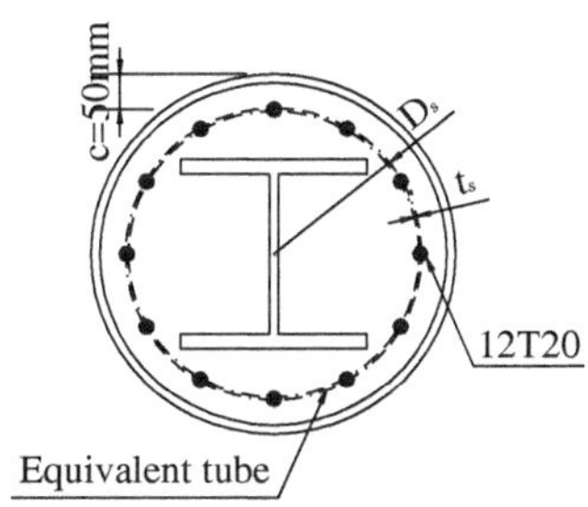

Fig. B.4 Equivalent tube section for reinforcing steel.

Concrete	C50/60, $f_{ck} = 50\ \text{N/mm}^2$
Steel tube	Grade S355, $f_y = 355\ \text{N/mm}^2$
Embedded steel section	Grade S355, $f_{ek} = 355\ \text{N/mm}^2$
Reinforcements	Grade 460, $f_{sk} = 460\ \text{N/mm}^2$

- Design strength and modulus of material

Unless otherwise stated, the subscript "e" stands for the encased steel section, and the "s" stands for the reinforcing steel.

$$f_{yd} = f_y/\gamma_a = 355/1.0 = 355\,\text{N/mm}^2$$
$$f_{sd} = f_{sk}/\gamma_s = 460/1.15 = 400\,\text{N/mm}^2$$
$$f_{ed} = f_{ek}/\gamma_a = 355/1.0 = 355\,\text{N/mm}^2$$
$$f_{cd} = f_{ck}/\gamma_c = 50/1.5 = 33.3\,\text{N/mm}^2$$
$$f_{cm} = f_{ck} + 8 = 50 + 8 = 58\,\text{N/mm}^2$$
$$E_a = E_s = E_e = 210\,\text{GPa}$$
$$E_{cm} = 22(f_{cm}/10)^{0.3} = 22(58/10)^{0.3} = 37.3\,\text{GPa}$$

- Cross sectional areas

$$A = (\pi/4)D^2 = (\pi/4) \times 508^2 = 202683\,\text{mm}^2$$
$$A_a = (\pi/4)\left[D^2 - (D - 2t_a)^2\right] = (\pi/4)\left[508^2 - (508 - 2 \times 12.5)^2\right]$$
$$= 19458\,\text{mm}^2$$
$$A_e = bh - (b - t_w)(h - 2t_f)$$
$$= 258.8 \times 266.7 - (258.8 - 12.8)(266.7 - 2 \times 20.5) = 13500\,\text{mm}^2$$
$$A_s = 12(\pi/4)d^2 = 12 \times (\pi/4) \times 20^2 = 3770\,\text{mm}^2$$
$$A_c = A - A_a - A_e - A_s = 202683 - 19458 - 13500 - 3770 = 165955\,\text{mm}^2$$

- Second moment of areas

For simplicity, the reinforcements are equivalently converted to a circular tube based on the same cross-sectional area and position of centerline, as shown in Fig. B.4 and Fig. B.5.

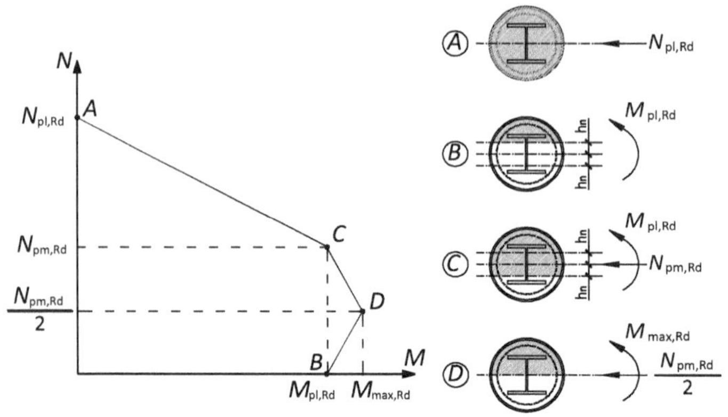

Fig. B.5 Simplified interaction curve for circular CFST column with encased steel section.

$$t_s = \frac{A_s}{\pi(D-2c)} = \frac{3770}{\pi \times (508 - 2 \times 50)} = 2.94\,\text{mm}$$
$$D_s = D - 2c + t_s = 508 - 2 \times 50 + 2.94 = 410.9\,\text{mm}$$
$$I = (\pi/64)D^4 = (\pi/64) \times 508^4 \times 10^{-4} = 326907\,\text{cm}^4$$
$$I_a = (\pi/64)\left[D^4 - (D - 2t_a)^4\right]$$
$$= (\pi/64)\left[508^4 - (508 - 2 \times 12.5)^4\right] \times 10^{-4} = 59755\,\text{cm}^4$$
$$I_s = (\pi/64)\left[D_s^4 - (D_s - 2t_s)^4\right]$$
$$= (\pi/64)\left[410.9^4 - (410.9 - 2 \times 2.94)^4\right] \times 10^{-4} = 7840\,\text{cm}^4$$
$$I_{ey} = \frac{1}{12}[bh^3 - (b - t_w)(h - 2t_f)^3]$$
$$= \frac{1}{12}\left[258.8 \times 266.7^3 - (258.8 - 12.8)(266.7 - 2 \times 20.5)^3\right] \times 10^{-4}$$
$$= 17343\,\text{cm}^4$$
$$I_{ez} = \frac{1}{12}\left[2t_f b^3 + (h - 2t_f)t_w^3\right]$$
$$= \frac{1}{12}[2 \times 20.5 \times 258.8^3 + (266.7 - 2 \times 20.5) \times 12.8^3] \times 10^{-4} = 5926\,\text{cm}^4$$
$$I_{cy} = I - I_a - I_s - I_{ey} = 326907 - 59755 - 7840 - 17343 = 241969\,\text{cm}^4$$
$$I_{cz} = I - I_a - I_s - I_{ez} = 326907 - 59755 - 7840 - 5926 = 253386\,\text{cm}^4$$

- Plastic modulus

$$W = D^3/6 = 508^3/6 \times 10^{-3} = 21850\,\text{cm}^3$$
$$W_a = \left[D^3 - (D - 2t_a)^3\right]/6 = \left[508^3 - (508 - 2 \times 12.5)^3\right]/6 \times 10^{-3} = 3070\,\text{cm}^3$$

$$W_s = \left[D_s^3 - (D_s - 2t_s)^3\right]/6 = \left[410.9^3 - (410.9 - 2 \times 2.94)^3\right]/6 \times 10^{-3} = 489\,\text{cm}^3$$

$$W_{ey} = \frac{1}{4}\left[bh^2 - (b - t_w)(h - 2t_f)^2\right]$$
$$= \frac{1}{4}\left[258.8 \times 266.7^2 - (258.8 - 12.8)(266.7 - 2 \times 20.5)^2\right] \times 10^{-3} = 1469\,\text{cm}^3$$

$$W_{ez} = \frac{1}{4}\left[2t_f b^2 + (h - 2t_f)t_w^2\right]$$
$$= \frac{1}{4}\left[2 \times 20.5 \times 258.8^2 + (266.7 - 2 \times 20.5) \times 12.8^2\right] \times 10^{-4} = 696\,\text{cm}^4$$

$$W_{cy} = W - W_a - W_s - W_{ey} = 21850 - 3070 - 489 - 1469 = 16821\,\text{cm}^4$$
$$W_{cz} = W - W_a - W_s - W_{ez} = 21850 - 3070 - 489 - 696 = 17594\,\text{cm}^4$$

- Check for local buckling

$$D/t_a = 508/12.5 = 40.6 < 90(235/f_y) = 90(235/355) = 59.6$$

Resistance against local buckling is adequate!

- Long-term effect

Age of concrete at loading in days: $t_0 = 30$
Age of concrete at moment considered in days: $t = \infty$
Relative humidity of ambient environment: RH = 50%
Perimeter of concrete section: $u = \pi(D - 2t_a) = \pi(508 - 2 \times 12.5) = 1517\,\text{mm}$
Notional size of concrete section: $h_0 = 2A_c/u = 2 \times 165955/1517 = 219\,\text{mm}$
Coefficient: $\alpha_1 = (35/f_{cm})^{0.7} = (35/58)^{0.7} = 0.70$
Coefficient: $\alpha_2 = (35/f_{cm})^{0.2} = (35/58)^{0.2} = 0.90$
Coefficient: $\alpha_3 = (35/f_{cm})^{0.5} = (35/58)^{0.5} = 0.78$

Factor: $\varphi_{\text{RH}} = \left(1 + \dfrac{1 - RH/100}{0.1\sqrt[3]{h_0}}\alpha_1\right)\alpha_2 = \left(1 + \dfrac{1 - 50/100}{0.1\sqrt[3]{219}} \times 0.70\right) \times 0.90 = 1.43$

Factor: $\beta(f_{cm}) = 16.8/\sqrt{f_{cm}} = 16.8/\sqrt{58} = 2.21$

Factor: $\beta(t_0) = 1/\left(0.1 + t_0^{0.2}\right) = 1/\left(0.1 + 30^{0.2}\right) = 0.48$

Factor: $\varphi_0 = \varphi_{\text{RH}}\beta(f_{cm})\beta(t_0) = 1.43 \times 2.21 \times 0.48 = 1.52$

Factor: $\beta_H = 1.5\left[1 + (0.012RH)^{18}\right]h_0 + 250\alpha_3$
$= 1.5 \times \left[1 + (0.012 \times 50)^{18}\right] \times 219 + 250 \times 0.78 = 522$

Factor: $\beta_c(t, t_0) = \left(\dfrac{t - t_0}{\beta_H + t - t_0}\right)^{0.3} = \left(\dfrac{\infty - 14}{522 + \infty - 14}\right)^{0.3} = 1.0$

Creep coefficient: $\varphi_t = \varphi_0\beta_c(t, t_0) = 1.52 \times 1.0 = 1.52$

- Elastic modulus of concrete considering long-term effect

$$E_{c,eff} = E_{cm}\frac{1}{1 + \left(N_{G,Ed}/N_{Ed}\right)\varphi_t} = \frac{37.3}{1 + (4000/10000) \times 1.52} = 23.2\,\text{GPa}$$

- Effective flexural stiffness of cross-section

$$\begin{aligned}(EI)_{eff,y} &= E_a I_a + E_s I_s + E_e I_{ey} + 0.6E_{c,eff} I_{cy}\\ &= [210 \times (59755 + 7840 + 17343) + 0.6 \times 23.2 \times 241969] \times 10^4\\ &= 2.12 \times 10^{11}\,\text{kN}\cdot\text{mm}^2\\ (EI)_{eff,z} &= E_a I_a + E_s I_s + E_e I_{ez} + 0.6E_{c,eff} I_{cz}\\ &= [210 \times (59755 + 7840 + 5926) + 0.6 \times 23.2 \times 253386] \times 10^4\\ &= 1.90 \times 10^{11}\,\text{kN}\cdot\text{mm}^2\end{aligned}$$

- Elastic critical Euler buckling resistance

$$N_{cr,y} = \frac{\pi^2 (EI)_{eff,y}}{L_{eff}^2} = \frac{\pi^2 \times 2.12 \times 10^{11}}{4000^2} = 130793\,\text{kN}$$

$$N_{cr,z} = \frac{\pi^2 (EI)_{eff,z}}{L_{eff}^2} = \frac{\pi^2 \times 1.90 \times 10^{11}}{4000^2} = 116984\,\text{kN}$$

- Characteristic plastic resistance of cross-section

$$\begin{aligned}N_{pl,Rk} &= A_a f_y + A_s f_{sk} + A_e f_{ek} + A_c f_{ck}\\ &= [(19458 + 13500) \times 355 + 3770 \times 460 + 165955 \times 50] \times 10^{-3}\\ &= 21475\,\text{kN}\end{aligned}$$

- Relative slenderness ratio

$$\overline{\lambda_y} = \sqrt{\frac{N_{pl,Rk}}{N_{cr,y}}} = \sqrt{\frac{21475}{130793}} = 0.408 < 0.5$$

$$\overline{\lambda_z} = \sqrt{\frac{N_{pl,Rk}}{N_{cr,z}}} = \sqrt{\frac{21475}{116984}} = 0.431 < 0.5$$

- Buckling curves and buckling reduction factors

Since a steel section is embedded in the CFST column, the buckling curves about both axes are "b". Thus, the imperfection factor is $\alpha = 0.34$.

$$\begin{aligned}\bar{\lambda} &= \max\left(\overline{\lambda_y}, \overline{\lambda_z}\right) = 0.431\\ \Phi &= 0.5\left[1 + \alpha(\bar{\lambda} - 0.2) + \bar{\lambda}^2\right]\\ &= 0.5\left[1 + 0.34 \times (0.431 - 0.2) + 0.431^2\right] = 0.632\end{aligned}$$

$$\chi = \min\left(\frac{1}{\Phi + \sqrt{\Phi^2 - \bar{\lambda}^2}}, 1.0\right)$$
$$= \min\left(\frac{1}{0.632 + \sqrt{0.632^2 - 0.431^2}}, 1.0\right) = 0.914$$

- Load eccentricity

$$e_{N,y} = \frac{\max\left(|M_{t,y}|, |M_{b,y}|\right)}{N_{Ed}} = \frac{700}{10000} \times 10^3 = 70\,\text{mm} > 0.1D = 50.8\,\text{mm}$$

Since the eccentricity $e_{\text{N,y}}$ is larger than 0.1D, the confinement effect is not taken into account.

- Simplified Interaction Curve

1) Point A (0, $N_{\text{pl,Rd}}$):
Full cross-section is under uniform compression. No bending moment is resultant from the compressive stresses on the cross-section.

$$N_{pl,Rd} = A_a f_{yd} + A_s f_{sd} + A_e f_{ed} + A_c f_{cd}$$
$$= [(19458 + 13500) \times 355 + 3770 \times 400 + 165955 \times 33.3] \times 10^{-3}$$
$$= 18753\,\text{kN}$$

2) Point B ($M_{\text{pl,Rd}}$, 0):
The cross-section is under partial compression and no axial force is formed. Assuming the neutral axis lies in the web of encased section ($h_n \leq h/2 - t_f$), the height of neutral axis is calculated where the areas of steel tube, equivalent tube for reinforcements, and concrete in the height of $2h_n$ are approximated as rectangles, and based on the force equilibrium between the tensile capacity of steel sections within the height of $2h_n$ is equal to the compression resistance of concrete in the compression zone. Unless otherwise stated, the tensile resistance of concrete in the tension zone is conservatively ignored.

$$h_n = \frac{A_c f_{cd}}{2Df_{cd} + 4t_a(2f_{yd} - f_{cd}) + 4t_s(2f_{sd} - f_{cd}) + 2t_w(2f_{ed} - f_{cd})}$$
$$= \frac{165955 \times 33.3}{2 \times 508 \times 33.3 + 4 \times 12.5 \times (2 \times 355 - 33.3) + 4 \times 2.94 \times (2 \times 400 - 33.3) + 2 \times 12.8 \times (2 \times 355 - 33.3)}$$
$$= 58.8\,\text{mm}$$
$$h_n = 58.8\text{ mm} < h/2 - t_f = 266.7/2 - 20.5 = 112.85\text{ mm}$$

Thus, the neutral axial lies in the web of the encased section. The plastic modulus of steel tube, equivalent tube of reinforcements, encased section, and concrete in the height of $2h_n$, bending about centreline of the cross-section are calculated as:

$$W_{a,n} = 2t_a h_n^2 = 2 \times 12.5 \times 58.8^2 \times 10^{-3} = 86.4\,\text{cm}^3$$
$$W_{s,n} = 2t_s h_n^2 = 2 \times 2.94 \times 58.8^2 \times 10^{-3} = 20.3\,\text{cm}^3$$

$$W_{ey,n} = t_w h_n^2 = 12.8 \times 58.8^2 \times 10^{-3} = 44.3\,\text{cm}^3$$
$$\begin{aligned} W_{cy,n} &= (D - 2t_a - 2t_s - t_w)h_n^2 \\ &= (508 - 2 \times 12.5 - 2 \times 2.94 - 12.8) \times 58.8^2 \times 10^{-3} = 1605\,\text{cm}^3 \end{aligned}$$

Taking moment about the centerline of the cross-section, the plastic bending resistance is determined from:

$$\begin{aligned} M_{pl,Rd} &= (W_a - W_{a,n})f_{yd} + (W_s - W_{s,n})f_{sd} + (W_{ey} - W_{ey,n})f_{ed} + 0.5(W_{cy} - W_{cy,n})f_{cd} \\ &= [(3070 - 86.4) \times 355 + (489 - 20.3) \times 400 + (1469 - 44.3) \\ &\quad \times 355 + 0.5 \times (16821 - 1605) \times 33.3] \times 10^{-3} = 2006\,\text{kN} \cdot \text{m} \end{aligned}$$

It should be noted that the plastic bending resistance can be calculated by taking moment about either line on the cross-section parallel to the y-y axis, as long as the aforementioned plastic modulus are determined according to the referred line.

3) Point C ($M_{\text{pl,Rd}}$, $N_{\text{pm,Rd}}$):

The cross-section is under partial compression but axial force is resultant from the compressive stresses. The axial force is equal to the compression capacities of concrete in the compression zone and steel sections within the height of $2h_n$. It is mentioned above that the compression capacity of steel sections within the height of $2h_n$ is equal to the compression capacity of concrete in the compression zone and out of the height of $2h_n$. Thus, the axial force is actually the full cross-sectional compression capacity of concrete and determined from:

$$N_{pm,Rd} = A_c f_{cd} = 165955 \times 33.3 \times 10^{-3} = 5526\,\text{kN}$$

4) Point D ($M_{\text{max,Rd}}$, $N_{\text{pm,Rd}}/2$):

The maximum plastic moment resistance $M_{\text{max,Rd}}$ is calculated when the h_n is equal to 0.

$$\begin{aligned} M_{\max,Rd} &= W_a f_{yd} + W_s f_{sd} + W_{ey} f_{ed} + 0.5 W_{cy} f_{cd} \\ &= [3070 \times 355 + 489 \times 400 + 1469 \times 355 + 0.5 \times 16821 \times 33.3] \times 10^{-3} \\ &= 2087\,\text{kN} \cdot \text{m} \end{aligned}$$

- Steel contribution ratio

$$\begin{aligned} \delta &= \left(A_a f_{yd} + A_e f_{ed}\right)/N_{pl,Rd} \\ &= (19458 + 13500) \times 355 \times 10^{-3}/18753 = 0.625 < 0.9 \end{aligned}$$

- Check for resistance of column in axial compression

$$\frac{N_{Ed}}{\chi N_{pl,Rd}} = \frac{10000}{0.914 \times 18753} = \frac{10000}{17140} = 0.583 < 1.0$$

Thus, the buckling resistance under axial compression is adequate!

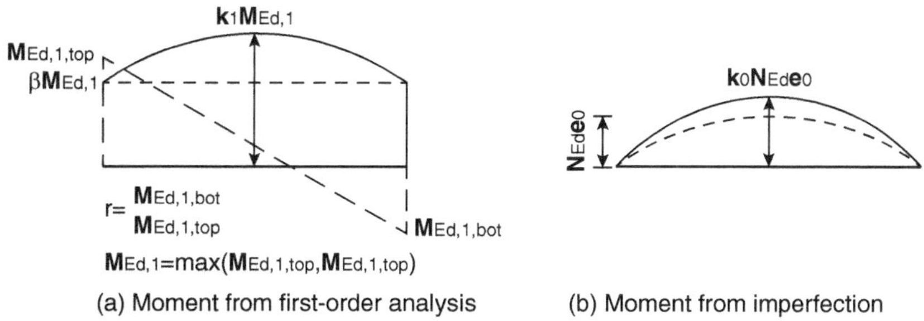

Fig. B.6 Consideration for second order effect.

- Effective flexural stiffness considering long-term effect

$$
\begin{aligned}
(EI)_{eff,II} &= K_0\left(E_a I_a + E_s I_s + E_e I_{ey} + K_{e,II} E_{c,m} I_{cy}\right) \\
&= 0.9 \times [210 \times (59755 + 7840 + 17343) + 0.5 \times 23.2 \times 241969] \times 10^4 \\
&= 1.86 \times 10^{11}\,\text{kN} \cdot \text{mm}^2
\end{aligned}
$$

- Critical normal force about y-y axis

The effective length is taken as the system length of column.

$$N_{cr,eff} = \frac{\pi^2 (EI)_{eff,II}}{L^2} = \frac{\pi^2 \times 1.86 \times 10^{11}}{4000^2} = 114601\,\text{kN}$$

- Second-order effect (refer to Chapter 4.4.2)

The second-order effect should be considered for both moments from first-order analysis and moment from imperfection as shown in Fig. B.6.

According to buckling curve "b", the initial imperfection about y-y axial is:

$$e_{0,y} = L/200 = 4000/200 = 20\,\text{mm}$$

Accordingly, the bending moment by the initial imperfection is determined as:

$$M_0 = N_{Ed} e_{0,y} = 10000 \times 20/1000 = 200\,\text{kN} \cdot \text{m}$$

According to the moment diagram by the initial imperfection, the factor β_0 for determination of moment to second-order effect is equal to 1.0. Thus, the amplification factor for the moment by the imperfection is calculated from:

$$k_0 = \frac{\beta_0}{1 - N_{Ed}/N_{cr,eff}} = \frac{1.0}{1 - 10000/114601} = 1.096$$

According to the first-order design moment diagram, the ratio of end moments is calculated as:

$$r = M_{t,y}/M_{b,y} = -500/700 = -0.714$$

Thus, the factor β_1 for determination of moment to second-order effect is determined:

$$\beta_1 = \max(0.66 + 0.44r, 0.44) = \max(0.66 + 0.44 \times (-0.714), 0.44) = 0.44$$

Thus, the amplification factor for the moment by the imperfection is calculated from:

$$k_1 = \frac{\beta_1}{1 - N_{Ed}/N_{cr,eff}} = \frac{0.44}{1 - 10000/114601} = 0.482$$

Thus, the design moment, considering second-order effect, is calculated as:

$$\begin{aligned} M_{Ed} &= Max[k_0 M_0 + k_1 Max(|M_{t,y}|, |M_{b,y}|), Max(|M_{t,y}|, |M_{b,y}|)] \\ &= Max[1.096 \times 200 + 0.482 \times Max(|-500|, |700|), Max(|-500|, |700|)] \\ &= 700\,\text{kN} \cdot \text{m} \end{aligned}$$

In this case, the second-order effect is not significant and the maximum end moment is taken as the design moment.

- Check for resistance of column in combined compression and uniaxial bending

For $N_{Ed} > N_{pm,Rd} = 5526\ kN$, the value for determining the plastic bending resistance $M_{pl,N,Rd}$ taking into account the normal force N_{Ed} is calculated from:

$$\mu_d = \frac{N_{pl,Rd} - N_{Ed}}{N_{pl,Rd} - N_{pm,Rd}} = \frac{18753 - 10000}{18753 - 5526} = 0.662 \quad \text{(refer to Eq.(4.30))}$$

$$\frac{M_{Ed}}{M_{pl,N,Rd}} = \frac{M_{Ed}}{\mu_d M_{pl,Rd}} = \frac{700}{0.662 \times 2006} = 0.527 < \alpha_M = 0.9$$

Thus, the resistance for combined axial compression and uniaxial bending is adequate. The external design force and bending moment, and the design M-N interaction curve are plotted in Fig. B.7 and Fig. B.8.

B.2.3 CHS 508 x 12.5 and UC 254 x 254 x 107 - S355 steel sections with C90/105 concrete and G460 reinforcements

In this section, the normal strength concrete (NSC) C40/50 is replaced by high strength concrete (HSC) C90/105. The steel grade is not changed.

- Design strength

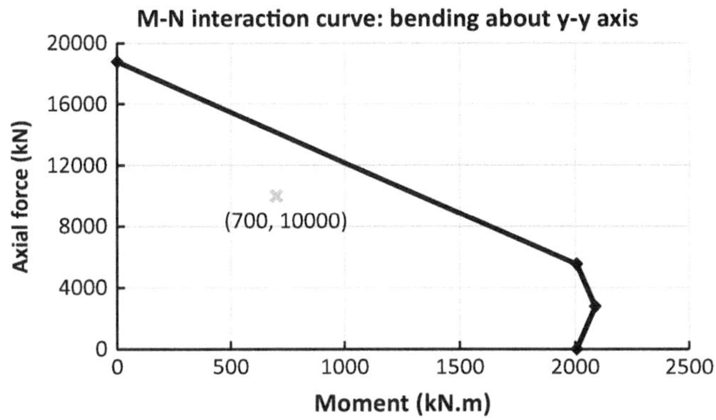

Fig. B.7 Design M–N interaction curve of circular CFST column.

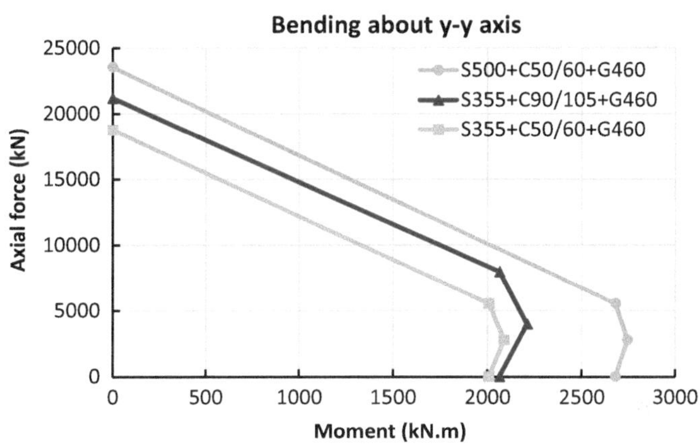

Fig. B.8 Design M–N interaction curves for CFST columns with various material grades.

Effective compressive strength and modulus of elasticity of the HSC are taken from Table 3.1, Eq. (3.1) and Eq. (3.2) of BC4.

$$f_{ck} = 72\,\text{N/mm}^2;\ E_{cm} = 41.1\,\text{GPa}$$

$$f_{cd} = f_{ck}/\gamma_c = 72/1.5 = 48\,\text{N/mm}^2$$

$$f_{cm} = f_{ck} + 8 = 72 + 8 = 80\,\text{N/mm}^2$$

- Effective flexural stiffness of cross-section

Creep coefficient could be similarly determined as $\phi_t = 1.12$

$$E_{c,eff} = \frac{E_{cm}}{1 + \left(N_{G,Ed}/N_{Ed}\right)\varphi_t} = \frac{41.1}{1 + (4000/10000) \times 1.12} = 28.3\,\text{GPa}$$

$$\begin{aligned}(EI)_{eff,y} &= E_a I_a + E_s I_s + E_e I_{ey} + 0.6 E_{c,eff} I_{cy} \\ &= [210 \times (59755 + 7840 + 17343) + 0.6 \times 28.3 \times 241969] \times 10^4 \\ &= 2.19 \times 10^{11}\ \text{kN} \cdot \text{mm}^2 \\ (EI)_{eff,z} &= E_a I_a + E_s I_s + E_e I_{ez} + 0.6 E_{c,eff} I_{cz} \\ &= [210 \times (59755 + 7840 + 5926) + 0.6 \times 28.3 \times 253386] \times 10^4 \\ &= 1.97 \times 10^{11}\ \text{kN} \cdot \text{mm}^2\end{aligned}$$

- Characteristic plastic resistance of cross-section

$$\begin{aligned}N_{cr,y} &= \frac{\pi^2 (EI)_{eff,y}}{L_{eff}^2} = \frac{\pi^2 \times 2.19 \times 10^{11}}{4000^2} = 135397\,\text{kN} \\ N_{cr,z} &= \frac{\pi^2 (EI)_{eff,z}}{L_{eff}^2} = \frac{\pi^2 \times 1.97 \times 10^{11}}{4000^2} = 121805\,\text{kN} \\ N_{pl,Rk} &= A_a f_y + A_s f_{sk} + A_e f_{ek} + A_c f_{ck} \\ &= [(19458 + 13500) \times 355 + 3770 \times 460 + 165955 \times 90] \times 10^{-3} = 25395\,\text{kN}\end{aligned}$$

- Relative slenderness ratio, buckling curves and buckling reduction factors

$$\begin{aligned}\overline{\lambda_y} &= \sqrt{\frac{N_{pl,Rk}}{N_{cr,y}}} = \sqrt{\frac{25395}{135397}} = 0.433 < 0.5; \\ \overline{\lambda_z} &= \sqrt{\frac{N_{pl,Rk}}{N_{cr,z}}} = \sqrt{\frac{25395}{121805}} = 0.457 < 0.5\end{aligned}$$

For buckling curves about both axis "b", the imperfection factor is $\alpha = 0.34$

$$\begin{aligned}\bar{\lambda} &= \max\left(\overline{\lambda_y}, \overline{\lambda_z}\right) = 0.457 \\ \Phi &= 0.5\left[1 + \alpha\left(\bar{\lambda} - 0.2\right) + \bar{\lambda}^2\right] \\ &= 0.5\left[1 + 0.34 \times (0.457 - 0.2) + 0.457^2\right] = 0.648 \\ \chi &= \min\left(\frac{1}{\Phi + \sqrt{\Phi^2 + \bar{\lambda}^2}}, 1.0\right) \\ &= \min\left(\frac{1}{0.648 + \sqrt{0.648^2 - 0.457^2}}, 1.0\right) = 0.903\end{aligned}$$

The confinement effect is also ignored since the eccentricity is larger than 0.1D.

- M-N interaction curve

Point A (0, $N_{\text{pl,Rd}}$):

$$N_{pl,Rd} = A_a f_{yd} + A_s f_{sd} + A_e f_{ed} + A_c f_{cd}$$

$$= [(19458 + 13500) \times 355 + 3770 \times 400 + 165955 \times 48] \times 10^{-3}$$
$$= 21187\,\text{kN}$$

Buckling resistance: $\chi N_{pl,Rd} = 0.903 \times 21187 = 19132$ kN
Point B ($M_{\text{pl,Rd}}$, 0):

$$h_n = \frac{A_c f_{cd}}{2Df_{cd} + 4t_a(2f_{yd} - f_{cd}) + 4t_s(2f_{sd} - f_{cd}) + 2t_w(2f_{ed} - f_{cd})}$$
$$= \frac{165955 \times 48}{2 \times 508 \times 48 + 4 \times 12.5 \times (2 \times 355 - 48) + 4 \times 2.94 \times (2 \times 400 - 48) + 2 \times 12.8 \times (2 \times 355 - 48)}$$
$$= 73.7\,\text{mm}$$

$h_n = 73.7$ mm $< h/2 - t_f = 266.7/2 - 20.5 = 112.85$ mm, thus, the neutral axial also lies in the web of the encased section.

$$W_{a,n} = 2t_a h_n^2 = 2 \times 12.5 \times 73.7^2 \times 10^{-3} = 135.8\,\text{cm}^3$$
$$W_{s,n} = 2t_s h_n^2 = 2 \times 2.94 \times 73.7^2 \times 10^{-3} = 31.9\,\text{cm}^3$$
$$W_{ey,n} = t_w h_n^2 = 12.8 \times 73.7^2 \times 10^{-3} = 69.5\,\text{cm}^3$$
$$W_{cy,n} = (D - 2t_a - 2t_s - t_w)h_n^2$$
$$= (508 - 2 \times 12.5 - 2 \times 2.94 - 12.8) \times 73.7^2 \times 10^{-3} = 2522\,\text{cm}^3$$
$$M_{pl,Rd} = (W_a - W_{a,n})f_{yd} + (W_s - W_{s,n})f_{sd} + (W_{ey} - W_{ey,n})f_{ed} + 0.5(W_{cy} - W_{cy,n})f_{cd}$$
$$= [(3070 - 135.8) \times 355 + (489 - 31.9) \times 400 + (1469 - 65.9) \times 355 + 0.5 \times (16821 - 2522) \times 33.3] \times 10^{-3} = 2065\,\text{kN} \cdot \text{m}$$

Point C ($M_{\text{pl,Rd}}$, $N_{\text{pm,Rd}}$):

$$N_{pm,Rd} = A_c f_{cd} = 165955 \times 48 \times 10^{-3} = 7964\,\text{kN}$$

Point D ($M_{\text{max,Rd}}$, $N_{\text{pm,Rd}}/2$):

$$M_{\max,Rd} = W_a f_{yd} + W_s f_{sd} + W_{ey} f_{ed} + 0.5 W_{cy} f_{cd}$$
$$= [3070 \times 355 + 489 \times 400 + 1469 \times 355 + 0.5 \times 16821 \times 48] \times 10^{-3}$$
$$= 2211\,\text{kN} \cdot \text{m}$$

- Steel contribution ratio

$$\delta = (A_a f_{yd} + A_e f_{ed})/N_{pl,Rd}$$
$$= (19458 + 13500) \times 355 \times 10^{-3}/21187 = 0.552 < 0.9$$

B.2.4 CHS 508 x 12.5 and UC 254 x 254 x 107 - S500 steel sections with C50/60 concrete and G460 reinforcements

In this section, the mild steel S355 is replaced by the high tensile steel (HTS) S500, the concrete grade is not changed.

- Characteristic plastic resistance of cross-section

$$N_{pl,Rk} = A_a f_y + A_s f_{sk} + A_e f_{ek} + A_c f_{ck}$$
$$= [(19458 + 13500) \times 500 + 3770 \times 460 + 165955 \times 50] \times 10^{-3} = 26530\,\text{kN}$$

- Relative slenderness ratio, buckling curves and buckling reduction factors

$$\overline{\lambda_y} = \sqrt{\frac{N_{pl,Rk}}{N_{cr,y}}} = \sqrt{\frac{26530}{130793}} = 0.450 < 0.5$$

$$\overline{\lambda_z} = \sqrt{\frac{N_{pl,Rk}}{N_{cr,z}}} = \sqrt{\frac{26530}{116984}} = 0.476 < 0.5$$

For buckling curves about both axis "b", the imperfection factor is $\alpha = 0.34$

$$\bar{\lambda} = \max\left(\overline{\lambda_y}, \overline{\lambda_z}\right) = 0.476$$
$$\Phi = 0.5\left[1 + \alpha\left(\bar{\lambda} - 0.2\right) + \bar{\lambda}^2\right]$$
$$= 0.5\left[1 + 0.34 \times (0.476 - 0.2) + 0.476^2\right] = 0.660$$
$$\chi = \min\left(\frac{1}{\Phi + \sqrt{\Phi^2 + \bar{\lambda}^2}}, 1.0\right)$$
$$= \min\left(\frac{1}{0.660 + \sqrt{0.660^2 - 0.476^2}}, 1.0\right) = 0.895$$

The confinement effect is also ignored since the eccentricity is larger than 0.1D.

- M–N interaction curve

Point A (0, $N_{\text{pl,Rd}}$):

$$N_{pl,Rd} = A_a f_{yd} + A_s f_{sd} + A_e f_{ed} + A_c f_{cd}$$
$$= [(19458 + 13500) \times 500 + 3770 \times 400 + 165955 \times 33.3] \times 10^{-3}$$
$$= 23538\,\text{kN}$$

Buckling resistance: $\chi N_{pl,Rd} = 0.895 \times 23538 = 21067$ kN

Point B ($M_{\text{pl,Rd}}$, 0):

$$h_n = \frac{A_c f_{cd}}{2D f_{cd} + 4t_a\left(2f_{yd} - f_{cd}\right) + 4t_s\left(2f_{sd} - f_{cd}\right) + 2t_w\left(2f_{ed} - f_{cd}\right)}$$
$$= \frac{165955 \times 33.3}{2 \times 508 \times 33.3 + 4 \times 12.5 \times (2 \times 500 - 33.3) + 4 \times 2.94 \times (2 \times 400 - 33.3) + 2 \times 12.8 \times (2 \times 500 - 33.3)}$$
$$= 47.3\,\text{mm}$$

Table B.1 Comparisons between circular CFST columns with various material strengths.

Material grades (Steel + Concrete + Rebars)	Steel contribution ratios	Design resistances			
		$\chi N_{pl,Rd}$	$N_{pm,Rd}$	$M_{pl,Rd}$	$M_{max,Rd}$
S355 + C50/60 + G460	0.625	0	0	0	0
S355 + C90/105 + G460	0.552	11.6%	44.0%	2.9%	5.9%
S500 + C50/60 + G460	0.701	22.9%	0	33.6%	31.5%

The neutral axial also lies in the web of the encased section.

$$
\begin{aligned}
W_{a,n} &= 2t_a h_n^2 = 2 \times 12.5 \times 47.3^2 \times 10^{-3} = 55.9\,\text{cm}^3 \\
W_{s,n} &= 2t_s h_n^2 = 2 \times 2.94 \times 47.3^2 \times 10^{-3} = 13.2\,\text{cm}^3 \\
W_{ey,n} &= t_w h_n^2 = 12.8 \times 47.3^2 \times 10^{-3} = 28.6\,\text{cm}^3 \\
W_{cy,n} &= (D - 2t_a - 2t_s - t_w)h_n^2 = (508 - 2 \times 12.5 - 2 \times 2.94 - 12.8) \\
&\quad \times 47.3^2 \times 10^{-3} = 1039\,\text{cm}^3 \\
M_{pl,Rd} &= (W_a - W_{a,n})f_{yd} + (W_s - W_{s,n})f_{sd} \\
&\quad + \left(W_{ey} - W_{ey,n}\right)f_{ed} + 0.5\left(W_{cy} - W_{cy,n}\right)f_{cd} \\
&= [(3070 - 55.9) \times 500 + (489 - 13.2) \times 400 + (1469 - 28.6) \\
&\quad \times 500 + 0.5 \times (16821 - 1039) \times 33.3] \times 10^{-3} = 2681\,\text{kN}\cdot\text{m}
\end{aligned}
$$

Point C ($M_{pl,Rd}$, $N_{pm,Rd}$):

$$
N_{pm,Rd} = A_c f_{cd} = 165955 \times 33.3 \times 10^{-3} = 5530\,\text{kN}
$$

Point D ($M_{max,Rd}$, $N_{pm,Rd}/2$):

$$
\begin{aligned}
M_{\max,Rd} &= W_a f_{yd} + W_s f_{sd} + W_{ey} f_{ed} + 0.5 W_{cy} f_{cd} \\
&= [3070 \times 500 + 489 \times 400 + 1469 \times 500 + 0.5 \times 16821 \times 133.3] \times 10^{-3} \\
&= 2746\ \text{kN}\cdot\text{m}
\end{aligned}
$$

- Steel contribution ratio:

$$
\begin{aligned}
\delta &= \left(A_a f_{yd} + A_e f_{ed}\right)/N_{pl,Rd} \\
&= (19458 + 13500) \times 500 \times 10^{-3}/23538 = 0.701 < 0.9
\end{aligned}
$$

B.2.5 Comparison and summary

The design resistances are compared for the aforementioned three composite sections. The composite section with steel tube of S355, encased steel section of S355, concrete of C50/60, and reinforcing steel of G460 is referred to for the comparison.

It can been seen in Table B.1, the axial buckling resistance ($\chi N_{pl,Rd}$) of the CFST column is improved by 11.6% by use of high strength concrete C90/105 replacing

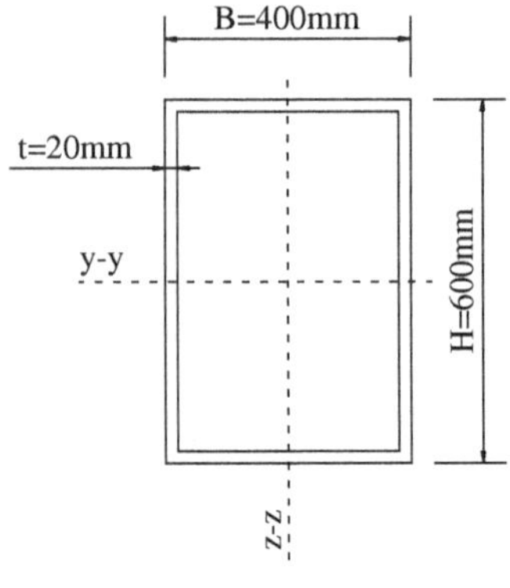

Fig. B.9 Cross-sectional dimensions of CFST column in Example 3.

normal strength concrete C50/60. However the increase of moment resistances ($M_{pl,Rd}$ and $M_{max,Rd}$) are smaller (less than 6%).

By use of steel S500 replacing S355, the axial buckling resistance is improved by 22.9%, and the increase of moment resistance is higher than 30%.

B.3 Rectangular concrete filled steel tubular column subject to axial compression and bi-axial bending

B.3.1 General

In Example 3, the design resistance of a rectangular concrete filled steel tubular column are checked against combined axial compression and bi-axial bending moments. The dimensions of the rectangular CFST column are shown in Fig. B.9 and Fig. B.10.

The column lengths and design loads are given as:

Column system length	$L = 6000$ mm
Effective length	$L_{eff} = 6000$ mm
Total design axial load	$N_{Ed} = 12{,}000$ kN
Design axial load that is permanent	$N_{G,Ed} = 5000$ kN
Design moment at bottom around y-y axis	$M_{b,y} = 900$ kN.m
Design moment at top around y-y axis	$M_{t,y} = -550$ kN.m
Design moment at bottom around y-y axis	$M_{b,z} = -600$ kN.m
Design moment at top around y-y axis	$M_{t,z} = 350$ kN.m

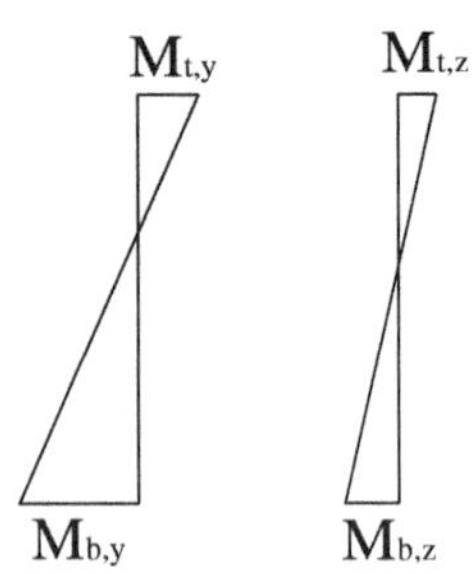

To evaluate and compare their resistance, the following steel and concrete material grades are taken into account:

a) RHS 400 × 600 × 20 - S355 steel with C50/60 concrete
b) RHS 400 × 600 × 20 - S355 steel with C90/105 concrete

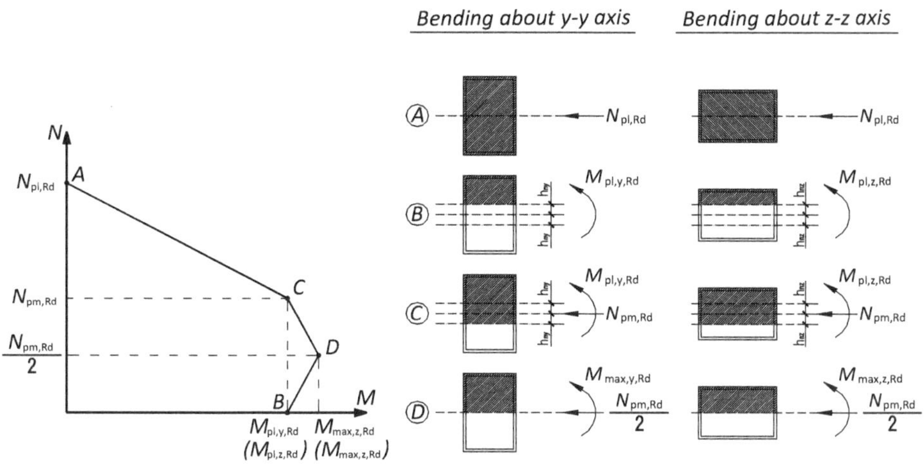

Fig. B.10 Simplified interaction curve for rectangular CFST columns.

c) RHS 400 × 600 × 20 - S500 steel with C50/60 concrete
d) RHS 400 × 600 × 20 - S500 steel with C90/105 concrete

B.3.2 RHS 400 x 600 x 20 - S355 steel tube infilled with C50/60 concrete

- Material

Concrete	C50/60, $f_{ck} = 50$ N/mm^2
Steel tube	Grade S355, $f_y = 355$ N/mm^2

- Design strengths and modulus of material

$$f_{yd} = f_y/\gamma_a = 355/1.0 = 355\,\text{N/mm}^2$$
$$f_{cd} = f_{ck}/\gamma_c = 50/1.5 = 33.3\,\text{N/mm}^2$$
$$f_{cm} = f_{ck} + 8 = 50 + 8 = 58\,\text{N/mm}^2$$
$$E_a = 210\text{ GPa}$$
$$E_{cm} = 22(f_{cm}/10)^{0.3} = 22(58/10)^{0.3} = 37.3\text{ GPa}$$

- Cross sectional areas

$$A = BH = 400 \times 600 \times 10^{-2} = 2400\,\text{cm}^2$$
$$A_a = [BH - (B - 2t_a)(H - 2t_a)]$$
$$= [400 \times 600 - (400 - 2 \times 20)(600 - 2 \times 20)] \times 10^{-2} = 384\,\text{cm}^2$$
$$A_c = A - A_a = 240000 - 38400 \times 10^{-2} = 2016\,\text{cm}^2$$

- Second moment of areas

$$
\begin{aligned}
I_y &= BH^3/12 = 400 \times 600^3/12 \times 10^{-4} = 720000\,\text{cm}^4 \\
I_z &= HB^3/12 = 600 \times 400^3/12 \times 10^{-4} = 320000\,\text{cm}^4 \\
I_{ay} &= \left[BH^3 - (B - 2t_a)(H - 2t_a)^3\right]/12 \\
&= \left[400 \times 600^3 - (400 - 2 \times 20)(600 - 2 \times 20)^3\right]/12 \times 10^{-4} \\
&= 193152\,\text{cm}^4 \\
I_{az} &= \left[HB^3 - (H - 2t_a)(B - 2t_a)^3\right]/12 \\
&= \left[600 \times 400^3 - (600 - 2 \times 20)(400 - 2 \times 20)^3\right]/12 \times 10^{-4} \\
&= 102272\,\text{cm}^4 \\
I_{cy} &= I_y - I_{ay} = 720000 - 193152 = 526848\,\text{cm}^4 \\
I_{cz} &= I_z - I_{az} = 320000 - 193152 = 217728\,\text{cm}^4
\end{aligned}
$$

- Plastic modulus

$$
\begin{aligned}
W_y &= BH^2/4 = 400 \times 600^2/4 \times 10^{-3} = 36000\,\text{cm}^3 \\
W_z &= HB^2/4 = 600 \times 400^2/4 \times 10^{-3} = 24000\,\text{cm}^3 \\
W_{ay} &= \left[BH^2 - (B - 2t_a)(H - 2t_a)^2\right]/4 \\
&= \left[400 \times 600^2 - (400 - 2 \times 20)(600 - 2 \times 20)^2\right]/4 \times 10^{-3} \\
&= 7776\,\text{cm}^3 \\
W_{az} &= \left[HB^2 - (H - 2t_a)(B - 2t_a)^2\right]/4 \\
&= \left[600 \times 400^2 - (600 - 2 \times 20)(400 - 2 \times 20)^2\right]/4 \times 10^{-3} \\
&= 5856\,\text{cm}^3 \\
W_{cy} &= W_y - W_{ay} = 36000 - 7776 = 28224\,\text{cm}^3 \\
W_{cz} &= W_z - W_{az} = 24000 - 5856 = 18144\,\text{cm}^3
\end{aligned}
$$

- Check for local buckling

$$H/t_a = 600/20 = 30 < 52\left(235/f_y\right) = 52(235/355) = 42.3$$

Resistance against local buckling is adequate!

- Long-term effect

Assuming age of concrete at loading in days: $t_0 = 30$

Age of concrete at moment considered in days: $t = \infty$

Relative humidity of ambient environment: RH = 50%

Perimeter of concrete section: $u = 2(B - 2t_a) + 2(H - 2t_a)$

$= 2(400 - 2 \times 20) + 2(600 - 2 \times 20) = 1840\,\text{mm}$

Notional size of concrete section: $h_0 = 2A_c/u = 2 \times 201600/1840 = 219\,\text{mm}$

Coefficient: $\alpha_1 = (35/f_{cm})^{0.7} = (35/58)^{0.7} = 0.70$

Coefficient: $\alpha_2 = (35/f_{cm})^{0.2} = (35/58)^{0.2} = 0.90$

Coefficient: $\alpha_3 = (35/f_{cm})^{0.5} = (35/58)^{0.5} = 0.78$

Factor: $\varphi_{\mathrm{RH}} = \left(1+\dfrac{1-RH/100}{0.1\sqrt[3]{h_0}}\alpha_1\right)$

$$\alpha_2 = \left(1+\frac{1-50/100}{0.1\sqrt[3]{219}} \times 0.70\right) \times 0.90 = 1.43$$

Factor: $\beta(f_{cm}) = 16.8/\sqrt{f_{cm}} = 16.8/\sqrt{58} = 2.21$

Factor: $\beta(t_0) = 1/\left(0.1 + t_0^{0.2}\right) = 1/\left(0.1 + 30^{0.2}\right) = 0.48$

Factor: $\varphi_0 = \varphi_{\mathrm{RH}}\beta(f_{cm})\beta(t_0) = 1.43 \times 2.21 \times 0.48 = 1.52$

Factor: $\beta_H = 1.5\left[1 + (0.012RH)^{18}\right]h_0 + 250\alpha_3$

$$= 1.5 \times \left[1 + (0.012 \times 50)^{18}\right] \times 219 + 250 \times 0.78 = 523$$

Factor: $\beta_c(t, t_0) = \left(\dfrac{t - t_0}{\beta_H + t - t_0}\right)^{0.3} = \left(\dfrac{\infty - 14}{522 + \infty - 14}\right)^{0.3} = 1.0$

Creep coefficient: $\varphi_t = \varphi_0\beta_c(t, t_0) = 1.52 \times 1.0 = 1.52$

- Elastic modulus of concrete considering long-term effect

$$E_{c,eff} = E_{cm}\frac{1}{1 + \left(N_{G,Ed}/N_{Ed}\right)\varphi_t} = \frac{37.3}{1 + (5000/12000) \times 1.52} = 22.8\,\mathrm{GPa}$$

- Effective flexural stiffness of cross-section

$$(EI)_{eff,y} = E_a I_{ay} + 0.6E_{c,eff}I_{cy}$$
$$= [210 \times 193152 + 0.6 \times 22.8 \times 526848] \times 10^4 = 4.78 \times 10^{11}\,\mathrm{kN \cdot mm^2}$$
$$(EI)_{eff,z} = E_a I_{az} + 0.6E_{c,eff}I_{cz}$$
$$= [210 \times 102272 + 0.6 \times 22.8 \times 217728] \times 10^4 = 2.45 \times 10^{11}\,\mathrm{kN \cdot mm^2}$$

- Elastic critical Euler buckling resistance

$$N_{cr,y} = \frac{\pi^2(EI)_{eff,y}}{L_{eff}^2} = \frac{\pi^2 \times 4.78 \times 10^{11}}{6000^2} = 130977\,\mathrm{kN}$$
$$N_{cr,z} = \frac{\pi^2(EI)_{eff,z}}{L_{eff}^2} = \frac{\pi^2 \times 2.45 \times 10^{11}}{6000^2} = 67053\,\mathrm{kN}$$

- Characteristic plastic resistance of cross-section

$$N_{pl,Rk} = A_a f_y + A_c f_{ck} = [384 \times 355 + 2016 \times 50] \times 10^{-1} = 23712\,\mathrm{kN}$$

- Relative slenderness ratio

$$\overline{\lambda_y} = \sqrt{\frac{N_{pl,Rk}}{N_{cr,y}}} = \sqrt{\frac{23712}{130977}} = 0.425$$

$$\overline{\lambda_z} = \sqrt{\frac{N_{pl,Rk}}{N_{cr,z}}} = \sqrt{\frac{23712}{67053}} = 0.595$$

- Buckling curves and buckling reduction factors

The buckling curves about both axes are "a". Thus, the imperfection factor is $\alpha = 0.21$.

$$\bar{\lambda} = \max\left(\overline{\lambda_y}, \overline{\lambda_z}\right) = \max(0.425, 0.595) = 0.595$$

$$\Phi = 0.5\left[1 + \alpha\left(\bar{\lambda} - 0.2\right) + \bar{\lambda}^2\right]$$

$$= 0.5\left[1 + 0.21 \times (0.595 - 0.2) + 0.595^2\right] = 0.718$$

$$\chi = \min\left(\frac{1}{\Phi + \sqrt{\Phi^2 - \bar{\lambda}^2}}, 1.0\right)$$

$$= \min\left(\frac{1}{0.718 + \sqrt{0.718^2 - 0.595^2}}, 1.0\right) = 0.892$$

- Simplified Interaction Curves

1) Point A $(0, N_{\mathrm{pl,Rd}})$:

Confinement effect is not taken into account for rectangular CFST column, thus

$$N_{pl,Rd} = A_a f_{yd} + A_c f_{cd} = [384 \times 355 + 2016 \times 33.3] \times 10^{-1} = 20352\,\mathrm{kN}$$

Buckling resistance: $\chi N_{pl,Rd} = 0.892 \times 20352 = 18154$ kN

2) Point B $(M_{\mathrm{pl,y,Rd}}, 0)$ & $(M_{\mathrm{pl,z,Rd}}, 0)$:

The position of neutral axis are determined from

$$h_{ny} = \frac{A_c f_{cd}}{2B f_{cd} + 4t_a\left(2f_{yd} - f_{cd}\right)}$$

$$= \frac{201600 \times 33.3}{2 \times 400 \times 33.3 + 4 \times 20 \times (2 \times 355 - 33.3)} = 83.2\,\mathrm{mm}$$

$$h_{nz} = \frac{A_c f_{cd}}{2H f_{cd} + 4t_a\left(2f_{yd} - f_{cd}\right)}$$

$$= \frac{201600 \times 33.3}{2 \times 600 \times 33.3 + 4 \times 20 \times (2 \times 355 - 33.3)} = 71.4\,\mathrm{mm}$$

The plastic modulus of steel tube and concrete in the height of $2h_n$, bending about centreline of the cross-section are calculated as:

$$W_{ay,n} = 2t_a h_{ny}^2 = 2 \times 20 \times 83.2^2 \times 10^{-3} = 276.9\,\text{cm}^3$$
$$W_{az,n} = 2t_a h_{nz}^2 = 2 \times 20 \times 71.4^2 \times 10^{-3} = 204\,\text{cm}^3$$
$$W_{cy,n} = (B - 2t_a)h_{ny}^2 = (400 - 2 \times 20) \times 83.2^2 \times 10^{-3} = 2492\,\text{cm}^3$$
$$W_{cz,n} = (H - 2t_a)h_{nz}^2 = (600 - 2 \times 20) \times 71.4^2 \times 10^{-3} = 2855\,\text{cm}^3$$

Taking moment about the centerline of the cross-section, the plastic bending resistance is determined from:

$$\begin{aligned} M_{pl,y,Rd} &= \left(W_{ay} - W_{ay,n}\right) f_{yd} + 0.5\left(W_{cy} - W_{cy,n}\right) f_{cd} \\ &= [(7776 - 276.9) \times 355 + 0.5 \times (28224 - 2492) \times 33.3] \times 10^{-3} \\ &= 3091\,\text{kN} \cdot \text{m} \end{aligned}$$

$$\begin{aligned} M_{pl,z,Rd} &= \left(W_{az} - W_{az,n}\right) f_{yd} + 0.5\left(W_{cz} - W_{cz,n}\right) f_{cd} \\ &= [(5856 - 204) \times 355 + 0.5 \times (18144 - 2855) \times 33.3] \times 10^{-3} \\ &= 2261\,\text{kN} \cdot \text{m} \end{aligned}$$

3) Point C $(M_{\text{pl,y,Rd}}, N_{\text{pm,Rd}})$ & $(M_{\text{pl,z,Rd}}, N_{\text{pm,Rd}})$:

$$N_{pm,Rd} = A_c f_{cd} = 201600 \times 33.3 \times 10^{-3} = 6720\,\text{kN}$$

4) Point D $(M_{\text{max,y,Rd}}, N_{\text{pm,Rd}}/2)$ & $(M_{\text{max,z,Rd}}, N_{\text{pm,Rd}}/2)$:

$$\begin{aligned} M_{\text{max},y,Rd} &= W_{ay} f_{yd} + 0.5 W_{cy} f_{cd} \\ &= [7776 \times 355 + 0.5 \times 28224 \times 33.3] \times 10^{-3} \\ &= 3231\,\text{kN} \cdot \text{m} \end{aligned}$$

$$\begin{aligned} M_{\text{max},z,Rd} &= W_{az} f_{yd} + 0.5 W_{cz} f_{cd} \\ &= [5856 \times 355 + 0.5 \times 18144 \times 33.3] \times 10^{-3} \\ &= 2381\,\text{kN} \cdot \text{m} \end{aligned}$$

- Steel contribution ratio

$$\begin{aligned} \delta &= A_a f_{yd} / N_{pl,Rd} = 384 \times 355 \times 10^{-1} / 20352 \\ &= 0.67 < 0.9 \end{aligned}$$

- Check for resistance of column in axial compression

$$\frac{N_{Ed}}{\chi N_{pl,Rd}} = \frac{12000}{0.892 \times 20352} = 0.661 < 1.0$$

Thus, the buckling resistance under axial compression is adequate!

- Check for resistance of column in combined compression and biaxial bending moments

The design value of effective flexural stiffness with long-term effect is calculated from:

$$\begin{aligned}
(EI)_{eff,II,y} &= K_0\left(E_a I_{ay} + K_{e,II} E_{c,m} I_{cy}\right) \\
&= 0.9 \times [210 \times 193152 + 0.5 \times 22.8 \times 526848] \times 10^4 \\
&= 4.19 \times 10^{11}\,\text{kN} \cdot \text{mm}^2 \\
(EI)_{eff,II,z} &= K_0\left(E_a I_{az} + K_{e,II} E_{c,m} I_{cz}\right) \\
&= 0.9 \times [210 \times 102272 + 0.5 \times 22.8 \times 217728] \times 10^4 \\
&= 2.16 \times 10^{11}\,\text{kN} \cdot \text{mm}^2
\end{aligned}$$

Thus, the critical normal forces with effective length taken as the system length of column are determined from:

$$\begin{aligned}
N_{cr,eff,y} &= \frac{\pi^2 (EI)_{eff,II,y}}{L^2} = \frac{\pi^2 \times 4.19 \times 10^{11}}{6000^2} = 114913\,\text{kN} \\
N_{cr,eff,z} &= \frac{\pi^2 (EI)_{eff,II,z}}{L^2} = \frac{\pi^2 \times 2.16 \times 10^{11}}{6000^2} = 59122\,\text{kN}
\end{aligned}$$

The second-order effect should be considered for both moments from first-order analysis and moment from imperfection, as shown in Fig. B.6. According to the buckling curve "a" and refer to Table 4.7, the initial imperfections about y-y axis and z-z axis are:

$$\begin{aligned}
e_{0,y} &= L/300 = 6000/300 = 20\,\text{mm} \\
e_{0,z} &= L/300 = 6000/300 = 20\,\text{mm}
\end{aligned}$$

Accordingly, the bending moments by the initial imperfections are determined as:

$$\begin{aligned}
M_{0,y} &= N_{Ed} e_{0,y} = 12000 \times 20/1000 = 240\,\text{kN} \cdot \text{m} \\
M_{0,z} &= N_{Ed} e_{0,z} = 12000 \times 20/1000 = 240\,\text{kN} \cdot \text{m}
\end{aligned}$$

According to the moment diagram by the initial imperfection, the factor β_0 for determination of moment to second-order effect is equal to 1.0. Thus, the amplification factors for the moments by the imperfection are calculated from:

$$\begin{aligned}
k_{0,y} &= \frac{\beta_0}{1 - N_{Ed}/N_{cr,eff,y}} = \frac{1.0}{1 - 12000/114913} = 1.117 \\
k_{0,z} &= \frac{\beta_0}{1 - N_{Ed}/N_{cr,eff,z}} = \frac{1.0}{1 - 12000/59122} = 1.255
\end{aligned}$$

According to the first-order design moment diagram, the ratio of end moments is calculated as:

$$r_y = M_{t,y}/M_{b,y} = -550/900 = -0.611$$
$$r_z = M_{t,z}/M_{b,z} = 350/(-600) = -0.583$$

Thus, the factors for determination of moment to second-order effect are determined:

$$\beta_{1,y} = \max\left(0.66 + 0.44r_y, 0.44\right) = \max\left(0.66 + 0.44 \times (-0.611), 0.44\right) = 0.44$$
$$\beta_{1,z} = \max\left(0.66 + 0.44r_z, 0.44\right) = \max\left(0.66 + 0.44 \times (-0.583), 0.44\right) = 0.44$$

Thus, the amplification factors for the moment by the imperfection are calculated from:

$$k_{1,y} = \frac{\beta_{1,y}}{1 - N_{Ed}/N_{cr,eff,y}} = \frac{0.44}{1 - 12000/114913} = 0.491$$
$$k_{1,z} = \frac{\beta_{1,z}}{1 - N_{Ed}/N_{cr,eff,z}} = \frac{0.44}{1 - 12000/59122} = 0.552$$

Thus, the design moments, considering second-order effect, are calculated as:

$$\begin{aligned} M_{y,Ed} &= Max[k_{0,y}M_{0,y} + k_{1,y}Max(|M_{t,y}|, |M_{b,y}|), Max(|M_{t,y}|, |M_{b,y}|)] \\ &= Max[1.117 \times 240 + 0.491 \times Max(|-550|, |900|), Max(|-550|, |900|)] \\ &= 900\,\text{kN}\cdot\text{m} \end{aligned}$$

$$\begin{aligned} M_{z,Ed} &= Max[k_{0,z}M_{0,z} + k_{1,z}Max(|M_{t,z}|, |M_{b,z}|), Max(|M_{t,z}|, |M_{b,z}|)] \\ &= Max[1.255 \times 240 + 0.552 \times Max(|350|, |-600|), Max(|350|, |-600|)] \\ &= 632\,\text{kN}\cdot\text{m} \end{aligned}$$

For $N_{Ed} = 12{,}000\ kN > N_{pm,Rd} = 6720\ kN$, the values for determining the plastic bending resistances $M_{pl,N,y,Rd}$ and $M_{pl,N,z,Rd}$ taking into account the normal force N_{Ed} are calculated from:

$$\mu_{dy} = \mu_{dz} = \frac{N_{pl,Rd} - N_{Ed}}{N_{pl,Rd} - N_{pm,Rd}} = \frac{20352 - 12000}{20352 - 6720} = 0.613$$
$$\frac{M_{y,Ed}}{M_{pl,N,y,Rd}} = \frac{M_{y,Ed}}{\mu_{dy}M_{pl,y,Rd}} = \frac{900}{0.613 \times 3091} = 0.475 < \alpha_{M,y} = 0.9$$
$$\frac{M_{z,Ed}}{M_{pl,N,z,Rd}} = \frac{M_{z,Ed}}{\mu_{dz}M_{pl,z,Rd}} = \frac{632}{0.613 \times 2261} = 0.456 < \alpha_{M,z} = 0.9$$
$$\frac{M_{y,Ed}}{\mu_{dy}M_{pl,y,Rd}} + \frac{M_{z,Ed}}{\mu_{dz}M_{pl,z,Rd}} = \frac{900}{0.613 \times 3091} + \frac{632}{0.613 \times 2261} = 0.932 < 1.0$$

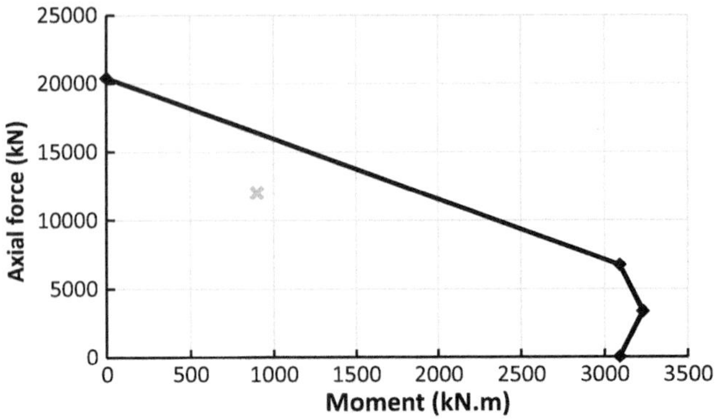

Fig. B.11 Design M–N curve for bending about y-y axis.

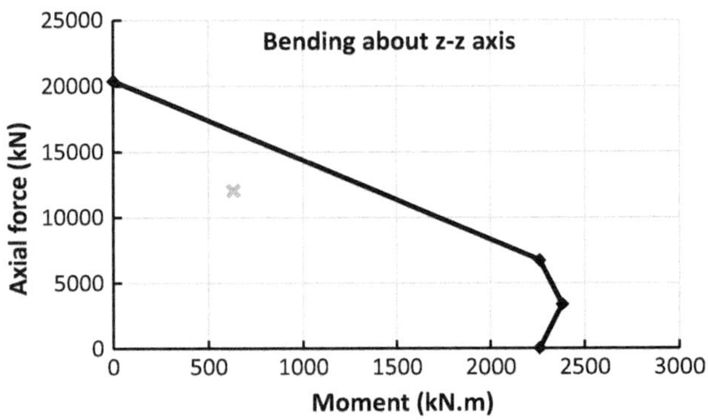

Fig. B.12 Design M–N curve for bending about z-z axis.

Thus, the resistance for combined axial compression and biaxial bending is adequate. The external design force and bending moment, and M-N interaction curves are plotted in Fig. B.11, Fig. B.12 and Fig. B.13.

B.3.3 RHS 400 x 600 x 20 - S355 steel tube infilled with C90/105 concrete

In this section, the high strength concrete C50/60 is replaced by higher strength concrete C90/105. The steel grade is not changed.

- Design strength

Effective compressive strength and modulus of elasticity are taken from to Table 3.1, Eq. (3.1) and Eq. (3.2).

$$f_{ck} = 72\,\text{N/mm}^2;\; E_{cm} = 41.1\,\text{GPa}$$

$$f_{cd} = f_{ck}/\gamma_c = 72/1.5 = 48\,\text{N/mm}^2;\; f_{cm} = f_{ck} + 8 = 72 + 8 = 80\,\text{N/mm}^2$$

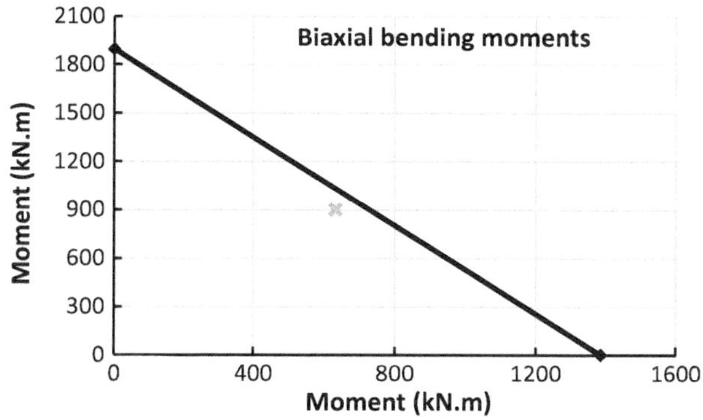

Fig. B.13 Check for bi-axial bending.

- Elastic modulus of concrete considering long-term effect

  Creep coefficient could be similarly determined as $\phi_t = 1.12$

$$E_{c,eff} = \frac{E_{cm}}{1 + (N_{G,Ed}/N_{Ed})\varphi_t} = \frac{41.1}{1 + (5000/12000) \times 1.12} = 28\,\text{GPa}$$

- Effective flexural stiffness of cross-section

$$(EI)_{eff,y} = E_a I_{ay} + 0.6 E_{c,eff} I_{cy} = [210 \times 193152 + 0.6 \times 28 \times 526848] \times 10^4 = 4.94 \times 10^{11}\,\text{kN} \cdot \text{mm}^2$$

$$(EI)_{eff,z} = E_a I_{az} + 0.6 E_{c,eff} I_{cz} = [210 \times 102272 + 0.6 \times 28 \times 217728] \times 10^4 = 2.51 \times 10^{11}\,\text{kN} \cdot \text{mm}^2$$

- Characteristic plastic resistance of cross-section

$$N_{cr,y} = \frac{\pi^2 (EI)_{eff,y}}{L_{eff}^2} = \frac{\pi^2 \times 4.94 \times 10^{11}}{6000^2} = 135431\,\text{kN}$$

$$N_{cr,z} = \frac{\pi^2 (EI)_{eff,z}}{L_{eff}^2} = \frac{\pi^2 \times 2.51 \times 10^{11}}{6000^2} = 68893\,\text{kN}$$

$$N_{pl,Rk} = A_a f_y + A_c f_{ck} = [384 \times 355 + 2016 \times 90] \times 10^{-1} = 28147\,\text{kN}$$

- Relative slenderness ratio, buckling curves and buckling reduction factors

$$\overline{\lambda_y} = \sqrt{\frac{N_{pl,Rk}}{N_{cr,y}}} = \sqrt{\frac{28147}{135431}} = 0.456; \ \overline{\lambda_z} = \sqrt{\frac{N_{pl,Rk}}{N_{cr,z}}} = \sqrt{\frac{28147}{68893}} = 0.639$$

$$\bar{\lambda} = \max\left(\overline{\lambda_y}, \overline{\lambda_z}\right) = 0.639$$

$$\Phi = 0.5[1 + \alpha(\bar{\lambda} - 0.2) + \bar{\lambda}^2]$$

$$= 0.5[1 + 0.21 \times (0.639 - 0.2) + 0.639^2] = 0.75$$

$$\chi = \min\left(\frac{1}{\Phi + \sqrt{\Phi^2 - \bar{\lambda}^2}}, 1.0\right)$$

$$= \min\left(\frac{1}{0.75 + \sqrt{0.75^2 - 0.639^2}}, 1.0\right) = 0.875$$

$$N_{pl,Rd} = A_a f_{yd} + A_c f_{cd} = [384 \times 355 + 2016 \times 41.1] \times 10^{-1} = 23309\,\text{kN}$$

Buckling resistance: $\chi N_{pl,Rd} = 0.875 \times 23309 = 20395\text{kN}$

- Steel contribution ratio

$$\delta = A_a f_{yd} / N_{pl,Rd} = 384 \times 355 \times 10^{-1}/23309 = 0.585 < 0.9$$

- M–N interaction curve

$$h_{ny} = \frac{A_c f_{cd}}{2B f_{cd} + 4t_a(2f_{yd} - f_{cd})}$$

$$= \frac{201600 \times 41.1}{2 \times 400 \times 41.1 + 4 \times 20 \times (2 \times 355 - 41.1)} = 105.9\,\text{mm}$$

$$h_{nz} = \frac{A_c f_{cd}}{2H f_{cd} + 4t_a(2f_{yd} - f_{cd})}$$

$$= \frac{201600 \times 41.1}{2 \times 600 \times 41.1 + 4 \times 20 \times (2 \times 355 - 41.1)} = 87.5\,\text{mm}$$

$$W_{ay,n} = 2t_a h_{ny}^2 = 2 \times 20 \times 105.9^2 \times 10^{-3} = 448.6\,\text{cm}^3$$

$$W_{az,n} = 2t_a h_{nz}^2 = 2 \times 20 \times 87.5^2 \times 10^{-3} = 306.3\,\text{cm}^3$$

$$W_{cy,n} = (B - 2t_a)h_{ny}^2 = (400 - 2 \times 20) \times 105.9^2 \times 10^{-3} = 4037.3\,\text{cm}^3$$

$$W_{cz,n} = (H - 2t_a)h_{nz}^2 = (600 - 2 \times 20) \times 87.5^2 \times 10^{-3} = 4287.5\,\text{cm}^3$$

$$M_{pl,y,Rd} = (W_{ay} - W_{ay,n})f_{yd} + 0.5(W_{cy} - W_{cy,n})f_{cd}$$

$$= [(7776 - 448.6) \times 355 + 0.5 \times (28224 - 4037.3) \times 41.1] \times 10^{-3}$$

$$= 3182\,\text{kN} \cdot \text{m}$$

$$M_{pl,z,Rd} = (W_{az} - W_{az,n})f_{yd} + 0.5(W_{cz} - W_{cz,n})f_{cd}$$

$$= [(5856 - 306.3) \times 355 + 0.5 \times (18144 - 4287.5) \times 41.1] \times 10^{-3}$$

$$= 2303\,\text{kN} \cdot \text{m}$$

$$N_{pm,Rd} = A_c f_{cd} = 201600 \times 41.1 \times 10^{-3} = 9677\,\text{kN}$$

$$M_{\max,y,Rd} = W_{ay} f_{yd} + 0.5 W_{cy} f_{cd}$$

$$= [7776 \times 355 + 0.5 \times 28224 \times 41.1] \times 10^{-3} = 3438\,\text{kN} \cdot \text{m}$$

$$M_{\max,z,Rd} = W_{az} f_{yd} + 0.5 W_{cz} f_{cd}$$

$$= [5856 \times 355 + 0.5 \times 18144 \times 41.1] \times 10^{-3} = 2514\,\text{kN} \cdot \text{m}$$

B.3.4 RHS 400 x 600 x 20 - S500 steel tube infilled with C50/60 concrete

In this section, the mild steel S355 is replaced by the high tensile steel (HTS) S500, the concrete grade is not changed.

- Plastic resistance of cross-section

$$N_{pl,Rk} = A_a f_y + A_c f_{ck} = [384 \times 500 + 2016 \times 50] \times 10^{-1} = 29280\,\text{kN}$$

$$\overline{\lambda_y} = \sqrt{\frac{N_{pl,Rk}}{N_{cr,y}}} = \sqrt{\frac{29280}{130977}} = 0.473;\ \overline{\lambda_z} = \sqrt{\frac{N_{pl,Rk}}{N_{cr,z}}} = \sqrt{\frac{29280}{67053}} = 0.661$$

$$\bar{\lambda} = \max\left(\overline{\lambda_y}, \overline{\lambda_z}\right) = 0.661$$

$$\Phi = 0.5[1 + \alpha(\bar{\lambda} - 0.2) + \bar{\lambda}^2]$$
$$= 0.5[1 + 0.21 \times (0.661 - 0.2) + 0.661^2] = 0.767$$

$$\chi = \min\left(\frac{1}{\Phi + \sqrt{\Phi^2 - \bar{\lambda}^2}}, 1.0\right) = \min\left(\frac{1}{0.767 + \sqrt{0.767^2 - 0.661^2}}, 1.0\right)$$
$$= 0.865$$

$$N_{pl,Rd} = A_a f_{yd} + A_c f_{cd} = [384 \times 500 + 2016 \times 33.3] \times 10^{-1} = 25920\,\text{kN}$$

Buckling resistance: $\chi N_{pl,Rd} = 0.865 \times 25920 = 22421\text{kN}$

- Steel contribution ratio

$$\delta = A_a f_{yd}/N_{pl,Rd} = 384 \times 500 \times 10^{-1}/22421 = 0.741 < 0.9$$

- M-N interaction curve

$$h_{ny} = \frac{A_c f_{cd}}{2B f_{cd} + 4t_a\left(2f_{yd} - f_{cd}\right)}$$
$$= \frac{201600 \times 33.3}{2 \times 400 \times 33.3 + 4 \times 20 \times (2 \times 500 - 33.3)} = 64.6\,\text{mm}$$

$$h_{nz} = \frac{A_c f_{cd}}{2H f_{cd} + 4t_a\left(2f_{yd} - f_{cd}\right)}$$
$$= \frac{201600 \times 33.3}{2 \times 600 \times 33.3 + 4 \times 20 \times (2 \times 500 - 33.3)} = 57.3\,\text{mm}$$

$$W_{ay,n} = 2t_a h_{ny}^2 = 2 \times 20 \times 64.6^2 \times 10^{-3} = 170\,\text{cm}^3$$

$$W_{az,n} = 2t_a h_{nz}^2 = 2 \times 20 \times 57.3^2 \times 10^{-3} = 131.3\,\text{cm}^3$$

$$W_{cy,n} = (B - 2t_a)h_{ny}^2 = (400 - 2 \times 20) \times 64.6^2 \times 10^{-3} = 1502.3\,\text{cm}^3$$

$$W_{cz,n} = (H - 2t_a)h_{nz}^2 = (600 - 2 \times 20) \times 57.3^2 \times 10^{-3} = 1838.6\,\text{cm}^3$$

$$M_{pl,y,Rd} = \left(W_{ay} - W_{ay,n}\right)f_{yd} + 0.5\left(W_{cy} - W_{cy,n}\right)f_{cd}$$
$$= [(7776 - 170) \times 500 + 0.5 \times (28224 - 1502.3) \times 33.3] \times 10^{-3}$$
$$= 4250\,\text{kN} \cdot \text{m}$$

$$M_{pl,z,Rd} = \left(W_{az} - W_{az,n}\right)f_{yd} + 0.5\left(W_{cz} - W_{cz,n}\right)f_{cd}$$

$$= [(5856 - 131.3) \times 500 + 0.5 \times (18144 - 1838.6) \times 33.3] \times 10^{-3}$$
$$= 3134\,\text{kN} \cdot \text{m}$$

$$N_{pm,Rd} = A_c f_{cd} = 201600 \times 33.3 \times 10^{-3} = 6720\,\text{kN}$$

$$M_{\max,y,Rd} = W_{ay} f_{yd} + 0.5 W_{cy} f_{cd}$$
$$= [7776 \times 500 + 0.5 \times 28224 \times 33.3] \times 10^{-3} = 4358\,\text{kN} \cdot \text{m}$$

$$M_{\max,z,Rd} = W_{az} f_{yd} + 0.5 W_{cz} f_{cd}$$
$$= [5856 \times 500 + 0.5 \times 18144 \times 33.3] \times 10^{-3} = 3230\,\text{kN} \cdot \text{m}$$

B.3.5 RHS 400 x 600 x 20 - S500 steel tube infilled with C90/105 concrete

In this section, the mild steel S355 is replaced by the high tensile steel (HTS) S500, and the normal strength concrete C50/60 is replaced by high strength concrete C90/105.

- Plastic resistance of cross-section

$$N_{pl,Rk} = A_a f_y + A_c f_{ck} = [384 \times 500 + 2016 \times 90] \times 10^{-1} = 33715\,\text{kN}$$

$$\overline{\lambda_y} = \sqrt{\frac{N_{pl,Rk}}{N_{cr,y}}} = \sqrt{\frac{33715}{135431}} = 0.499;\ \overline{\lambda_z} = \sqrt{\frac{N_{pl,Rk}}{N_{cr,z}}} = \sqrt{\frac{33715}{68893}} = 0.7$$

$$\bar{\lambda} = \max\left(\overline{\lambda_y}, \overline{\lambda_z}\right) = 0.7$$

$$\Phi = 0.5\left[1 + \alpha\left(\bar{\lambda} - 0.2\right) + \bar{\lambda}^2\right] = 0.5\left[1 + 0.21 \times (0.7 - 0.2) + 0.7^2\right] = 0.797$$

$$\chi = \min\left(\frac{1}{\Phi + \sqrt{\Phi^2 - \bar{\lambda}^2}}, 1.0\right)$$
$$= \min\left(\frac{1}{0.797 + \sqrt{0.797^2 - 0.7^2}}, 1.0\right) = 0.848$$

$$N_{pl,Rd} = A_a f_{yd} + A_c f_{cd} = [384 \times 500 + 2016 \times 41.1] \times 10^{-1} = 28877\,\text{kN}$$

Buckling resistance: $\chi N_{pl,Rd} = 0.848 \times 28877 = 24488\text{kN}$

- Steel contribution ratio

$$\delta = A_a f_{yd} / N_{pl,Rd} = 384 \times 500 \times 10^{-1} / 24488 = 0.665 < 0.9$$

- M–N interaction curve

$$h_{ny} = \frac{A_c f_{cd}}{2B f_{cd} + 4t_a\left(2f_{yd} - f_{cd}\right)}$$
$$= \frac{201600 \times 41.1}{2 \times 400 \times 41.1 + 4 \times 20 \times (2 \times 500 - 41.1)} = 84.5\,\text{mm}$$

$$h_{nz} = \frac{A_c f_{cd}}{2H f_{cd} + 4t_a\left(2f_{yd} - f_{cd}\right)}$$
$$= \frac{201600 \times 41.1}{2 \times 600 \times 41.1 + 4 \times 20 \times (2 \times 500 - 41.1)} = 72.3\,\text{mm}$$

Table B.2 Comparison between rectangular CFST columns with various material strengths.

Sections	Steel contribution ratios	Design resistances				
		$\chi N_{\mathrm{pl,Rd}}$	$M_{\mathrm{pl,y,Rd}}$	$M_{\mathrm{pl,z,Rd}}$	$M_{\mathrm{max,y,Rd}}$	$M_{\mathrm{max,z,Rd}}$
S355 + C50/60	0.670	0	0	0	0	0
S355 + C90/105	0.585	12.3%	2.9%	1.8%	6.4%	5.6%
S500 + C50/60	0.741	23.5%	37.5%	38.6%	34.9%	35.6%
S500 + C90/105	0.665	34.9%	41.1%	41.0%	41.3%	41.2%

$$W_{ay,n} = 2t_a h_{ny}^2 = 2 \times 20 \times 84.5^2 \times 10^{-3} = 285.6\,\mathrm{cm}^3$$

$$W_{az,n} = 2t_a h_{nz}^2 = 2 \times 20 \times 72.3^2 \times 10^{-3} = 209.1\,\mathrm{cm}^3$$

$$W_{cy,n} = (B - 2t_a)h_{ny}^2 = (400 - 2 \times 20) \times 84.5^2 \times 10^{-3} = 2570.5\,\mathrm{cm}^3$$

$$W_{cz,n} = (H - 2t_a)h_{nz}^2 = (600 - 2 \times 20) \times 72.3^2 \times 10^{-3} = 2927.3\,\mathrm{cm}^3$$

$$\begin{aligned} M_{pl,y,Rd} &= \left(W_{ay} - W_{ay,n}\right) f_{yd} + 0.5\left(W_{cy} - W_{cy,n}\right) f_{cd} \\ &= [(7776 - 285.6) \times 500 + 0.5 \times (28224 - 2570.5) \times 41.1] \times 10^{-3} \\ &= 4361\,\mathrm{kN \cdot m} \end{aligned}$$

$$\begin{aligned} M_{pl,z,Rd} &= \left(W_{az} - W_{az,n}\right) f_{yd} + 0.5\left(W_{cz} - W_{cz,n}\right) f_{cd} \\ &= [(5856 - 209.1) \times 500 + 0.5 \times (18144 - 2927.3) \times 41.1] \times 10^{-3} \\ &= 3188\,\mathrm{kN \cdot m} \end{aligned}$$

$$N_{pm,Rd} = A_c f_{cd} = 201600 \times 41.1 \times 10^{-3} = 9677\,\mathrm{kN}$$

$$\begin{aligned} M_{\max,y,Rd} &= W_{ay} f_{yd} + 0.5 W_{cy} f_{cd} \\ &= [7776 \times 500 + 0.5 \times 28224 \times 41.1] \times 10^{-3} = 4565\,\mathrm{kN \cdot m} \end{aligned}$$

$$\begin{aligned} M_{\max,z,Rd} &= W_{az} f_{yd} + 0.5 W_{cz} f_{cd} \\ &= [5856 \times 500 + 0.5 \times 18144 \times 41.1] \times 10^{-3} = 3363\,\mathrm{kN \cdot m} \end{aligned}$$

B.3.6 Comparison and summary

The design resistances are compared for the aforementioned four rectangular composite sections as shown in Table B.2, Fig. B.14 and Fig. B.15. The composite section with steel tube of S355 and concrete of C50/60 is referred to for comparison.

By using high strength concrete C90/105 replacing normal strength concrete C50/60, the axial buckling resistance ($\chi N_{\mathrm{pl,Rd}}$) of the CFST column is improved by 12.3%, and the increase of moment resistances ($M_{\mathrm{pl,y,Rd}}$, $M_{\mathrm{pl,z,Rd}}$, $M_{\mathrm{max,y,Rd}}$, $M_{\mathrm{max,z,Rd}}$) are smaller (less than 7%).

By using high strength steel S500 to replace S355 steel, the axial buckling resistance is improved by 23.5%, and the increase of moment resistances are larger (higher than 34.9%). By using high strength concrete C90/105 to replace normal strength concrete C50/60 and use of S500 steel to replace S355 steel, the axial buckling resistance is further improved to 34.9%, and the increase of moment resistances are more than 40%.

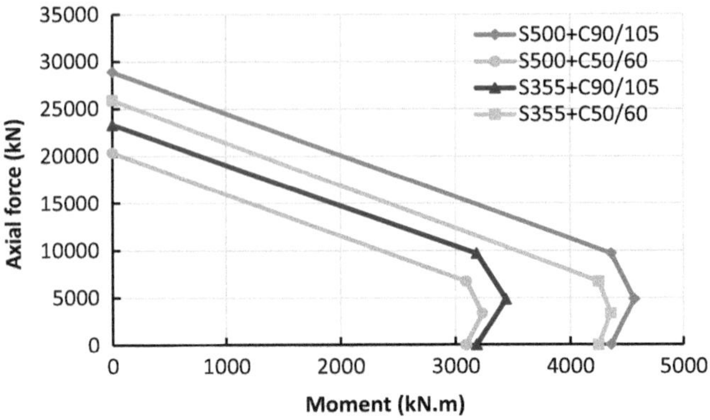

Fig. B.14 Design M–N interaction curves about y-y axis.

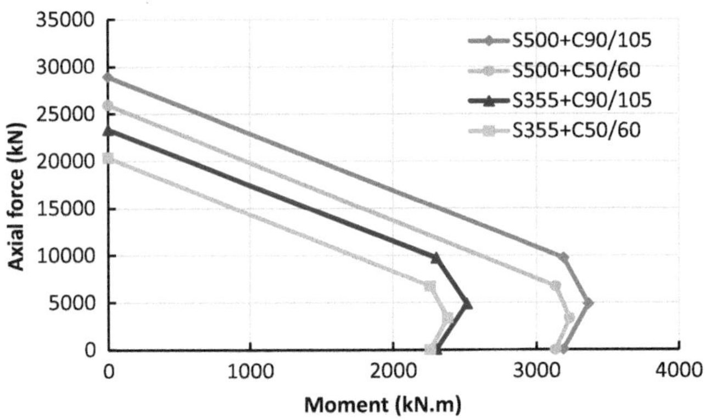

Fig. B.15 Design M–N interaction curves about z-z axis.

B.4 Concrete encased steel member subject to axial compression and bending

B.4.1 General

A concrete encased steel member is subject to axial compression and bending about the major axis. The following steel, concrete and reinforcing materials are used:

a) UC 254 × 254 × 107 S355 sections with C50/60 concrete and G5000 reinforcements
b) UC 254 × 254 × 107 S355 sections with C90/105 concrete and G500 reinforcements
c) UC 254 × 254 × 107 S500 sections with C50/60 concrete and G500 reinforcements

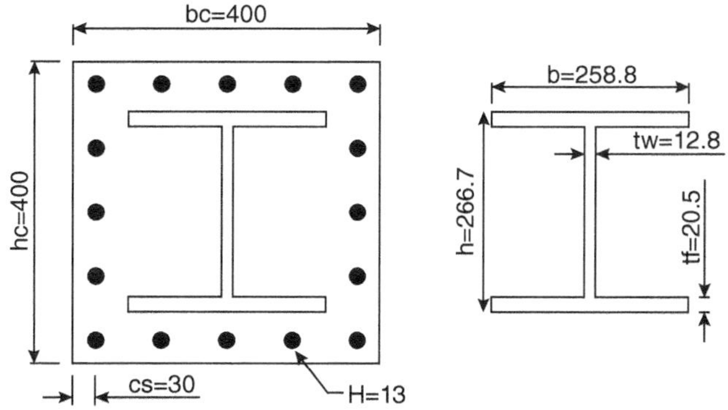

B.4.2 UC 254 x 254 x 107 S355 sections with C50/60 concrete and G500 reinforcements

- Design parameters

Concrete	C50/60, $f_{ck} = 50$ N/mm^2
Embedded steel section	Grade S355, $f_{ek} = 355$ N/mm^2
Reinforcements	Grade 500, $f_{sk} = 500$ N/mm^2
Column system length	$L = 4000$ mm
Effective length	$L_{eff} = 4000$ mm
Total design axial load	$N_{Ed} = 6000$ kN
Design axial load that is permanent	$N_{G,Ed} = 4000$ kN
Design moment at bottom around y-y axis	$M_{b,y} = 400$ kN.m
Design moment at top around y-y axis	$M_{t,y} = -200$ kN.m

- Design strength and modulus

$$f_{sd} = f_{sk}/\gamma_s = 500/1.15 = 435 \text{ N/mm}^2$$
$$f_{ed} = f_{ek}/\gamma_a = 355/1.0 = 355 \text{ N/mm}^2$$
$$f_{cd} = f_{ck}/\gamma_c = 50/1.5 = 33.3 \text{ N/mm}^2$$
$$f_{cm} = f_{ck} + 8 = 50 + 8 = 58 \text{ N/mm}^2$$
$$E_s = E_e = 210 \text{ GPa}$$
$$E_{cm} = 22(f_{cm}/10)^{0.3} = 22(58/10)^{0.3} = 37.3 \text{ GPa}$$

- Cross sectional areas

$$A = b_c h_c = 400 \times 400 = 160000 \text{ mm}^2$$
$$A_e = bh - (b - t_w)(h - 2t_f)$$
$$= 258.8 \times 266.7 - (258.8 - 12.8)(266.7 - 2 \times 20.5) = 13500 \text{ mm}^2$$
$$A_s = 16(\pi/4)d^2 = 16 \times (\pi/4) \times 13^2 = 2120 \text{ mm}^2$$
$$A_c = A - A_e - A_s = 160000 - 13500 - 2120 = 144380 \text{ mm}^2$$

- Second moment of areas

For simplicity, the reinforcements are equivalently converted to a square tube based on the same cross-sectional area and position of centerline.

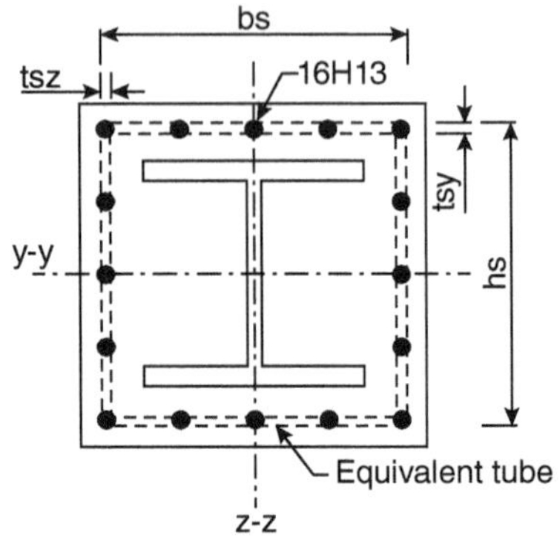

$$t_{sy} = 5(\pi/4)d^2/(b_c - 2cs) = 5(\pi/4) \times 13^2/(400 - 2 \times 30) = 2 \text{ mm}$$
$$t_{sz} = 5(\pi/4)d^2/(h_c - 2cs) = 5(\pi/4) \times 13^2/(400 - 2 \times 30) = 2 \text{ mm}$$
$$h_s = (h_c - 2cs) + t_{sy} = 342 \text{ mm}$$
$$b_s = (b_c - 2cs) + t_{sz} = 342 \text{ mm}$$
$$I_{sy} = \frac{1}{12}\left[b_s h_s^3 - (b_s - 2t_{sz})(h_s - 2t_{sy})^3\right] = 5115 \text{ cm}^4$$
$$I_{sz} = \frac{1}{12}\left[h_s b_s^3 - (h_s - 2t_{sy})(b_s - 2t_{sz})^3\right] = 5115 \text{ cm}^4$$
$$I_{ey} = \frac{1}{12}\left[bh^3 - (b - t_w)(h - 2t_f)^3\right]$$
$$= \frac{1}{12}\left[258.8 \times 266.7^3 - (258.8 - 12.8)(266.7 - 2 \times 20.5)^3\right]$$
$$= 17343 \text{ cm}^4$$
$$I_{ez} = \frac{1}{12}\left[2t_f b^3 + (h - 2t_f){t_w}^3\right]$$
$$= \frac{1}{12}\left[2 \times 20.5 \times 258.8^3 + (266.7 - 2 \times 20.5) \times 12.8^3\right] = 5926 \text{ cm}^4$$
$$I_{cy} = I - I_{sy} - I_{ey} = 213333 - 5115 - 17343 = 190876 \text{ cm}^4$$
$$I_{cz} = I - I_{sz} - I_{ez} = 213333 - 5115 - 5926 = 202292 \text{ cm}^4$$

Plastic modulus

$$W_y = b_c h_c^2/4 = 400^3/4 = 16000 \text{ cm}^3$$
$$W_z = h_c b_c^2/4 = 400^3/4 = 16000 \text{ cm}^3$$
$$W_{sy} = \frac{1}{4}\left[b_s h_s^2 - (b_s - 2t_{sz})(h_s - 2t_{sy})^2\right]$$

$$= \frac{1}{4}\left[342^3 - (342 - 2 \times 2)^3\right] = 338 \text{ cm}^3$$

$$W_{sz} = \frac{1}{4}\left[h_s b_s^2 - (h_s - 2t_{sy})(b_s - 2t_{sz})^2\right]$$

$$= \frac{1}{4}\left[342^3 - (342 - 2 \times 2)^3\right] = 338 \text{ cm}^3$$

$$W_{ey} = \frac{1}{4}\left[bh^2 - (b - t_w)(h - 2t_f)^2\right]$$

$$= \frac{1}{4}\left[258.8 \times 266.7^2 - (258.8 - 12.8)(266.7 - 2 \times 20.5)^2\right] = 1469 \text{ cm}^3$$

$$W_{ez} = \frac{1}{4}\left[2t_f b^2 + (h - 2t_f)t_w^2\right]$$

$$= \frac{1}{4}\left[2 \times 20.5 \times 258.8^2 + (266.7 - 2 \times 20.5) \times 12.8^2\right] = 696 \text{ cm}^3$$

$$W_{cy} = W_y - W_{sy} - W_{ey} = 16000 - 338 - 1469 = 14192 \text{ cm}^3$$

$$W_{cz} = W_z - W_{sz} - W_{ez} = 16000 - 338 - 696 = 14966 \text{ cm}^3$$

- Long-term effect

  Age of concrete at loading in day: $t_0 = 30$
  Age of concrete at the moment considered in days: $t = \infty$
  Relative humidity of ambient environment: RH = 80%
  Perimeter of concrete section: $u = 2b_c + 2h_c = 1600$ mm
  Notional size of concrete section: $h_0 = 2A_c/u = 2 \times 144400/1600 = 180$ mm
  Coefficient: $\alpha_1 = (35/f_{cm})^{0.7} = (35/58)^{0.7} = 0.70$
  Coefficient: $\alpha_2 = (35/f_{cm})^{0.2} = (35/58)^{0.2} = 0.90$
  Coefficient: $\alpha_3 = (35/f_{cm})^{0.5} = (35/58)^{0.5} = 0.78$
  Factor: $\phi_{RH} = \left(1 + \frac{1-RH/100}{0.1\sqrt[3]{h_0}}\alpha_1\right)\alpha_2 = \left(1 + \frac{1-80/100}{0.1\sqrt[3]{180.5}}0.70\right) \times 0.90 = 1.13$
  Factor: $\beta(f_{cm}) = 16.8/\sqrt{f_{cm}} = 16.8/\sqrt{58} = 2.21$
  Factor: $\beta(t_0) = 1/(0.1 + t_0^{0.2}) = 1/(0.1 + 30^{0.2}) = 0.48$
  Factor: $\phi_0 = \phi_{RH}\beta(f_{cm})\beta(t_0) = 1.13 \times 2.21 \times 0.48 = 1.20$
  Factor: $\beta_H = 1.5\left[1 + (0.012RH)^{18}\right]h_0 + 250\alpha_3$
  $= 1.5\left[1 + (0.012 \times 80)^{18}\right] \times 180 + 250 \times 0.78 = 595$
  Factor: $\beta_c(t, t_0) = \left(\frac{t-t_0}{\beta_H+t-t_0}\right)^{0.3} = 1$
  Creep coefficient: $\phi_t = \phi_0\beta_c(t,t_0) = 1.20 \times 1 = 1.20$

- Elastic modulus of concrete considering long-term effect

$$E_{c,eff} = E_{cm}\frac{1}{1 + (N_{G,Ed}/N_{Ed})\phi_t} = \frac{37.3}{1 + (4000/6000) \times 1.20} = 20.7\,\text{GPa}$$

- Effective flexural stiffness of cross-section

$$(EI)_{eff,y} = E_s I_{sy} + E_e I_{ey} + 0.6E_{c,eff} I_{cy}$$

$$
\begin{aligned}
&= [210 \times (5115 + 17343) + 0.6 \times 20.7 \times 190876] \times 10^4 \\
&= 7.09 \times 10^{10}\ \text{kN} \cdot \text{mm}^2 \\
(EI)_{eff,z} &= E_s I_{sz} + E_e I_{ez} + 0.6 E_{c,eff} I_{cz} \\
&= [210 \times (5115 + 5926) + 0.6 \times 20.7 \times 202292] \times 10^4 \\
&= 4.83 \times 10^{10}\ \text{kN} \cdot \text{mm}^2
\end{aligned}
$$

- Elastic critical Euler buckling resistance

$$
N_{cr,y} = \frac{\pi^2 (EI)_{eff,y}}{L_{eff}^2} = \frac{\pi^2 \times 7.09 \times 10^{10}}{4000^2} = 43721\ \text{kN}
$$

$$
N_{cr,z} = \frac{\pi^2 (EI)_{eff,z}}{L_{eff}^2} = \frac{\pi^2 \times 4.83 \times 10^{10}}{4000^2} = 29807\ \text{kN}
$$

- Characteristic plastic resistance of cross-section

$$
\begin{aligned}
N_{pl,Rk} &= A_s f_{sk} + A_e f_{ek} + 0.85 A_c f_{ck} \\
&= 2120 \times 500 + 13500 \times 355 + 0.85 \times 144400 \times 50 = 11990\ \text{kN}
\end{aligned}
$$

- Relative slenderness ratio

$$
\overline{\lambda_y} = \sqrt{\frac{N_{pl,Rk}}{N_{cr,y}}} = \sqrt{\frac{11990}{43721}} = 0.522
$$

$$
\overline{\lambda_z} = \sqrt{\frac{N_{pl,Rk}}{N_{cr,z}}} = \sqrt{\frac{11990}{29807}} = 0.632
$$

- Buckling curves and buckling reduction factors

For a fully encased steel column, the buckling curve about major axis is "b", and about minor axis is "c". Thus, the imperfection factor is $\alpha = 0.34$ and $\alpha = 0.49$, respectively.

$$
\begin{aligned}
\Phi_y &= 0.5\left[1 + \alpha_y(\overline{\lambda_y} - 0.2) + \overline{\lambda_y}^2\right] \\
&= 0.5\left[1 + 0.34 \times (0.522 - 0.2) + 0.522^2\right] = 0.691 \\
\chi_y &= \min\left(\frac{1}{\Phi_y + \sqrt{\Phi_y^2 - \overline{\lambda_y^2}}}, 1.0\right) \\
&= \min\left(\frac{1}{0.691 + \sqrt{0.691^2 - 0.522^2}}, 1.0\right) = 0.874 \\
\Phi_z &= 0.5\left[1 + \alpha_z(\overline{\lambda_z} - 0.2) + \overline{\lambda_z}^2\right] \\
&= 0.5\left[1 + 0.49 \times (0.632 - 0.2) + 0.632^2\right] = 0.806
\end{aligned}
$$

$$\chi_z = \min\left(\frac{1}{\Phi_z + \sqrt{\Phi_z^2 - \overline{\lambda}_z^2}}, 1.0\right)$$
$$= \min\left(\frac{1}{0.806 + \sqrt{0.806^2 - 0.632^2}}, 1.0\right) = 0.766$$
$$\chi = \min\left(\chi_y, \chi_z\right) = \min(0.874, 0.766) = 0.766$$

- Simplified Interaction Curve

1) Point A (0, $N_{\mathrm{pl,Rd}}$):

Full cross-section is under uniform compression. No bending moment is resultant from the compressive stresses on the cross-section.

$$N_{pl,Rd} = A_s f_{sd} + A_e f_{ed} + 0.85 A_c f_{cd}$$
$$= 2120 \times 435 + 13500 \times 355 + 0.85 \times 144400 \times 33.3 = 9806 \text{ kN}$$

2) Point B ($M_{\mathrm{pl,Rd}}$, 0):

The cross-section is under partial compression and no axial force is formed. Assuming the neutral axis lies in the web of encased section ($h_n \leq h/2 - t_f$), the height of neutral axis is calculated where the areas of the equivalent tube for reinforcements, and concrete in the height of $2h_n$ are approximated as rectangles, and based on the force equilibrium between the tensile capacity of steel sections within the height of $2h_n$ is equal to the compression resistance of concrete in the compression zone. Unless other stated, the tensile resistance of concrete in the tension zone is conservatively ignored.

$$h_n = \frac{0.85 A_c f_{cd}}{2 \times 0.85 f_{cd} b_c + 4 t_{sz}(2 f_{sd} - 0.85 f_{cd}) + 2 t_w (2 f_{ed} - 0.85 f_{cd})}$$
$$= \frac{0.85 \times 144400 \times 33.3}{2 \times 0.85 \times 33.3 \times 400 + 4 \times 2 \times (2 \times 400 - 0.85 \times 33.3) + 2 \times 12.8 \times (2 \times 355 - 0.85 \times 33.3)}$$
$$= 88.7 \text{ mm}$$

$h_n = 88.7$ mm $< h/2 - t_f = 266.7/2 - 20.5 = 112.85$ mm, thus, the neutral axial lies in the web of the encased section.

The plastic modulus of the equivalent tube of reinforcements, encased section, and concrete in the height of $2h_n$, bending about centreline of the cross-section are calculated as:

$$W_{sy,n} = 2 t_{sz} h_n^2 = 2 \times 2 \times 88.7^2 \times 10^{-3} = 31.5 \text{ cm}^3$$
$$W_{ey,n} = t_w h_n^2 = 12.8 \times 88.7^2 \times 10^{-3} = 100.7 \text{ cm}^3$$
$$W_{cy,n} = (b_c - 2 t_{sz} - t_w) h_n^2 = (400 - 2 \times 2 - 12.8) \times 88.7^2 \times 10^{-3} = 3015 \text{ cm}^3$$

Taking moment about the centerline of the cross-section, the plastic bending resistance is determined from:

$$\begin{aligned} M_{pl,Rd} &= (W_{sy} - W_{sy,n})f_{sd} + (W_{ey} - W_{ey,n})f_{ed} + 0.5(W_{cy} - W_{cy,n})f_{cd} \\ &= [(338 - 31.5) \times 400 + (1469 - 100.7) \times 355 + 0.5 \\ &\quad \times (14192 - 3015) \times 33.3] \times 10^{-3} \\ &= 767 \text{ kN} \cdot \text{m} \end{aligned}$$

It should be noted that the plastic bending resistance can be calculated by taking moment about either line on the cross-section parallel to the y-y axis, as long as the aforementioned plastic modulus are determined according to the referred line.

3) Point C ($M_{\text{pl,Rd}}$, $N_{\text{pm,Rd}}$):

The cross-section is under partial compression but axial force is resultant from the compressive stresses. The axial force is equal to the compression capacities of concrete in the compression zone and steel sections within the height of $2h_n$. It is mentioned above that the compression capacity of steel sections within the height of $2h_n$ is equal to the compression capacity of concrete in the compression zone and out of the height of $2h_n$. Thus, the axial force is actually the full cross-sectional compression capacity of concrete and determined from:

$$N_{pm,Rd} = 0.85 A_c f_{cd} = 0.85 \times 144400 \times 33.3 \times 10^{-3} = 4091 \text{ kN}$$

4) Point D ($M_{\text{max,Rd}}$, $N_{\text{pm,Rd}}/2$):

The maximum plastic moment resistance $M_{\text{max,Rd}}$ is calculated when the h_n is equal to 0.

$$\begin{aligned} M_{\text{max},Rd} &= W_{sy} f_{sd} + W_{ey} f_{ed} + 0.5 \times 0.85 W_{cy} f_{cd} \\ &= [338 \times 435 + 1469 \times 355 + 0.5 \times 0.85 \times 14192 \times 33.3] \times 10^{-3} \\ &= 870 \text{ kN} \cdot \text{m} \end{aligned}$$

- Steel contribution ratio

$$\begin{aligned} \delta &= (A_e f_{ed})/N_{pl,Rd} \\ &= (13500 \times 355) \times 10^{-3}/9733 = 0.492 < 0.9 \end{aligned}$$

- Check for resistance of column in axial compression

$$\frac{N_{Ed}}{\chi N_{pl,Rd}} = \frac{6000}{0.766 \times 9806} = 0.805 < 1.0$$

Thus, the buckling resistance under axial compression is adequate!

- Check for resistance of column in combined compression and uniaxial bending

For the determination of the internal forces considering second-order effect, the design value of effective flexural stiffness should be calculated as following with long-term effect included.

$$
\begin{aligned}
(EI)_{eff,II} &= K_0(E_s I_{sy} + E_e I_{ey} + K_{e,II} E_{cm} I_{cy}) \\
&= 0.9 \times [210 \times (5115 + 17343) + 0.5 \times 20.7 \times 190876] \times 10^4 \\
&= 6.02 \times 10^{10}\ \text{kN} \cdot \text{mm}^2
\end{aligned}
$$

Thus, the critical normal force about y-y axis with effective length taken as the system length of column is determined from:

$$
N_{cr,eff} = \frac{\pi^2 (EI)_{eff,II}}{L^2} = \frac{\pi^2 \times 6.02 \times 10^{10}}{4000^2} = 37154\ \text{kN}
$$

The second-order effect should be considered for both moments from first-order analysis and moment from imperfection as following:

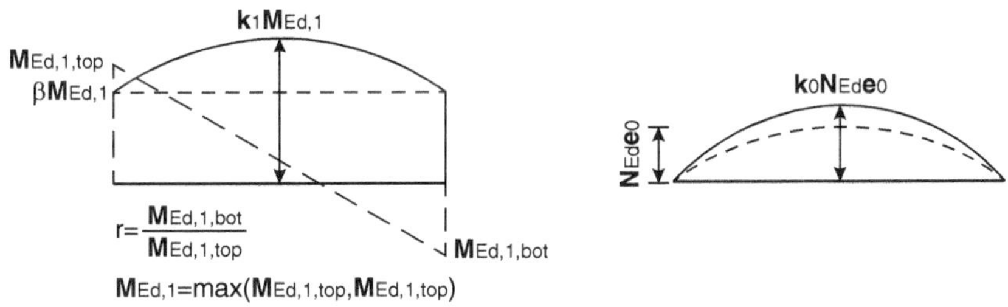

(a) Moment from first-order analysis (b) Moment from imperfection

According to buckling curve “b”, the initial imperfection about y–y axial is:

$$
e_{0,y} = L/200 = 4000/200 = 20\,\text{mm}
$$

Accordingly, the bending moment by the initial imperfection is determined as:

$$
M_0 = N_{Ed} e_{0,y} = 6000 \times 20/1000 = 120\ \text{kN} \cdot \text{m}
$$

According to the moment diagram by the initial imperfection, the factor β_0 for determination of moment to second-order effect is equal to 1.0. Thus, the amplification factor for the moment by the imperfection is calculated from:

$$
k_0 = \frac{\beta_0}{1 - N_{Ed}/N_{cr,eff}} = \frac{1.0}{1 - 6000/37154} = 1.193
$$

According to the first-order design moment diagram, the ratio of end moments is calculated as:

$$
r = M_{t,y}/M_{b,y} = -200/400 = -0.5
$$

Thus, the factor β_1 for determination of moment to second-order effect is determined:

$$\beta_1 = \max(0.66 + 0.44r, 0.44) = \max(0.66 + 0.44 \times (-0.5), 0.44) = 0.44$$

Thus, the amplification factor for the moment by the imperfection is calculated from:

$$k_1 = \frac{\beta_1}{1 - N_{Ed}/N_{cr,eff}} = \frac{0.44}{1 - 6000/37154} = 0.525$$

Thus, the design moment, considering second-order effect, is calculated as:

$$\begin{aligned} M_{Ed} &= \max[k_0 M_0 + k_1 \max(|M_{t,y}|, |M_{b,y}|), \max(|M_{t,y}|, |M_{b,y}|)] \\ &= \max[1.193 \times 120 + 0.525 \times \max(|-200|, |400|), \max(|-200|, |400|)] \\ &= 400 \text{ kN} \cdot \text{m} \end{aligned}$$

In this case, the second-order effect is not significant and the maximum end moment is taken as the design moment. For $N_{Ed} > N_{pm,Rd} = 4091$ kN, the value for determining the plastic bending resistance $M_{pl,N,Rd}$ taking into account the normal force N_{Ed} is calculated from:

$$\mu_d = \frac{N_{pl,Rd} - N_{Ed}}{N_{pl,Rd} - N_{pm,Rd}} = \frac{9806 - 6000}{9806 - 4091} = 0.662$$

$$\frac{M_{Ed}}{M_{pl,N,Rd}} = \frac{M_{Ed}}{\mu_d M_{pl,Rd}} = \frac{400}{0.662 \times 767} = 0.788 < \alpha_M = 0.9$$

Thus, the resistance for combined axial compression and uniaxial bending is adequate. The external design force and bending moment, and M-N interaction curve are plotted in Fig. B.16:

- Fire resistance check

The fire resistance of this column can be determined according to the tabulated data in Table 4.4 of EN 1994–1–2. Given cross-sectional dimension $h_c = b_c = 400$ mm, concrete cover of steel section $c = 67$ mm and axial distance of reinforcing bars $u_s = 30$ mm, this CES column can achieve a fire

UC 254×254×107 S355 sections with C90/105 concrete and G500 reinforcements

Effective compressive strength and modulus of elasticity are taken from Table 3.1 and Table 3.2.

$$\begin{aligned} f_{ck} &= 72\,\text{N/mm}^2;\ E_{cm} = 41.1\,\text{GPa} \\ f_{cd} &= f_{ck}/\gamma_c = 72/1.5 = 48\,\text{N/mm}^2 \\ f_{cm} &= f_{ck} + 8 = 72 + 8 = 80\,\text{N/mm}^2 \end{aligned}$$

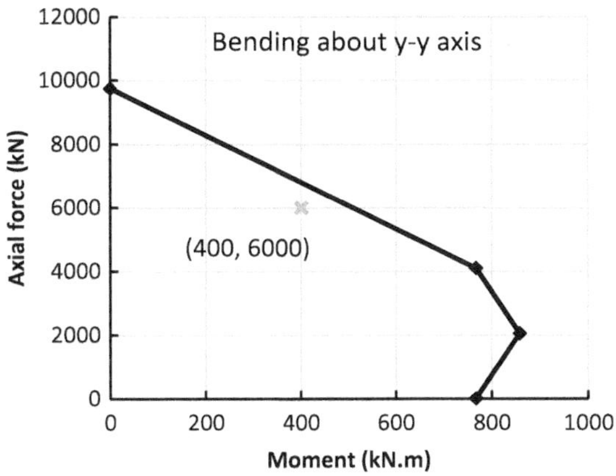

Fig. B.16 Design M–N curve for bending about y-y axis.

Creep coefficient could be similarly determined: $\phi_t = 1.20$

$$E_{c,eff} = E_{cm}\frac{1}{1+(N_{G,Ed}/N_{Ed})\phi_t} = \frac{41.1}{1+(4000/6000)\times 1.20} = 25.4\ \text{GPa}$$

$$\begin{aligned}(EI)_{eff,y} &= E_s I_{sy} + E_e I_{ey} + 0.6E_{c,eff}I_{cy}\\ &= [210\times(5115+17343)+0.6\times 25.4\times 190876]\times 10^4\\ &= 7.63\times 10^{10}\ \text{kN}\cdot\text{mm}^2\end{aligned}$$

$$\begin{aligned}(EI)_{eff,z} &= E_s I_{sz} + E_e I_{ez} + 0.6E_{c,eff}I_{cz}\\ &= [210\times(5115+5926)+0.6\times 25.4\times 202292]\times 10^4\\ &= 5.41\times 10^{10}\ \text{kN}\cdot\text{mm}^2\end{aligned}$$

$$N_{cr,y} = \frac{\pi^2(EI)_{eff,y}}{L_{eff}^2} = \frac{\pi^2\times 7.63\times 10^{10}}{4000^2} = 47069\ \text{kN}$$

$$N_{cr,z} = \frac{\pi^2(EI)_{eff,z}}{L_{eff}^2} = \frac{\pi^2\times 5.41\times 10^{10}}{4000^2} = 33356\ \text{kN}$$

$$\begin{aligned}N_{pl,Rk} &= A_s f_{sk} + A_e f_{ek} + 0.85A_c f_{ck}\\ &= 2120\times 500 + 13500\times 355 + 0.85\times 144400\times 72 = 14690\ \text{kN}\end{aligned}$$

$$\overline{\lambda_y} = \sqrt{\frac{N_{pl,Rk}}{N_{cr,y}}} = \sqrt{\frac{14690}{47069}} = 0.557$$

$$\overline{\lambda_z} = \sqrt{\frac{N_{pl,Rk}}{N_{cr,z}}} = \sqrt{\frac{14690}{33356}} = 0.662$$

For a fully encased steel column, the buckling curve about major axis is "b", and about minor axis is "c". Thus, the imperfection factor is $\alpha = 0.34$ and $\alpha = 0.49$, respectively.

$$\Phi_y = 0.5\left[1 + \alpha_y(\overline{\lambda_y} - 0.2) + \overline{\lambda_y}^2\right]$$
$$= 0.5\left[1 + 0.34 \times (0.557 - 0.2) + 0.557^2\right] = 0.716$$

$$\chi_y = \min\left(\frac{1}{\Phi_y + \sqrt{\Phi_y^2 - \overline{\lambda}_y^2}}, 1.0\right)$$
$$= \min\left(\frac{1}{0.716 + \sqrt{0.716^2 - 0.557^2}}, 1.0\right) = 0.858$$

$$\Phi_z = 0.5\left[1 + \alpha_z(\overline{\lambda_z} - 0.2) + \overline{\lambda_z}^2\right]$$
$$= 0.5\left[1 + 0.49 \times (0.662 - 0.2) + 0.662^2\right] = 0.832$$

$$\chi_z = \min\left(\frac{1}{\Phi_z + \sqrt{\Phi_z^2 - \overline{\lambda}_z^2}}, 1.0\right)$$
$$= \min\left(\frac{1}{0.832 + \sqrt{0.832^2 - 0.662^2}}, 1.0\right) = 0.748$$

$$\chi = \min\left(\chi_y, \chi_z\right) = \min(0.858, 0.748) = 0.748$$

Point A (0, $N_{\mathrm{pl,Rd}}$):

$$N_{pl,Rd} = A_s f_{sd} + A_e f_{ed} + 0.85 A_c f_{cd}$$
$$= 2120 \times 435 + 13500 \times 355 + 0.85 \times 144400 \times 48 = 11606 \text{ kN}$$

Buckling resistance: $\chi N_{pl,Rd} = 0.748 \times 11606 = 8626$
Point B ($M_{\mathrm{pl,Rd}}$, 0):

$$h_n = \frac{0.85 A_c f_{cd}}{2 \times 0.85 f_{cd} b_c + 4t_{sz}(2f_{sd} - 0.85 f_{cd}) + 2t_w(2f_{ed} - 0.85 f_{cd})}$$
$$= \frac{0.85 \times 144400 \times 48}{2 \times 0.85 \times 48 \times 400 + 4 \times 2 \times (2 \times 400 - 0.85 \times 48) + 2 \times 12.8 \times (2 \times 355 - 0.85 \times 48)}$$
$$= 105.8 \,\text{mm}$$

$h_n = 105.8$ mm $< h/2 - t_f = 266.7/2 - 20.5 = 112.85$ mm, thus, the neutral axial also lies in the web of the encased section.

$$W_{sy,n} = 2t_{sz}h_n^2 = 2 \times 2 \times 105.8^2 \times 10^{-3} = 44.8 \text{ cm}^3$$
$$W_{ey,n} = t_w h_n^2 = 12.8 \times 105.8^2 \times 10^{-3} = 143.3 \text{ cm}^3$$
$$W_{cy,n} = (b_c - 2t_{sz} - t_w)h_n^2 = (400 - 2 \times 2 - 12.8) \times 105.8^2 \times 10^{-3}$$
$$= 4289 \text{ cm}^3$$
$$M_{pl,Rd} = (W_{sy} - W_{sy,n})f_{sd} + (W_{ey} - W_{ey,n})f_{ed} + 0.5(W_{cy} - W_{cy,n})f_{cd}$$

$$= [(338 - 44.8) \times 400 + (1469 - 143.3) \times 355 + 0.5 \times (14192 - 4289) \times 48] \times 10^{-3} = 791 \text{ kN} \cdot \text{m}$$

Point C ($M_{\text{pl,Rd}}$, $N_{\text{pm,Rd}}$):

$$N_{pm,Rd} = 0.85A_c f_{cd} = 0.85 \times 144400 \times 48 \times 10^{-3} = 5891 \text{ kN}$$

Point D ($M_{\text{max,Rd}}$, $N_{\text{pm,Rd}}/2$):

$$M_{\text{max},Rd} = W_{sy} f_{sd} + W_{ey} f_{ed} + 0.5 \times 0.85 W_{cy} f_{cd}$$
$$= [338 \times 435 + 1469 \times 355 + 0.5 \times 0.85 \times 14192 \times 48] \times 10^{-3}$$
$$= 958 \text{ kN} \cdot \text{m}$$

Steel contribution ratio:

$$\delta = (A_e f_{ed})/N_{pl,Rd}$$
$$= (13500 \times 355) \times 10^{-3}/11606 = 0.416 < 0.9$$

- Fire resistance check

Since this CES column is made of high-strength concrete, the load level needs to be determined before applying the tabulated data in Table 4.4 of EN 1994–1–2. The recommended load reduction factor $\eta_{fi} = 0.65$ in EN1994-1–2 is adopted to calculate the design actions under fire conditions.

Design axial load under fire conditions:

$$N_{fi,d,t} = \eta_{fi} N_{Ed} = 0.65 \times 6000 = 3900 \text{ kN}$$
$$M_{fi,d,t} = \eta_{fi} M_{Ed} = 0.65 \times 400 = 260 \text{ kN} \cdot \text{m}$$

The plastic bending resistance $M_{pl,N,Rd}$ taking into account the normal force $N_{fi,d,t}$ can be determined from the M-N interaction curve.

$$M_{pl,N,Rd} = 776 \text{ kN} \cdot \text{m}$$

The load level under fire conditions can be calculated as:

$$\eta_{fi,t} = \frac{M_{fi,d,t}}{\alpha_M \times M_{pl,N,Rd}} = \frac{260}{0.9 \times 776} = 0.372 < 0.5$$

Therefore, the fire resistance of this column can be determined according to the tabulated data in Table 4.4 of EN 1994–1–2. Given cross-sectional dimension $h_c = b_c = 400$ mm, concrete cover of steel section $c = 67$ mm and axial distance of reinforcing bars $u_s = 30$ mm, this CES column can achieve a fire rating of R120.

B.4.3 UC 254 x 254 x 107 S500 sections with C50/60 concrete and G500 reinforcements

$$N_{pl,Rk} = A_s f_{sk} + A_e f_{ek} + 0.85 A_c f_{ck}$$
$$= 2120 \times 500 + 13500 \times 500 + 0.85 \times 144400 \times 33.3 = 13948 \text{ kN}$$

$$\overline{\lambda_y} = \sqrt{\frac{N_{pl,Rk}}{N_{cr,y}}} = \sqrt{\frac{13948}{43721}} = 0.563$$

$$\overline{\lambda_z} = \sqrt{\frac{N_{pl,Rk}}{N_{cr,z}}} = \sqrt{\frac{13948}{29807}} = 0.682$$

For a fully encased steel column, the buckling curve about major axis is "b", and about minor axis is "c". Thus, the imperfection factor is $\alpha = 0.34$ and $\alpha = 0.49$, respectively.

$$\Phi_y = 0.5\left[1 + \alpha_y(\overline{\lambda_y} - 0.2) + \overline{\lambda_y}^2\right]$$
$$= 0.5\left[1 + 0.34 \times (0.563 - 0.2) + 0.563^2\right] = 0.720$$

$$\chi_y = \min\left(\frac{1}{\Phi_y + \sqrt{\Phi_y^2 - \overline{\lambda_y^2}}}, 1.0\right)$$

$$= \min\left(\frac{1}{0.720 + \sqrt{0.720^2 - 0.563^2}}, 1.0\right) = 0.855$$

$$\Phi_z = 0.5\left[1 + \alpha_z(\overline{\lambda_z} - 0.2) + \overline{\lambda_z}^2\right]$$
$$= 0.5\left[1 + 0.49 \times (0.682 - 0.2) + 0.682^2\right] = 0.851$$

$$\chi_z = \min\left(\frac{1}{\Phi_z + \sqrt{\Phi_z^2 - \overline{\lambda_z^2}}}, 1.0\right)$$

$$= \min\left(\frac{1}{0.851 + \sqrt{0.851^2 - 0.682^2}}, 1.0\right) = 0.736$$

$$\chi = \min\left(\chi_y, \chi_z\right) = \min(0.855, 0.736) = 0.736$$

Point A ($0, N_{\text{pl,Rd}}$):

$$N_{pl,Rd} = A_s f_{sd} + A_e f_{ed} + 0.85 A_c f_{cd}$$
$$= 2120 \times 435 + 13500 \times 500 + 0.85 \times 144400 \times 33.3 = 11764 \text{ kN}$$

Buckling resistance: $\chi N_{pl,Rd} = 0.736 \times 11764 = 8604$ kN

Point B ($M_{\mathrm{pl,Rd}}$, 0):

$$h_n = \frac{0.85A_c f_{cd}}{2 \times 0.85 f_{cd} b_c + 4t_{sz}(2f_{sd} - 0.85f_{cd}) + 2t_w(2f_{ed} - 0.85f_{cd})}$$
$$= \frac{0.85 \times 144400 \times 33.3}{2 \times 0.85 \times 33.3 \times 400 + 4 \times 2 \times (2 \times 435 - 0.85 \times 33.3) + 2 \times 12.8 \times (2 \times 500 - 0.85 \times 33.3)}$$
$$= 76.4\,\mathrm{mm}$$

The neutral axial also lies in the web of the encased section.

$$W_{sy,n} = 2t_{sz}h_n^2 = 2 \times 2 \times 76.4^2 \times 10^{-3} = 23.3\ \mathrm{cm}^3$$
$$W_{ey,n} = t_w h_n^2 = 12.8 \times 76.4^2 \times 10^{-3} = 74.7\ \mathrm{cm}^3$$
$$W_{cy,n} = (b_c - 2t_{sz} - t_w)h_n^2 = (400 - 2\times2 - 12.8) \times 76.4^2 \times 10^{-3} = 2237\ \mathrm{cm}^3$$
$$M_{pl,Rd} = (W_{sy} - W_{sy,n})f_{sd} + (W_{ey} - W_{ey,n})f_{ed} + 0.5(W_{cy} - W_{cy,n})f_{cd}$$
$$= [(338 - 23.3) \times 435 + (1469 - 74.7) \times 500 + 0.5 \times (14192 - 2237) \times 33.3] \times 10^{-3}$$
$$= 1006\ \mathrm{kN \cdot m}$$

Point C ($M_{\mathrm{pl,Rd}}$, $N_{\mathrm{pm,Rd}}$):

$$N_{pm,Rd} = 0.85A_c f_{cd} = 0.85 \times 144400 \times 33.3 \times 10^{-3} = 4091\ \mathrm{kN}$$

Point D ($M_{\mathrm{max,Rd}}$, $N_{\mathrm{pm,Rd}}/2$):

$$M_{\mathrm{max},Rd} = W_{sy}f_{sd} + W_{ey}f_{ed} + 0.5 \times 0.85 W_{cy} f_{cd}$$
$$= [338 \times 435 + 1469 \times 500 + 0.5 \times 0.85 \times 14192 \times 33.3] \times 10^{-3}$$
$$= 1083\ \mathrm{kN \cdot m}$$

Steel contribution ratio:

$$\delta = (A_e f_{ed})/N_{pl,Rd}$$
$$= (13500 \times 500) \times 10^{-3}/11764 = 0.577 < 0.9$$

- Fire resistance check

Table B.3 Comparison between CES columns with various material strengths.

Material grades (Steel + Concrete) + Rebars)	Steel contribution ratios	Design resistances			
		$\chi N_{pl,Rd}$	$N_{pm,Rd}$	$M_{pl,Rd}$	$M_{max,Rd}$
S355 + C50/60 + G500	0.492	0	0	0	0
S355 + C90/105 + G500	0.416	15.7%	44.0%	3.1%	10.3%
S500 + C50/60 + G500	0.577	15.4%	0	29.5%	24.8%

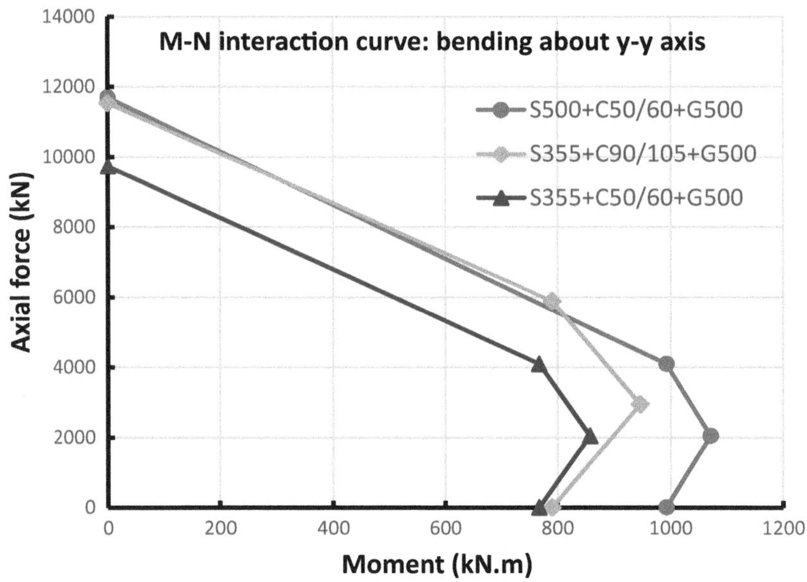

Through this study, it could be concluded that for CES columns with small eccentricities (large axial load and small bending moments), it is beneficial to use high strength concrete; whereas for CES columns with smaller axial load and higher bending moments, the use of high strength steel is beneficial.

The fire resistance of this column can be determined according to the tabulated data in Table 4.4 of EN 1994–1–2. Given cross-sectional dimension $h_c = b_c = 400$ mm, concrete cover of steel section $c = 67$ mm and axial distance of reinforcing bars $u_s = 30$ mm, this CES column can achieve a fire rating of R120.

B.4.4 Comparison and summary

The design resistances are compared for the aforementioned three composite sections. The composite section with encased steel section of S355, concrete of C50/60, and reinforcing steel of G500 is referred to for comparison.

As tabulated in Table B.3, by use of high strength concrete C90/105 replacing normal strength concrete C50/60, the axial buckling resistance ($\chi N_{pl,Rd}$) of the CES column is improved by 15.7%, and the increase of moment resistances ($M_{pl,Rd}$ and $M_{max,Rd}$) are 3.1% and 10.3%, respectively.

By use of steel S500 replacing S355, the axial buckling resistance is improved by 15.4%, and the increase of moment resistance is as high as 29.5% and 24.8%.

Through this study, it could be concluded that for CES columns with small eccentricities (large axial load and small bending moments), it is beneficial to use high strength concrete; whereas for CES columns with smaller axial load and higher bending moments, the use of high strength steel is beneficial.

Design spreadsheets for composite columns

C

C.1 General

The design spreadsheets for composite columns are developed for both Concrete Filled Steel Tubular (CFST) members and Concrete Encased Steel (CES) members. The spreadsheets, in excel file format, are made available to the readers of this book. The spreadsheets are applicable to composite members with steel section grade within S235 to S550 and concrete class within C20/25 to C90/105. Beyond the material range, advanced calculation methods should be sought. This appendix provides general guidance on the use of spreadsheets for design purpose, while the detailed instructions are provided in the main program.

The spreadsheets for CFST and CES members are isolated in two folders, both of which consist of excel files named "Database for steel sections" and "Main program" as shown in Fig. C.1. The users must open these two files in the folder simultaneously to commence the operation in "Main program". In view of the similarity of the two folders, only the spreadsheet for CFST member is used for illustration purpose.

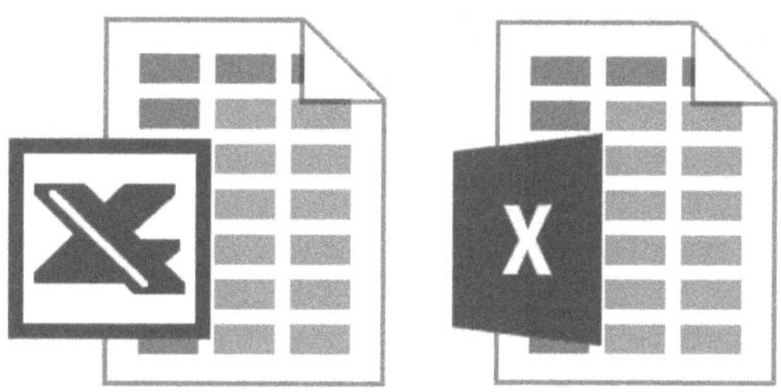

Fig. C.1 Constituents of spreadsheets (using CFST as example).

C.2 Database for steel sections

The database for steel sections includes all the standard hot-finished sections and the custom sections to be defined by the users. As shown in Fig. C.2(a), "UB", "UC", "CHS" and "RHS" refer to the hot-rolled Universal Beam, Universal Column, Circular Hollow Section and Rectangular Hollow Section, respectively, and they can be supplemented with the sections from other countries such as America, China, Japan, India, etc. Besides that, user-defined section is also allowed in the sheet named

Design of Steel-Concrete Composite Structures Using High Strength Materials. DOI: 10.1016/C2019-0-05474-X

"Custom_H" and "Custom_RHS". As highlighted in Fig. C.2(b), the users only need to input the dimensions of the section, the other sectional properties are automatically calculated.

Universal Beams to BS4 Part1 1993 - Dimensions & Properties

Designation	Mass Per metre	Depth of section	Width of section	Thickness: Web	Thickness: Flange	Root radius	Depth between fillets	Ratios for Local Buckling: Flange	Ratios for Local Buckling: Web	Second Moment of Area: Axis y-y	Second Moment of Area: Axis z-z	Radius of Gyration: Axis y-y	Radius of Gyration: Axis z-z	Elastic Modulus: Axis y-y	Elastic Modulus: Axis z-z	Plastic Modulus: Axis y-y	Plastic Modulus: Axis z-z	Bu Par
		h	b	t_w	t_f	r	d	b/2t	d/s	I_y	I_z	r_y	r_z	$W_{e,y}$	$W_{e,z}$	$W_{p,y}$	$W_{p,z}$	
	kg/m	mm	mm	mm	mm	mm	mm			cm^4	cm^4	cm	cm	cm^3	cm^3	cm^3	cm^3	
0	0	0	0	0	0	0	0	0	0	0	0	0	0	0	0	0	0	
1016x305x487	487	1036.1	308.5	30	54.1	30	867.9	2.85	28.9	1021400	26720	41	6.6	19720	1732	23200	2800	0
1016x305x437	437	1025.9	305.4	26.9	49	30	867.9	3.12	32.3	909900	23450	40	6.5	17740	1535	20760	2469	0
1016x305x393	393	1016	303	24.4	43.9	30	868.2	3.45	35.6	807700	20500	40	6.4	15900	1353	18540	2168	0
1016x305x349	349	1008.1	302	21.1	40	30	868.1	3.77	41.1	723100	18460	40	6.4	14350	1223	16590	1941	0
1016x305x314	314	1000	300	19.1	35.9	30	868.2	4.18	45.5	644200	16230	40	6.4	12880	1082	14850	1713	0
1016x305x272	272	990.1	300	16.5	31	30	868.1	4.84	52.6	554000	14000	40	6.4	11190	934	12830	1470	0
1016x305x249	249	980.2	300	16.5	26	30	868.2	5.77	52.6	481300	11750	39	6.1	9821	784	11350	1245	0
1016x305x222	222	970.3	300	16	21.1	30	868.1	7.11	54.3	408000	9546	38	5.8	8409	636	9807	1020	[illegible]
914x419x388	388	921	420.5	21.4	36.6	24.1	799.6	5.74	37.4	719600	45440	38	9.6	15630	2161	17670	3341	0
914x419x343	343	911.8	418.5	19.4	32	24.1	799.6	6.54	41.2	625800	39160	38	9.5	13730	1871	15480	2890	0
914x305x289	289	926.6	307.7	19.5	32	19.1	824.4	4.81	42.3	504200	15600	37	6.5	10880	1014	12570	1601	0
914x305x253	253	918.4	305.5	17.3	27.9	19.1	824.4	5.47	47.7	436300	13300	37	6.4	9501	871	10940	1371	0
914x305x224	224	910.4	304.1	15.9	23.9	19.1	824.4	6.36	51.8	376400	11240	36	6.3	8269	739	9535	1163	0
914x305x201	201	903	303.3	15.1	20.2	19.1	824.4	7.51	54.6	325300	9423	36	6.1	7204	621	8351	982	0
838x292x226	227	850.9	293.8	16.1	26.8	17.8	761.7	5.48	47.3	339700	11360	34	6.3	7985	773	9155	1212	[illegible]
838x292x194	194	840.7	292.4	14.7	21.7	17.8	761.7	6.74	51.8	279200	9066	34	6.1	6641	620	7640	974	0
838x292x176	176	834.9	291.7	14	18.8	17.8	761.7	7.76	54.4	246000	7799	33	5.9	5893	535	6808	842	0
762x267x197	197	769.8	268	15.6	25.4	16.5	686	5.28	44	240000	8175	31	5.7	6234	610	7167	959	0
762x267x173	173	762.2	266.7	14.3	21.6	16.5	686	6.17	48	205300	6850	31	5.6	5387	514	6198	807	0
762x267x147	147	754	265.2	12.8	17.5	16.5	686	7.58	53.6	168500	5455	30	5.4	4470	411	5156	647	0
762x267x134	134	750	264.4	12	15.5	16.5	686	8.53	57.2	150700	4788	30	5.3	4018	362	4644	570	0
686x356x217	217	695.5	354.7	15.4	24.8	15.2	615.5	7.15	39.97	234200	18440	29	8.2	6735	1040	7558	1598	0
686x356x240	240	700.8	356.1	16.8	27.4	15.2	615.6	6.50	36.64	261200	20670	29	8.2	7455	1161	8389	1787	0
686x356x265	265	706.4	357.8	18.4	30.2	15.2	615.6	5.92	33.46	290800	23120	29	8.3	8233	1292	9297	1992	0
686x356x289	289	714	356.4	19.1	34	15.2	615.6	5.24	32.23	325500	25720	30	8.4	9118	1444	10300	2223	0
686x356x384	384	736.1	362.5	24.9	45	15.2	615.7	4.03	24.73	447800	35780	30	8.6	12170	1974	13930	3057	0
686x254x170	170	692.9	255.8	14.5	23.7	15.2	615.1	5.4	42.4	170300	6630	28	5.5	4916	518	5631	811	0
686x254x152	152	687.5	254.5	13.2	21	15.2	615.1	6.06	46.6	150400	5784	28	5.5	4374	455	5000	710	0
686x254x140	140	683.5	253.7	12.4	19	15.2	615.1	6.68	49.6	136300	5183	28	5.4	3987	409	4558	638	0
686x254x125	125	677.9	253	11.7	16.2	15.2	615.1	7.81	52.6	118000	4383	27	5.2	3481	346	3994	542	0
610x324x155	155	611.1	323.8	12.7	19.1	12.7	547.5	8.48	43.11	129200	10790	26	7.4	4227	666	4734	1023	0
610x324x174	174	616.2	325.1	14	21.6	12.7	547.6	7.53	39.11	147200	12380	26	7.5	4776	762	5360	1170	0
610x324x195	195	621.8	326.5	15.4	24.4	12.7	547.6	6.69	35.56	166381	14172	26	7.6	5352	868	6023	1335	0
610x324x217	217	628.4	327.7	16.5	27.7	12.7	547.6	5.92	33.19	190800	16260	26	7.7	6074	993	6845	1527	0

UB | UC | Custom_H | CHS | RHS | Custom_RHS

(a)

Costom H Sections

Designation	Mass Per metre	Depth of section	Width of section	Thickness: Web	Thickness: Flange	Root radius	Depth between fillets	Ratios for Local Buckling: Flange	Ratios for Local Buckling: Web	Second Moment of Area: Axis y-y	Second Moment of Area: Axis z-z	Radius of Gyration: Axis y-y	Radius of Gyration: Axis z-z	Elastic Modulus: Axis y-y	Elastic Modulus: Axis z-z	Plastic Modulus: Axis y-y	Plastic Modulus: Axis z-z
		h	b	t_w	t_f	r	d	b/2t	d/s	I_y	I_z	r_y	r_z	$W_{e,y}$	$W_{e,z}$	$W_{p,y}$	$W_{p,z}$
	kg/m	mm	mm	mm	mm	mm	mm			cm^4	cm^4	cm	cm	cm^3	cm^3	cm^3	cm^3
0	0	0	0	0	0	0	0	0	0	0	0	0	0	0	0	0	0
550x550x335	336	550	550	20	30	0	490	9.17	24.5	242936	83220.2	24	14	8834	3026	9781	4587
HEA220	51	210	220	7	11	0	188	10.00	26.86	5184	1952.67	9	5.5	493.7	177.5	543.4	269
210x220x48	48	210	220	7	11	0	188	10.00	26.86	5184	1952.67	9.2	5.6	493.7	177.5	543.4	269
254x254x80	80	256.3	254.8	9.4	15.6	0	225.1	8.17	23.95	12590	4304	11	6.5	983	338	1091	512
75x50x6.7	6.7	75	50	4	6	0	63	4.17	15.75	80	12.5336	3.1	1.2	21.31	5.013	24.67	7.75
60x40x5.3	5.3	60	40	4	6	0	48	3.33	12	39	6.4256	2.4	1	12.94	3.213	15.26	4.99

(b)

Fig. C.2 Database of steel sections.

C.3 Main program

The main program is based on the design procedure described in Chapter 4, which includes three steps, namely the data input, calculations, and checks. Users need to key in the design loads, section size and material properties. As highlighted in Fig. C.3, only the highlighted cells need to be manually input by the users.

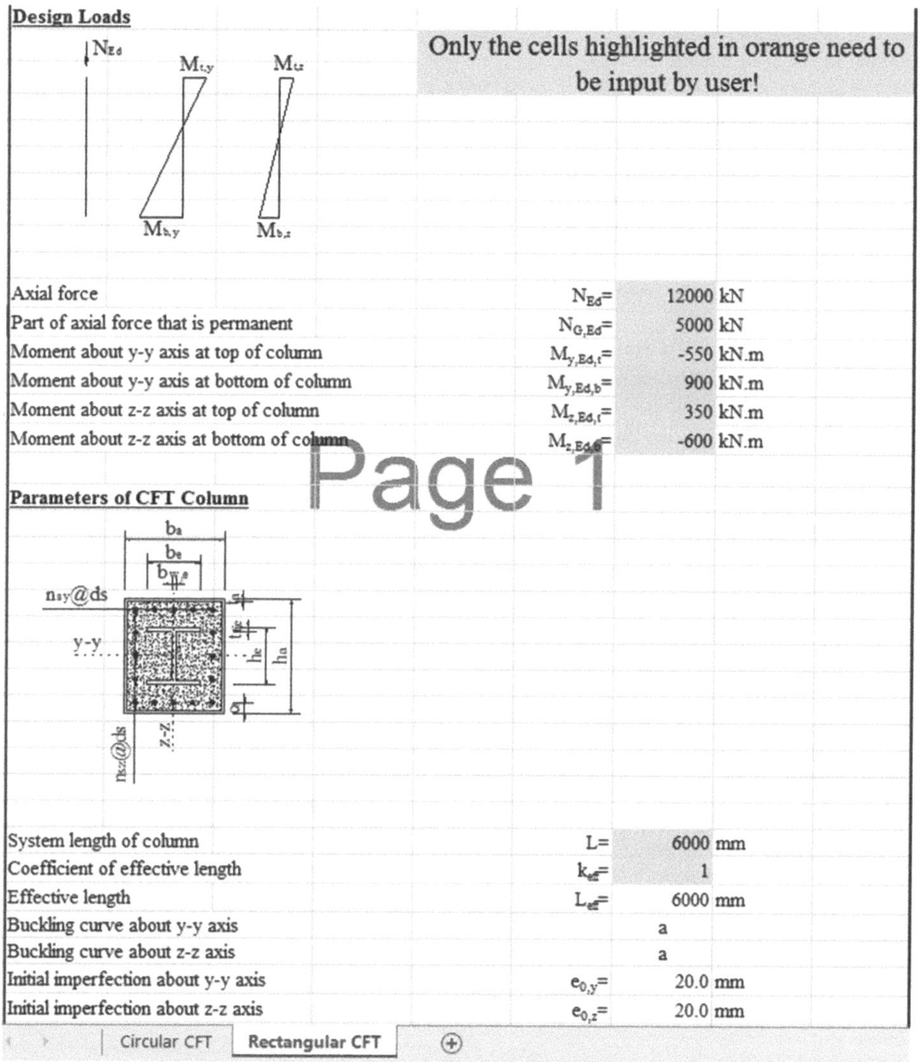

Fig. C.3 Data input in the main program.

In the calculation process, the main program considers the long-term effect, second-order effect and the construction of N-M interaction diagram. Only a few data need to be input as shown in Fig. C.4. For the ease of calculation, all the reinforcements are equivalently converted into a tube section with the same center position and sectional area as that of isolated bars. Details of this conversion are given in the comments of the excel sheet.

Long-term effect			
Age of concrete at first loading in days	t_0=	30	day
Age of concrete at the moment considered in days	t=	∞	days
Relative humidity of ambient environment	RH=	50	%
Notional size of concrete section	h_0=	219	mm
Coefficient	α_1=	0.70	mm
Coefficient	α_2=	0.90	mm
Coefficient	α_3=	0.78	mm
Factor to allow for effect of RH on notional creep	ϕ_{RH}=	1.43	
Factor to allow for effect of f_{cm} on notional creep	$\beta(f_{cm})$=	2.21	
Factor to allow for effect of t_0 on notional creep	$\beta(t_0)$=	0.48	
Notional creep coefficient	ϕ_0=	1.52	
Factor to allow for effect of RH and h_0	β_H=	523	
Coefficient to describe development of creep	$\beta_c(t,t_0)$=	1.00	
Creep coefficient	ϕ_t=	1.52	
Elastic modulus of concrete under long-term effect	$E_{c,eff}$=	22.8	GPa
Second-order Effect			
Effective flexural stiffness about y-y axis	$(EI)_{eff,II,y}$=	4.19E+11	$kN.mm^2$
Effective flexural stiffness about z-z axis	$(EI)_{eff,II,z}$=	2.16E+11	$kN.mm^2$
Elastic critical normal force about y-y axis	$N_{cr,eff,y}$=	114913	kN
Elastic critical normal force about z-z axis	$N_{cr,eff,z}$=	59122	kN
Factor for initial imperfection about y-y axis	$\beta_{e0,y}$=	1.0	
Factor for initial imperfection about z-z axis	$\beta_{e0,z}$=	1.0	
Ratio between top and bottom moments, y-y axis	r_y=	-0.611	
Ratio between top and bottom moments, z-z axis	r_z=	-0.583	
Factor for first-order moment about y-y axis	$\beta_{eN,y}$=	0.44	
Factor for first-order moment about z-z axis	$\beta_{eN,z}$=	0.44	
Amplification factor about y-y axis	$k_{e0,y}$=	1.117	
Amplification factor about z-z axis	$k_{e0,z}$=	1.255	
Amplification factor about y-y axis	$k_{eN,y}$=	0.491	
Amplification factor about z-z axis	$k_{eN,z}$=	0.552	
Design second-order moment about y-y axis	$M_{y,Ed}$=	900	kN.m
Design second-order moment about z-z axis	$M_{z,Ed}$=	632	kN.m

Page 3

(a)

Fig. C.4 Calculation in the main program.

Simplified Interaction Curve: Point A

Characteristic value of plastic resistance	$N_{pl,Rk}$=	23712	kN
Effective flexural stiffness about y-y axis	$(EI)_{eff,y}$=	4.78E+11	kN.mm^2
Effective flexural stiffness about z-z axis	$(EI)_{eff,z}$=	2.45E+11	kN.mm^2
Elastic critical normal force about y-y axis	$N_{cr,y}$=	130977	kN
Elastic critical normal force about z-z axis	$N_{cr,z}$=	67053	kN
Relative slenderness about y-y axis	λ_y=	0.425	
Relative slenderness about z-z axis	λ_z=	0.595	
Load eccentricity about y-y axis	$e_{N,y}$=	75	mm
Load eccentricity about z-z axis	$e_{N,z}$=	50	mm
Plastic resistance of cross-section	$N_{pl,Rd}$=	20352	kN
Steel contribution ratio	δ=	0.670	

Simplified Interaction Curve: Point B

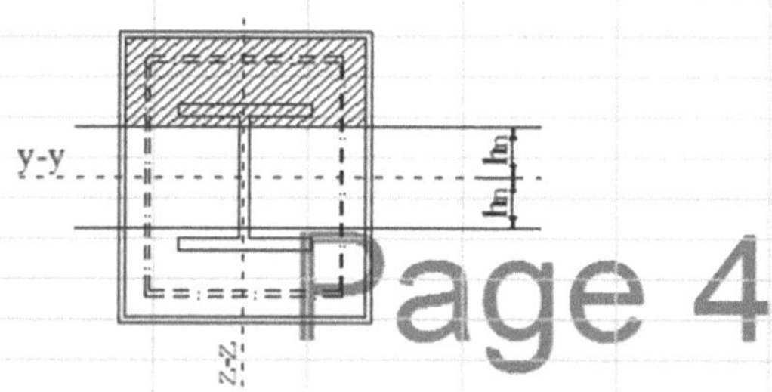

Bending about y-y axis

Assuming height of neutral axis	h_{ny}=	83.2	mm
Area of steel tube within $2h_n$	$A_{a,ny}$=	67	cm^2
Area of reinforcements within $2h_n$	$A_{s,ny}$=	0	cm^2
Area of encased section within $2h_n$	$A_{e,ny}$=	0	cm^2
Area of concrete beyond $2h_n$ under compression	$A_{c,ny}$=	709	cm^2
Design tensile resistance of steel within $2h_n$	N_{ny}=	2362	kN
Design normal resistance of concrete beyond $2h_n$	$N_{c,ny}$=	2362	kN
	N_{ny}-$N_{c,ny}$=	0	
Plastic modulus of steel tube beyond $2h_n$ about neutral axis	$W_{a,ny}$=	7499	cm^3
Plastic modulus of reinforcements beyond $2h_n$ about neutral axis	$W_{s,ny}$=	0	cm^3

(b)

Fig. C.4 *Continued*

The final check is performed for the scenarios under axial compression and under combined compression and bending. As shown in Fig. C.5, when the calculated design capacity exceeds the design forces, the cells turn green. Otherwise the users need to change the input data by increasing the cross-section size or material grade.

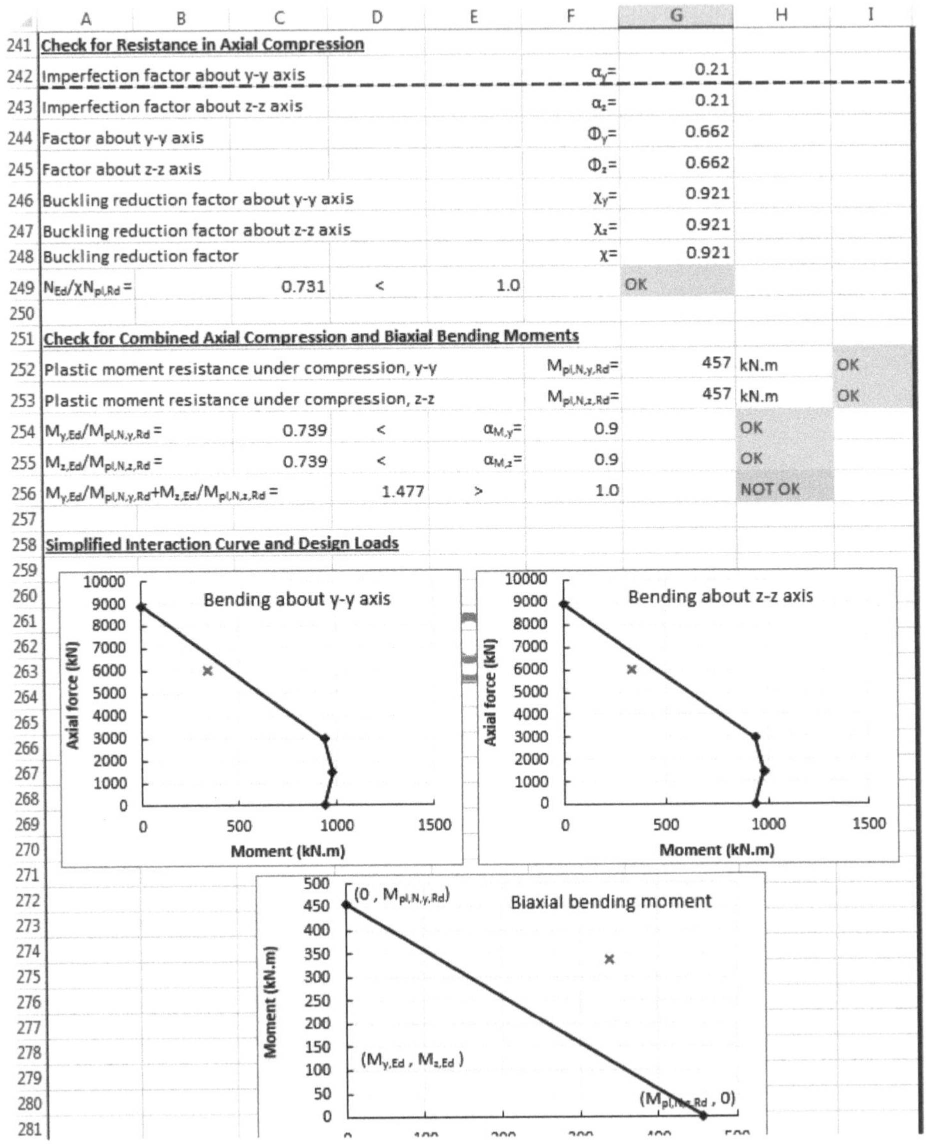

Fig. C.5 Final check in the main program.

Readers can use the work examples given in Appendix B to test the spreadsheets and see if the output is same as those shown in the work examples.

The spreadsheets can be downloaded from: https://www.elsevier.com/books-and-journals/book-companion/9780128233962.

References

Abe, F., 2004. Bainitic and martensitic creep-resistant steels. Curr. Opin. Solid State Mater. Sci. 8 (3–4), 305–311.

Abrams, M.S., 1971. Compressive Strength of Concrete at Temperatures to 1600 F. Portland Cement Association, pp. 33–58.

ACI 231R-10, 2010. Report on Early-Age Cracking: Causes, Measurement and Mitigation. American Concrete Institute, USA.

ACI 239R-18, 2018. Committee – Ultra High Performance Concrete: An Emerging Technology Report. American Concrete Institute, USA.

ACI 318-19, 2019. Building Code Requirements For Structural Concrete And Commentary, 318.

ACI Committee, 2003. Guide for Measuring, Mixing, Transporting, and Placing Concrete. ACI 304R-00.

ACI Committee, 2010. Report on High Strength Concrete. ACI 363R-10.

Aghajani, A., Somsen, C., Pesicka, J., Bendick, W., Hahn, B., Eggeler, G., 2009. Microstructural evolution in T24, a modified 2 (1/4) Cr–1Mo steel during creep after different heat treatments. Mater. Sci. Eng.: A 510, 130–135.

AIJ, 1997. Standards for Structural Calculation of Steel Reinforced Concrete Structures. Architectural Institute of Japan, Tokyo, Japan.

AISC-LRFD, 1999. Load and Resistance Factor Design Specification for Structural Steel Buildings. American Institute of Steel Construction.

Ansari, F., Li, Q., 1998. High-strength concrete subjected to triaxial compression. Mater. J. 95 (6), 747–755.

ANSI/AISC 360:2005, 2005. Specification for Structural Steel Buildings. American Institute of Steel Construction.

ANSI/AISC 360:2018, 2018. Specification for Structural Steel Buildings. American Institute of Steel Construction.

ANSI/AISC 360-10, 2010. Specification for Structural Steel Buildings. American Institute of Steel Construction.

ANSI/AISC 360-16, 2016. Specification for structural steel buildings. American Institute of Steel Construction (AISC), Chicago, USA.

Architectural Institute of Japan, 1997. Recommendations for Design and Construction of Concrete Filled Steel Tubular Structures. AIJ, Japan.

AS/NZS 2327, 2017. Australian/New Zealand Standard. Composite structures–Composite steel-concrete construction in buildings.

ASCE/SEI/SFPE 29-05, 2007. Standard Calculation Methods for Structural Fire Protection. American Society of Civil Engineers.

ASTM C1856/C1856M-17, 2017. Standard Practice for Fabricating and Testing Specimens of Ultra-High Performance Concrete. *ASTM International, West ConshohockenPA*.

ASTM E 1269, 2018. Standard Test Method for Determining Specific Heat Capacity by Differential Scanning Calorimetry. ASTM International, West Conshohocken, PA, USA.

ASTM E 21-20, 2020. Standard Test Methods for Elevated Temperature Tension Tests of Metallic Materials. ASTM Internaltional, West Conshohocken, PA.

ASTM E1461-13, 2013. Standard Test Method for Thermal Diffusivity by the Flash Method. ASTM International, West Conshohocken, PA, USA.

Bahr, O., Schaumann, P., Bollen, B., Bracke, J., 2013. Young's modulus and Poisson's ratio of concrete at high temperatures: experimental investigations. Mater. Des. 45, 421–429.

Design of Steel-Concrete Composite Structures Using High Strength Materials. DOI: 10.1016/C2019-0-05474-X

Balmer, G.G., 1944. Shear Strength of Concrete Under High Triaxial Stress – Computation of Mohr's Envelope as a Curve. SP-23. Structural Research Laboratory, U.S. Bureau of Reclamation.

Bamonte, PATRICK, Gambarova, P.G., 2010. Thermal and mechanical properties at high temperature of a very high-strength durable concrete. J. Mater. Civ. Eng. 22 (6), 545–555.

BC4, 2021. Design guide for composite steel and concrete members with high strength materials - An Extension of EN1994-1-1 Method to C90/105 Concrete and S550 Steel Section. Building and Construction Authority, 40pp.

Behnood, A., Ziari, H., 2008. Effects of silica fume addition and water to cement ratio on the properties of high-strength concrete after exposure to high temperatures. Cem. Concr. Compos. 30 (2), 106–112.

Bhadeshia, H., Honeycombe, R., 2017. Steels: Microstructure and Properties. Butterworth-Heinemann.

BS 5400, 2005. Steel, Concrete and Composite Bridges – Part 5: Code of Practice for Design of Composite Bridges. British Standards Institution, London.

BS 5950-8, 2003. Structural Use of Steelwork in Building-Part 8: Code of Practice of Fire Resistant Design. British Standard Institution.

BS 7419, 2012. Specification for Holding Down Bolts. British Standards Institution.

BS EN 13381-2, 2014. Test Methods for Determining the Contribution to the fire resistance of structural members – Part 2: Vertical protective membranes. British Standard Institution.

BS EN 13381-6, 2012. Test Methods for Determining the Contribution to the Fire Resistance of Structural Members – Part 2: Applied Protection to Concrete Filled Hollow Steel Columns. British Standard Institution.

Castilo, C., Durrani, A.J., 1990. Effect of transient high temperature on high strength concrete. ACI Mater. J. 87 (1), 47–53.

Chen, B., Liu, J., 2004. Residual strength of hybrid-fiber-reinforced high-strength concrete after exposure to high temperatures. Cem. Concr. Res. 34 (6), 1065–1069.

Chen, S., Wu, P., 2017. Analytical model for predicting axial compressive behavior of steel reinforced concrete column. J. Constr. Steel Res. 128, 649–660.

Cheng, F.P., Kodur, V.K.R., Wang, T.C., 2004. Stress-strain curves for high strength concrete at elevated temperatures. J. Mater. Civ. Eng. 16 (1), 84–90.

Choi, I.R., Chung, K.S., Kim, D.H., 2014. Thermal and mechanical properties of high-strength structural steel HSA800 at elevated temperatures. Mater. Des. 63, 544–551.

Chung, K.S., Kim, J.H., Yoo, J.H., 2013. Experimental and analytical investigation of high-strength concrete-filled steel tube square columns subjected to flexural loading. Steel Compos. Struct. 14 (2), 133–153.

Collins, M.P., Mitchell, D., Macgregor, J.G., 1993. Structural design considerations for high-strength concrete. Concr. Int. 15 (5), 27–34.

Contro, R., Poggi, C., Cazzani, A., 1988. Numerical analysis of fire effects on beam structures. Eng Comput (Swansea) 5, 53–58.

Corradi, L., Poggi, C., Setti, P., 1990. Interaction domains for steel beam-columns in fire conditions. J. Constr. Steel Res. 17 (3), 217–235.

Corus Construction & Industrial. RQT fabrication guide; 2006.

Craveiro, H.D., Rodrigues, J.P.C., Santiago, A., Laím, L., 2016. Review of the high temperature mechanical and thermal properties of the steels used in cold formed steel structures–The case of the S280 Gd+ Z steel. Thin-Walled Structures 98, 154–168.

Cree, D., Pliya, P., Green, M.F., Noumowé, A., 2017. Thermal behaviour of unstressed and stressed high strength concrete containing polypropylene fibers at elevated temperature. J. Struct. Fire Eng. 8 (4), 402–417.

Curbach, M., Spek, K., 2008. Ultra high performance concrete under biaxial compression. In: Fehling, E, Schmidt, M, Sturwald, S (Eds.), Proceedings of the Second International

Symposium on Ultra High Performance Concrete. Kassel. Kassel University Press, Germany, pp. 477–484 March 05-072008.

Dai, X., Lam, D., 2010. Numerical modelling of the axial compressive behaviour of short concrete-filled elliptical steel columns. J. Constr. Steel Res. 66 (7), 931–942.

Davison, B., Owens, G.W., 2012. Steel Designers' Manual, 7th Edition. The Steel Construction Institute. John Wiley & Sons Ltd.

Diederichs, U., Jumppanen, U.M., Penttala, V., 1988. Material properties of high strength concrete at elevated temperatures. IABSE 13th congress, Helsinki, 489–494.

Ding, J., Wang, Y.C., 2008. Realistic modelling of thermal and structural behaviour of unprotected concrete filled tubular columns in fire. J. Constr. Steel Res. 64 (10), 1086–1102.

Ellobody, E., Young, B., Lam, D., 2006. Behaviour of normal and high strength concrete-filled compact steel tube circular stub columns. J. Constr. Steel Res. 62 (7), 706–715.

El-Tawil, S., Deierlein, G.G., 1999. Strength and ductility of concrete encased composite columns. J. Struct. Eng. 125 (9), 1009–1019.

EN 10025-2, 2004. Hot-Rolled Products of Structural Steels – Part 2. Technical delivery conditions for non-alloy structural steels.

EN 10025-3, 2004. Hot-Rolled Products of Structural Steels – Part 3. Technical delivery conditions for normalized/normalized rolled weldable fine grain structural steels.

EN 10025-4, 2004. Hot-Rolled Products of Structural Steels – Part 4. Technical delivery conditions for thermomechanical rolled weldable fine grain structural steels.

EN 10025-6, 2004. Hot-Rolled Products of Structural Steels – Part 6. Technical delivery conditions for flat products of high strength structural steels in the quenched and tempered condition.

EN 10080, 2005. Steel for the Reinforcement of Concrete – Weldable Reinforcing Steel – General.

EN 1011-2, 2001. Welding – Recommendations for Welding of Metallic Materials – Part 2: Arc Welding of Ferritic Steels.

EN 10149-2, 1996. Hot-Rolled Flat Products Made of High Strength Steels for Cold Forming – Part 2. Delivery conditions for thermomechanically rolled steels.

EN 1090-2, 2008. Execution of Steel Structures and Aluminium Structures. Part 2. Technical requirements for steel structures.

EN 1990, 2002. Eurocode – Basis of Structural Design.

EN 1991-1-1, 2002. Eurocode 1: Actions on Structures – Part 1-1. General actions – Densities, self-weight, imposed loads for buildings.

EN 1991-1-2, 2002. Eurocode 1: Actions on Structures – Part 1-2. General actions – Actions on structures exposed to fire.

EN 1992-1-1, 2004. Eurocode 2: Design of Concrete Structures – Part 1-1. General rules and rules for buildings.

EN 1992-1-2, 2004. Eurocode 2: Design of Concrete Structures – Part 1-2. General rules – Structural fire design.

EN 1993-1-1, 2005. Eurocode 3 - Design of Steel Structures – Part 1-1. General rules and rules for buildings.

EN 1993-1-10, 2005. Eurocode 3 - Design of Steel Structures – Part 1-10. Materials toughness and through-thickness properties.

EN 1993-1-12, 2007. Eurocode 3 – Design of Steel Structures – Part 1-12. Additional rules for the extension of EN 1993 up to steel grades S700.

EN 1993-1-2, 2005. Eurocode 3: Design of Steel Structures – Part 1-2. General rules – Structural fire design.

EN 1993-1-8, 2005. Eurocode 3 - Design of Steel Structures – Part 1-8. Design of joints.

EN 1994-1-1, 2004. Eurocode 4: Design of Composite Steel And Concrete Structures – Part 1-1. General rules and rules for buildings.

EN 1994-1-2, 2005. Eurocode 4: Design of Composite Steel and Concrete Structures – Part 1-2. General rules – Structural fire design.

EN 22553 (1995), ISO 2553, 1992. Welded, Brazed and Soldered Joints – Symbolic Representation on Drawings.

EN ISO 1043-1, 2011. Plastics - Symbols and Abbreviated Terms. Part 1. Basic polymers and their special characteristics.

EN ISO 17636-1, 2013. Non-Destructive Testing of Welds – Radiographic Testing. X- and gamma-ray techniques with film.

EN ISO 17636-2, 2013. Non-Destructive Testing of Welds – Radiographic Testing. X- and gamma-ray techniques with digital detectors.

EN ISO 17637, 2011. Non-Destructive Testing Welds – Visual Testing of Fusion-Welded Joints.

EN ISO 17640, 2010. Non-Destructive Testing Welds – Ultrasonic Testing. Techniques, testing levels, and assessment.

EN ISO 6507-1 (2005). Metallic materials - Vickers hardness test. Part 1: test method.

EN ISO 9712 (2012). Non-destructive testing of welds – Qualification and certification of NDT personnel.

Endo, F., Yamanaka, M., Watnabe, T., Kageyama, M., Yoshida, O., Katsumata, H., Sano, T., 2011 Advanced Technologies Applied at the New "Techno Station" Building in Tokyo21. *Structural engineering international*, Japan, pp. 508–513.

Endo, S., Nakata, N., 2015. Development of Thermo-Mechanical Control Process (TMCP) and high performance steel in JFE Steel. JFE Tech. Rep *20*, 1–7.

Espinos, A., Romero, M.L., Hospitaler, A., 2010. Advanced model for predicting the fire response of concrete filled tubular columns. J. Constr. Steel Res. 66 (8–9), 1030–1046.

Felicetti R., Gambarova P.G., Rosati G.P., Corsi F., Giannuzzi G., 1996. Risidual mechanical properties of high strength concretes subjected to high-temperature cycles. Proceedings of 4th International Symposium on Utilization of High-Strength/High Performance Concrete, Paris, France: 579-588.

Foster, S.J., 2001. On behavior of high-strength concrete columns: cover spalling, steel fibers, and ductility. Struct. 98 (4), 583–589.

Foster, S.J., Liu, J., Sheikh, S.A., 1998. Cover spalling in HSC columns loaded in concentric compression. J. Struct. Eng. 124 (12), 1431–1437.

GB 50936, 2014. Technical Code for Concrete-Filled Steel Tubular Structures. Ministry of Housing and Urban-Rural Development of the People's Republic of China in Chinese.

Gho, W.M., Liu, D., 2004. Flexural behaviour of high-strength rectangular concrete-filled steel hollow sections. J. Constr. Steel Res. 60 (11), 1681–1696.

Goode, C.D., 2008. Composite columns-1819 tests on concrete-filled steel tube columns compared with Eurocode 4. The Structural Engineer 86 (16), 33–38.

Graybeal, B.A., 2006. Material Property Characterization of Ultra-High Performance Concrete. Federal Highway Administration. Office of Infrastructure Research and Development, United States No. FHWA-HRT-06-103.

Guler, S., Copur, A., Aydogan, M., 2012. Flexural behaviour of square UHPC-filled hollow steel section beams. Struct. Eng. Mech. 43 (2), 225–237.

Haber, Z.B., De la Varga, I., Graybeal, B.A., Nakashoji, B., El-Helou, R., 2018. Properties and Behavior of UHPC-Class Materials. Federal Highway Administration. Office of Infrastructure Research and Development, United States No. FHWA-HRT-18-036.

Hadi, M.N., 2007. Using fibres to enhance the properties of concrete columns. Constr. Build. Mater. 21 (1), 118–125.

Hammer, T.A., 1995. High-Strength Concrete Phase 3, Compressive Strength and Elastic Modulus at Elevated Temperatures. SP6 Fire resistance, Report 6.1, SINTEF Structure and concrete STF70 A95023.

Harmathy, T.Z., 1970. Thermal Properties of Concrete at Elevated Temperatures. Research paper No.426 of the Division of Building Research, Ottawa.

Hu, H.T., Huang, C.S., Wu, M.H., Wu, Y.M., 2003. Nonlinear analysis of axially loaded concrete-filled tube columns with confinement effect. J. Struct. Eng. 129 (10), 1322–1329.

Huang, S.S., Burgess, I.W., 2012. Effect of transient strain on strength of concrete and CFT columns in fire–part 2: simplified and numerical modelling. Eng. Struct. 44, 389–399.

Huang, Z.F., Tan, K.H., Ting, S.K., 2006. Heating rate and boundary restraint effects on fire resistance of steel columns with creep. Eng. Struct. 28 (6), 805–817.

Ianizzi, R., Schleich, J.B., 1991. Mechanical Properties of Structural Steel at Elevated Temperatures-Comparisons by Numerical Simulations. ARBED Researches, Luxembourg RPS Rep.No.15/90.

Imran, I., Pantazopoulou, S.J., 1996. Experimental study of plain concrete under triaxial stress. ACI Materials Journal-American Concrete Institute 93 (6), 589–601.

Incropera, F.P., DeWitt, D.P., Bergman, T.L., 2007. Lavine A.S. Introduction to Heat Transfer, 5th edition. John Wiley & Sons, Inc.

ISO 13918., 2017. Welding – Studs and Ceramic Ferrules for Arc Stud Welding. ISO/TC 44/SC 10 Quality management in the field of welding.

ISO 17660-1., 2006. Welding — Welding of Reinforcing Steel — Part 1: Load-bearing Welded Joints. ISO/TC 44/SC 10 Quality management in the field of welding.

ISO 6892-2., 2018. Metallic Materials - Tensile Testing - Part 2: Method of Test at Elevated Temperature. ISO/TC 164/SC 1 Uniaxial testing.

ISO 7091., 2000. Plain Washers, Normal Series, Product Grade C. ISO/TC 2 Fasteners.

ISO, 2015. Plastics—Determination of Thermal Conductivity and Thermal Diffusivity—Part 2: Transient Plane Heat Source (Hot Disc) Method. ISO 22007-2, Geneva ISO.

Jamkhaneh, M.E., Ahmadi, M., Sadeghian, P., 2020. Simplified relations for confinement factors of partially and highly confined areas of concrete in partially encased composite columns. Eng. Struct. 208, 110303.

Jeanes, D.C., 1985. Application of the computer in modeling fire endurance of structural steel floor systems. Fire Saf. J. 9 (1), 119–135.

JGJ 138-2016, 2016. Code for Design of Composite Structures 2016. China: Ministry of Housing and Urban-Rural Development of the People's Republic of China In Chinese.

Johnson, R.P., Anderson, D., 2004. Designers' Guide to EN 1994-1-1: Eurocode 4: Design of Composite Steel and Concrete Structures. *General Rules and Rules for Buildings*, Thomas Telford.

Karthik, M.M., Mander, J.B., 2011. Stress-block parameters for unconfined and confined concrete based on a unified stress-strain model. J. Struct. Eng. 137 (2), 270–273.

Khaliq, W., Kodur, V., 2011. Thermal and mechanical properties of fiber reinforced high performance self-consolidating concrete at elevated temperatures. Cem. Concr. Res. 41 (11), 1112–1122.

Khaliq, W., 2018. Mechanical and physical response of recycled aggregates high-strength concrete at elevated temperatures. Fire Saf. J. 96, 203–214.

Kirby, B.R., Preston, R.R., 1988. High temperature properties of hot-rolled, structural steels for use in fire engineering design studies. Fire Saf. J. 13 (1), 27–37.

Kodur, V., Khaliq, W., 2011. Effect of temperature on thermal properties of different types of high-strength concrete. J. Mater. Civ. Eng. 23 (6), 793–801.

Kodur, V., Banerji, S., Solhmirzaei, R., 2020. Effect of temperature on thermal properties of ultrahigh-performance concrete. J. Mater. Civ. Eng. 32 (8), 04020210.

Kodur, V.K.R., Sultan, M.A., 2003. Effect of temperature on thermal properties of high-strength concrete. J. Mater. Civ. Eng. 15 (2), 101–107.

Lai, B., Liew, J.R., Wang, T., 2019a. Buckling behaviour of high strength concrete encased steel composite columns. J. Constr. Steel Res. 154, 27–42.

Lai, B., Liew, J.R., Xiong, M., 2019b. Experimental study on high strength concrete encased steel composite short columns. Constr. Build. Mater. 228, 116640.

Lai, B., Liew, J.R., Le Hoang, A., 2019c. Behavior of high strength concrete encased steel composite stub columns with C130 concrete and S690 steel. Eng. Struct. 200, 109743.

Lai, B., Liew, J.R., Le Hoang, A., Xiong, M., 2019d. A unified approach to evaluate axial force-moment interaction curves of concrete encased steel composite columns. Eng. Struct. 201, 109841.

Lai, B., Liew, J.R., 2020a. Design and testing of concrete encased steel composite beam-columns with C90 concrete and S690 steel section. Eng. Struct. 220, 110995.

Lai, B., Liew, J.R., 2020b. Axial-moment interaction of high strength concrete encased steel composite columns: experimental investigation. J. Constr. Steel Res. 175, 106370.

Lai, B., Liew, J.R., 2020c. Axial-moment interaction of high strength concrete encased steel composite columns: design recommendation. J. Constr. Steel Res. 170, 106136.

Lai, B., Liew, J.R., 2021. Investigation on axial load-shorting behaviour of high strength concrete encased steel composite section. Eng. Struct. 227, 111401.

Laneyrie, C., Beaucour, A.L., Green, M.F., Hebert, R.L., Ledesert, B., Noumowe, A., 2016. Influence of recycled coarse aggregates on normal and high performance concrete subjected to elevated temperatures. Constr. Build. Mater. 111, 368–378.

Li, G.C., Qiu, Z.M., Yang, Z.J., Chen, B.W., Feng, Y.H., 2020. Seismic performance of high strength concrete filled high strength square steel tubes under cyclic pure bending. Adv. Steel Constr. 16 (2), 112–123.

Li, G., Liu, D., Yang, Z., Zhang, C., 2017. Flexural behavior of high strength concrete filled high strength square steel tube. J. Constr. Steel Res. 128, 732–744.

Li, H., Hao, X., Qiao, Q., Zhang, B., Li, H., 2020. Thermal properties of hybrid fiber-reinforced reactive powder concrete at high temperature. J. Mater. Civ. Eng. 32 (3), 04020022.

Liang, Q.Q., 2011a. High strength circular concrete-filled steel tubular slender beam–columns, part I: numerical analysis. J. Constr. Steel Res. 67 (2), 164–171.

Liang, Q.Q., 2011b. High strength circular concrete-filled steel tubular slender beam–columns, part II: fundamental behavior. J. Constr. Steel Res. 67 (2), 172–180.

Lie, T.T., Chabot, M., 1992. Experimental Studies on the Fire Resistance of Hollow Steel Columns Filled with Plain Concrete. National Research Council Canada Internal Report No.

Lie, T.T., Chabot, M., 1990. A method to predict the fire resistance of circular concrete filled hollow steel columns. J. Fire Prot. Eng. 2 (4), 111–124.

Liew, J.Y.R., 2019. Design Guide for Buildable Steel Connections. Singapore Structural Steel Society, pp. 671pp Download from https://ssss.org.sg/~ssssorgs/images/stories/docs/Design_guide_for_buildable_steel_connections_Final_Version_20190327.pdf.

Lim, J.C., Ozbakkaloglu, T., 2014a. Lateral strain-to-axial strain relationship of confined concrete. J. Struct. Eng. 141 (5), 04014141.

Lim, J.C., Ozbakkaloglu, T., 2014b. Stress–strain model for normal-and light-weight concretes under uniaxial and triaxial compression. Constr. Build. Mater. 71, 492–509.

Llewellyn, D.T., Hudd, R.C., 1998. Steel: Metallurgy & Applications, 3rd edition. Elsevier Science.

Lu, H., Zhao, X.L., Han, L.H., 2011. FE modelling and fire resistance design of concrete filled double skin tubular columns. J. Constr. Steel Res. 67 (11), 1733–1748.

Lu, X., Hsu, C.T.T., 2006. Behavior of high strength concrete with and without steel fiber reinforcement in triaxial compression. Cem. Concr. Res. 36 (9), 1679–1685.

Lyassami, M., Shahriari, D., Ben Fredj, E., Morin, J.B., Jahazi, M., 2018. Numerical simulation of water quenching of large size steel forgings: effects of macrosegregation and grain size on phase distribution. J. Manuf. Mater. Proces. 2 (2), 34.

Maekawa, K., Okamura, H., Pimanmas, A., 2003. Non-Linear Mechanics of Reinforced Concrete. CRC Press.

Mander, J.B., Priestley, M.J., Park, R., 1988. Theoretical stress-strain model for confined concrete. J. Struct. Eng. 114 (8), 1804–1826.

Mehta, P.K., Monteiro., P.J.M., 2006. Materials. McGrawHill, PJM, United States.

Mittemeijer, E.J., 2010. Recovery, recrystallization and grain growth Fundamentals of Materials Science. Springer, Berlin, Heidelberg.

Miyao, T., Nakamura, N., Sugiyama, T., 1997. Evaluation of bearing strength of backing rings for concrete filled tube. J. Archit. Build. Sci 4, 51–56.

Naus, D.J., 2006. *The Effect of Elevated Temperature on Concrete Materials and Structures-A Literature Review* (No. ORNL/TM–2005/553). Oak Ridge National Laboratory (United States). Funding organisation: ORNL work for others, United States.

Outinen, J., Kaitila, O., Mäkeläinen, P., 2001. High-Temperature Testing of Structural Steel and Modeling of Structures at Fire Temperatures. Helsinki University of Technology Laboratory of Steel Structures, Helsinki, Finland Research Report. No. TKK-TER-23.

Ozbakkaloglu, T., Saatcioglu, M., 2004. Rectangular stress block for high-strength concrete. ACI Struct. J. 101 (4), 475–483.

Oztekin, E., Pul, S., Husem, M., 2003. Determination of rectangular stress block parameters for high performance concrete. Eng. Struct. 25 (3), 371–376.

Phan, L.T., Carino, N.J., 2002. Effects of test conditions and mixture proportions on behavior of high-strength concrete exposed to high temperatures. ACI Mater. J. 99 (1), 54–66.

Poh, K.W., 2001. Stress-strain-temperature relationship for structural steel. J. Mater. Civ. Eng. 13 (5), 371–379.

Purkiss, J.A., 1988. Developments in the fire safety design of structural steelwork. J. Constr. Steel Res. 11 (3), 149–173.

Renaud, C., Joyeux, D., Kruppa, J., 2004. Improvement and extension of the simple calculation method for fire resistance of unprotected concrete filled hollow columns *CIDECT Research Project 15Q-12/03. Saint-Rémy-lès-Chevreuse. Centre Technique Industriel de la Construction Métallique (CTICM), France.*

Richart, F.E., Brandtzæg, A., Brown, R.L., 1928. A Study of the Failure of Concrete Under Combined Compressive Stresses. University of Illinois at Urbana Champaign, College of Engineering. Engineering Experiment Station.

RILEM Technical Committee, 2007. Recommendation of RILEM TC 200-HTC: mechanical concrete properties at high temperatures – modeling and applications, Part 2: stress-strain relation. Mater. Struct. 40, 855–864.

Rubert, A., Schaumann, P., 1988. Critical temperatures of steel columns exposed to fire. Fire Saf. J. 13 (1), 39–44.

Sawada, K., Ohba, T., Kushima, H., Kimura, K., 2005. Effect of microstructure on elastic property at high temperatures in ferritic heat resistant steels. Mater. Sci. Eng.: A 394 (1–2), 36–42.

Schaupp, T., Schroepfer, D., Kromm, A., Kannengiesser, T., 2017. Welding residual stresses in 960 MPa grade QT and TMCP high-strength steels. J. Manuf. Process. 27, 226–232.

Schneider, U., Diederichs, U., Ehm, C., 1982. Effect of temperature on steel and concrete for PCRV's. Nucl. Eng. Des. 67 (2), 245–258.

Shah, S.N.R., Akashah, F.W., Shafigh, P., 2019. Performance of high strength concrete subjected to elevated temperatures: a review. Fire. Technol. 55 (5), 1571–1597.

Shikanai, N., Mitao, S., Endo, S., 2008. Recent development in microstructural control technologies through the thermo-mechanical control process (TMCP) with JFE steel's high-performance plates. JFE technical report 11 (6), 1–6.

Shin, K.Y., Kim, S.B., Kim, J.H., Chung, M., Jung, P.S., 2002. Thermo-physical properties and transient heat transfer of concrete at elevated temperatures. Nucl. Eng. Des. 212 (1–3), 233–241.

Singapore Civil Defence Force (SCDF, 2018. Code of practice for fire precautions in buildings, 8th Edition.

Steau, E., Mahendran, M., Poologanathan, K., 2020. Elevated temperature thermal properties of carbon steels used in cold-formed light gauge steel frame systems. J. Build. Eng. 28, 101074.

Syuichi, M A T S U M O T O, Kazumasa, G O T O, Shusuke, K U R O I W A, Yoichi, T A K A S E, 2013. Application of concrete filled tubular column with fc 150n/mm2 high strength concrete and 780n/mm2 class high strength steel to high-rise building. REPORT OF TAISEI TECHNOLOGY CENTER, 46 No.

Takaki, J.I., Nakamura, N., Hirota, M., Miyao, T., Inosako, T., Sugiyama, T., 1997. Experimental study on bearing strength between tubular column with welded line inside surface, and filled concrete. AIJ Proceedings, 22445 No.

Takaki, J.I., Nakamura, N., Miyao, T., Inosako, T., Sugiyama, T., Nanba, T., 1999. Bearing strength of concrete for circular tube column. AIJ Proceedings, 22566 No.

Tao, Z., Wang, Z.B., Yu, Q., 2013. Finite element modelling of concrete-filled steel stub columns under axial compression. J. Constr. Steel Res. 89, 121–131.

The Steel Construction Institute and The British Constructional Steelwork Association Limited, 2014. Joints in steel construction. Simple joints to Eurocode 3, P358.

The Steel Construction Institute, 2005. Non-contradictory complementary information: shear resistance of a fin plate connection. SN017a-EN-EU NCCI.

Thomas, M., Green, B., O'Neal, E., Perry, V., Hayman, S., Hossack, A., 2012. Marine performance of UHPC at Treat Island, Proceedings of the 3rd International Symposium on UHPC and Nanotechnology for High Performance Construction Materials. *Kassel, Germany*, pp. 365–370.

Toman, J., Černý, R., 2001. Thermal conductivity of high performance concrete in wide temperature and moisture ranges. Acta Polytechnica 41 (1), 8–10.

Varma, A.H., Ricles, J.M., Sause, R., Lu, L.W., 2002. Seismic behavior and modeling of high-strength composite concrete-filled steel tube (CFT) beam–columns. J. Constr. Steel Res. 58 (5–8), 725–758.

Wang, Y.C., 2002. Steel and Composite Structures – Behaviour and Design for Fire Safety. Spon Press, London and New York.

Wang, J., Sun, Q., Li, J., 2019. Experimental study on seismic behavior of high-strength circular concrete-filled thin-walled steel tubular columns. Eng. Struct. 182, 403–415.

Wang, Y.B., Liew, J.R., 2016. Constitutive model for confined ultra-high strength concrete in steel tube. Constr. Build. Mater. 126, 812–822.

Wang, Y.B., Liew, J.Y., Lee, S.C., Xiong, D.X., 2016. Experimental study of ultra-high-strength concrete under triaxial compression. ACI Mater. J. 113 (1), 105–112.

WES 3001, 2012. Weldable High Strength Steel plates. The Japan Welding Engineering Society.

Williams, B., Heard, W., Graham, S., Nie, X., 2020. Effect of specimen geometry on triaxial compressive response of high-strength concrete. Constr. Build. Mater. 244, 118348.

Wilzer, J., Lüdtke, F., Weber, S., Theisen, W., 2013. The influence of heat treatment and resulting microstructures on the thermophysical properties of martensitic steels. J. Mater. Sci. 48 (24), 8483–8492.

Wurmbauer, H., Leitner, H., Panzenböck, M., Scheu, C., Clemens, H., 2010. Short-term creep behavior of an X 37 Cr Mo V 5-1 hot-work tool steel with almost bainitic and fully martensitic microstructures. Steel Res. Int. 81 (7), 569–575.

Xiong, M.X., Liew, J.Y.R., 2015 Spalling Behavior and Residual Resistance of Fibre Reinforced Ultra-High Performance Concrete After Elevated Temperatures65. Materiales de Construcción, pp. 1–10.

Xiong, M.X., Liew, J.R., 2016a. Mechanical properties of heat-treated high tensile structural steel at elevated temperatures. Thin-Walled Structures 98, 169–176.

Xiong, M.X., Liew, J.R., 2016b. Mechanical behaviour of ultra-high strength concrete at elevated temperatures and fire resistance of ultra-high strength concrete filled steel tubes. Mater. Des. 104, 414–427.

Xiong, M.X., Liew, J.R., 2018. Discussion on the applicability of the MN interaction curve for the fire resistance design of CFT members. Thin-Walled Structures 125, 172–186.

Xiong, M.X., Liew, J.R., 2020a. Experimental study to differentiate mechanical behaviours of TMCP and QT high strength steel at elevated temperatures. Constr. Build. Mater. 242, 118105.

Xiong, M.X., Liew, J.R., 2020b. Buckling behavior of circular steel tubes infilled with C170/185 ultra-high-strength concrete under fire. Eng. Struct. 212, 110523.

Xiong, M.X., Liew, J.R., 2021. Fire resistance of high-strength steel tubes infilled with ultra-high-strength concrete under compression. J. Constr. Steel Res. 176, 106410.

Xiong, M.X., Yan, J.B., 2016. Buckling length determination of concrete filled steel tubular column under axial compression in standard fire test. Mater. Struct. 49 (4), 1201–1212.

Xiong, M.X., Liew, J.R., Wang, Y.B., Xiong, D.X., Lai, B.L., 2020. Effects of coarse aggregates on physical and mechanical properties of C170/185 ultra-high strength concrete and compressive behaviour of CFST columns. Constr. Build. Mater. 240, 117967.

Xiong, M.X., Wang, Y., Liew, J.R., 2016. Evaluation on thermal behavior of concrete-filled steel tubular columns based on modified finite difference method. Adv. Struct. Eng. 19 (5), 746–761.

Xiong, M.X., Xiong, D.X., Liew, J.R., 2017a. Axial performance of short concrete filled steel tubes with high-and ultra-high-strength materials. Eng. Struct. 136, 494–510.

Xiong, M.X., Xiong, D.X., Liew, J.R., 2017b. Flexural performance of concrete filled tubes with high tensile steel and ultra-high strength concrete. J. Constr. Steel Res. 132, 191–202.

Xiong, M.X., Xiong, D.X., Liew, J.R., 2017c. Behaviour of steel tubular members infilled with ultra high strength concrete. J. Constr. Steel Res. 138, 168–183.

Yan, J.B., Liew, J.R., Zhang, M.H., Wang, J.Y., 2014. Mechanical properties of normal strength mild steel and high strength steel S690 in low temperature relevant to Arctic environment. Mater. Des. 61, 150–159.

Yang, I.H., Park, J., 2020. A study on the thermal properties of high-strength concrete containing CBA fine aggregates. Materials (Basel) 13 (7), 1493.

Yu, W., Qian, Y.J., Wu, H.B., Yang, Y.H., 2011. Effect of heat treatment process on properties of 1000 MPa ultra-high strength steel. J. Iron. Steel Res. Int. 18 (2), 64–69.

Zeiml, M., Leithner, D., Lackner, R., Mang, H.A., 2006. How do polypropylene fibers improve the spalling behavior of in-situ concrete?. Cem. Concr. Res. 36 (5), 929–942.

Zhang, J., Cai, R., Li, C., Liu, X., 2020. Seismic behavior of high-strength concrete columns reinforced with high-strength steel bars. Eng. Struct. 218, 110861.

Zhao, X., Qin, H., Chen, Y.Y., 2014. Experimental study on constitutive model of steel confined concrete in SRC columns with cruciform steel section. J. Build. Struct. 35 (4), 268–279 in Chinese.

Zhao, X., Wen, F., Chan, T.M., Cao, S., 2019. Theoretical stress–strain model for concrete in steel-reinforced concrete columns. J. Struct. Eng. 145 (4), 04019009.

Index

Page numbers followed by "*f*" and "*t*" indicate, figures and tables respectively.

M

N

O

P

Q

R

S

T

U

V

W

Y

FSC
www.fsc.org
MIX
Papier | Fördert
gute Waldnutzung
FSC® C083411

Kolibri 360 GmbH
Ferdinand-Jühlke-Straße 7
99095 Erfurt, Deutschland
produktsicherheit@kolibri360.de